Case Studies of Wildlife Ecology and Conservation in India

This volume brings together a collection of case studies examining wildlife ecology and conservation across India.

The book explores and examines a wide range of fauna across different terrains and habitats in India, revealing key issues and concerns for biodiversity conservation, with a particular emphasis on the impact of humans and climate change. Case-studies are as wide-ranging as tigers, leopards, sloth bears, pheasants, insects, and birds, across a diverse range of landscapes, including forests, wetlands, and nature reserves, and even a university campus. Split into three parts, Part I focuses on how the distribution of animals is influenced by the availability of resources such as food, water, and space. Chapters examine key determinants, such as diet and prey and habitat preferences, with habitat loss also being an important factor. In Part II, chapters examine human-wildlife interactions, dealing with issues such as the impact of urbanisation, the establishment of nature reserves, and competition for resources. The book concludes with an examination of landscape ecology and conservation, with chapters in Part III focusing on habitat degradation, changes in land-use patterns, and ecosystem management. Overall, the volume not only reflects the great breadth and depth of biodiversity in India but offers important insights into the challenges facing biodiversity conservation not only in this region but worldwide.

This volume will be of great interest to students and scholars of wildlife ecology, conservation biology, biodiversity conservation, and the environmental sciences more broadly.

Orus Ilyas is Associate Professor in the Department of Wildlife Sciences at Aligarh Muslim University, India. She has studied mammals in India since 1995 and has worked in the high-altitude Himalayas for more than 25 years. She has worked on various projects funded by government and non-government organisations, such as WWF India and the Council for Advancement of People's Action and Rural Technology (CAPART).

Afifullah Khan is Professor in the Department of Wildlife Sciences at Aligarh Muslim University, India. He started his research career by investigating the ecology of Asian elephants and continued to work on mammalian and avian species and communities. His current research explores ecological patterns and processes at a landscape level and the effects of EMF on plants and birds.

Routledge Studies in Conservation and the Environment

This series includes a wide range of inter-disciplinary approaches to conservation and the environment, integrating perspectives from both social and natural sciences. Topics include, but are not limited to, development, environmental policy and politics, ecosystem change, natural resources (including land, water, oceans and forests), security, wildlife, protected areas, tourism, human-wildlife conflict, agriculture, economics, law and climate change.

Women and Wildlife Trafficking
Participants, Perpetrators and Victims
Edited by Helen U. Agu and Meredith L. Gore

Conservation, Land Conflicts and Sustainable Tourism in Southern Africa
Contemporary Issues and Approaches
Edited by Regis Musavengane and Llewellyn Leonard

Threatened Freshwater Animals of Tropical East Asia
Ecology and Conservation in a Rapidly Changing Environment
David Dudgeon

Conservation Effectiveness and Concurrent Green Initiatives
Li An, Conghe Song, Qi Zhang, and Eve Bohnett

Religion and Nature Conservation
Global Case Studies
Edited by Radhika Borde, Alison A. Ormsby, Stephen M. Awoyemi, and Andrew G. Gosler

Jackals, Golden Wolves, and Honey Badgers
Cunning, Courage, and Conflict with Humans
Keith Somerville

Case Studies of Wildlife Ecology and Conservation in India
Edited by Orus Ilyas and Afifullah Khan

For more information about this series, please visit: www.routledge.com/Routledge -Studies-in-Conservation-and-the-Environment/book-series/RSICE

Case Studies of Wildlife Ecology and Conservation in India

Edited by Orus Ilyas and Afifullah Khan

Routledge
Taylor & Francis Group
LONDON AND NEW YORK

earthscan
from Routledge

First published 2023
by Routledge
4 Park Square, Milton Park, Abingdon, Oxon OX14 4RN

and by Routledge
605 Third Avenue, New York, NY 10158

Routledge is an imprint of the Taylor & Francis Group, an informa business

British Library Cataloguing-in-Publication Data
A catalogue record for this book is available from the British Library

Library of Congress Cataloging-in-Publication Data
Names: Ilyas, Orus, 1972- editor. | Khan, Afifullah, editor.
Title: Case studies of wildlife ecology and conservation in India/edited by Orus Ilyas and Afifullah Khan.
Description: Milton Park, Abingdon, Oxon; New York, NY: Routledge, 2023. | Includes bibliographical references and index.
Identifiers: LCCN 2022022366 (print) | LCCN 2022022367 (ebook) | ISBN 9781032342986 (hardback) | ISBN 9781032342993 (paperback) | ISBN 9781003321422 (ebook)
Subjects: LCSH: Wildlife conservation–India. | Habitat conservation–India.
Classification: LCC QL84.5.I4 C37 2023 (print) | LCC QL84.5.I4 (ebook) | DDC 333.95/40954–dc23/eng/20220729
LC record available at https://lccn.loc.gov/2022022366
LC ebook record available at https://lccn.loc.gov/2022022367

ISBN: 978-1-032-34298-6 (hbk)
ISBN: 978-1-032-34299-3 (pbk)
ISBN: 978-1-003-32142-2 (ebk)

DOI: 10.4324/9781003321422

Typeset in Goudy
by Deanta Global Publishing Services, Chennai, India

Contents

PART III
Landscape Ecology and Conservation   197

Figures

Tables

Contributors

Riyaz Ahmad, Coordinator and project Head, Jammu and Kashmir, has a doctoral degree in wildlife sciences on mountain ungulates, and the work has been published in peer-reviewed journals. He has around 20 publications on the work done in ungulates and birds of Jammu and Kashmir. He has been involved in putting together the management of plans for several protected areas of Jammu and Kashmir. He works with local communities, government, and security agencies to conserve the threatened biodiversity of Jammu and Kashmir. In this direction, he has organised several workshops also.

Kaleem Ahmed, Assistant Professor, Aligarh Muslim University, India, is working as an assistant professor in the Wildlife Sciences Department at Aligarh Muslim University, Aligarh. He has been actively engaged in wildlife research for the last ten years and has handled many projects from endangered barasingha in terai forest to biodiversity conservation of Kumaon Himalayas. Dr. Kaleem teaches Animal Behavior, Ornithology, Wildlife Legislation, Wildlife Conservation, and other sources at the Department of Wildlife Sciences at postgraduate level.

Khursheed Ahmed, Scientist and head in centre for Mountain Wildlife Sciences, Sher-e-Kashmir University of Agricultural Sciences & Technology of Kashmir, India, with an MSc and PhD in wildlife sciences and a specialised training in wildlife management from Macaulay Institute, Aberdeen, Scotland, has wide experience in the field of wildlife management and conservation especially on the Himalayan ungulates and avifauna. He has undertaken more than 20 research projects funded by reputed national and international funding agencies, Indo-US Science and Technology Forum, MOEF & CC, DST, SERB, and J&K Govt. He has the distinction of creating history by successfully capturing and collaring hangul for the first time and resolving and documenting the phylogenetic status and taxonomic relationship of Kashmir red deer or hangul in family Cervidae, which has resulted in the species being accorded separate species status by IUCN. He has been serving in the cause of conservation and scientific research as a Member Wildlife Advisory Board, Jammu & Kashmir; Programme Advisory Committee on Animal Sciences, SERB, Govt. of India; and the Regional

Expert Committee on Management Effectiveness Evaluation (MEE) of National Parks and Sanctuaries, Ministry of Environment Forests & Climate Change (MOEF&CC), Govt. of India; Expert Member Biodiversity (Animal Sciences) National Mission on Himalayan Studies, MOEF&CC; State Coordinator, IBA-IBCN, BNHS-RSPB, Birdlife International, United Kingdom Programme; member, Monitoring Committee on Hangul Conservation Breeding and Reintroduction Programme, Central Zoo Authority, Ministry of Environment & Forests (MOEF & CC), Govt. of India; Member Conservation Breeding Specialist Group South Asia (CBSG-SA), IUCN; member of Deer Specialist Group, IUCN, Member Antelope Specialist Group, IUCN 8; member of the governing board, WWF India, J & K Chapter; Member, Technical Committee for Research in Wildlife, Jammu & Kashmir Government.

Cassandra Bugir, PhD, Research Scholar, University of Newcastle, Newcastle, Australia, has a background in anthro-zoology and predominantly focuses on human-animal relationships. Cassandra's PhD thesis focuses on the prey preferences of hominid groups—extinct and extant—as well as the genomic cousin, the chimpanzee. Determining hominids' dietary choices and reaffirming predatory status provides valuable information to transform into solutions for conservation efforts for both human hunter-gatherer communities and animal populations

Aimon Bushra, PhD, Research Scholar, Wildlife Institute of India, received M.Tech degree in Remote Sensing and GIS with specialisation in forest resources and ecosystem analysis from the Indian Institute of Remote Sensing, Dehradun. She worked in an environmental consultancy in Dubai as a GIS specialist. Presently, she is working on the project "Himalayan alpine biodiversity characterization and Information System Network" at WII and pursuing a PhD on the topic "Patterns of plant species diversity across environmental gradients in Indian Trans Himalaya" registered in Forest Research.

Pooja Chaurasia, Senior Manager, Environmental and Forestry Services of Skill Art and Beyond (SAB), completed her doctoral degree on the ecology of the golden jackal. She was also a team member in a tiger reintroduction programme in Sariska Tiger Reserve, India. She had worked with WWF India in Living Ganga Programme. As a researcher under training, she served Gharial Conservation Alliance in Katerniaghat gharial project. She has experience in both qualitative and quantitative research with advanced techniques. She has published a total of five scientific research articles.

Tabassum Choudhary, Ph.D., Research Scholar, Aligarh Muslim University, is a full-time research scholar at the Department of Law, Aligarh Muslim University. Her specialisation is in commercial law, and her research area lies in cybercrimes. She has 16 years of teaching experience and has several publications related to child pornography, cyber terrorism, drug abuse, sexual harassment, and human rights education. In addition, she has also been registered with Uttar Pradesh Bar Council for the last 18 years.

Shahid Ahmad Dar, PhD, Research Scholar, Wildlife Institute of India, is a Ph.D. Research Scholar, pursuing his degree on "Ecology, genetics and conservation of Himalayan brown bear in Western Himalaya". Before beginning his Ph.D., he worked with the Wildlife Trust of India and Wildlife Institute of India in several projects, "Ecological aspects of Markhor (*Capra falconeri*) in Pir Panjal range of the Himalaya" and "Population genetic structure and gene flow in brown bear (*Ursus arctos isabellinus*) populations in India and assess extent of gene flow between populations of India and Pakistan: Conservation and Forensic Implications" respectively.

Nishith Dharaiya, Associate Professor, Hemchandracharya North Gujarat University, Patan (Gujarat) India, is in charge of the Wildlife and Conservation Research Lab at the Department of Life Sciences. He has 15 years of experience in various aspects of the sloth bear in Gujrat, Maharashtra, north-eastern India. He is a co-chair of IUCN SSC Sloth Bear Expert Team and member of various teams of bear specialist groups of IUCN. Presently he is working on sloth bear conservation and outreach education in Gujarat in collaboration with Gujarat forest department and other international agencies.

Kedar Gore, Director, The Corbett Foundation, India, holds a master's degree in Zoology and a post-graduate diploma in management. He has been actively engaged in wildlife conservation, nature education, and environmental protection since 1996. He has extensive experience in implementing landscape-level conservation programmes across India, benefitting threatened species and important habitats. He is a member of three IUCN Commissions and the Indian National Committee of IUCN India. He has authored several scientific and popular articles. He is recipient of several awards including the IUCN CEC Excellence Award, South & Southeast Asia in 2019.

Jadumoni Goswami, Education Officer, The Corbett Foundation, is currently working in The Corbett Foundation as Education and Outreach Coordinator for Wildlife. His professional interests centre around environmental education and photographic documentation of wilderness. He has co-authored many papers based on field research in international journals.

Bilal Habib, Scientist-E, Wildlife Institute of India, is a conservation biologist interested in the integration of quantitative and interdisciplinary approaches to conservation challenges. His doctoral research was on the ecology of the Indian wolf in the Great Indian Bustard Sanctuary, Maharashtra, India. He also worked as Wildlife Survey Program Manager for the Wildlife Conservation Society's Afghanistan Biodiversity Project and was involved in the design and coordination of wildlife surveys in the Wakhan Corridor (Pamirs), capacity building in the environment sector, and development of the Red List.

Abdul Haleem, Aligarh Muslim University, India, completed his master's degree in wildlife sciences from Aligarh Muslim University. He joined DST-SERB Project "Ecology of Ungulates in Pench Tiger Reserve" and earned his Ph.D.. Presently, he is working as freelancer in the field of wildlife.

Matt W. Hayward, Faculty, Bangor University, North Wales, works as Associate Professor at Bangor University, North Wales, UK. He holds two post-doctoral degrees from Walter Sisulu University, Africa, and Nelson Mandela Metropolitan University, Africa on bush meat hunting in the coastal forests of the Transkei, and the reintroduction of lions, spotted- hyenas, and leopards to Addo Elephant National Park, respectively. Presently working on reintroducing red squirrels; ascertaining the impact of pine marten on squirrels; the context dependence of humans on wolf ecology; spatial ecology of peccaries and elephants; invasive snakes; and the impact of neonicotinoids on fossorial mammals. His research interests include the conservation ecology of threatened species, the factors that threaten them, and the methods he can use to conserve them effectively.

Munawwar Husain, retired Professor, Aligarh Muslim University, India, is an ex-medical superintendent at the Department of Forensic Medicine, Jawaharlal Nehru Medical College, Aligarh. He has vast experience in teaching several courses in forensic medicine and medico-legal issues.

Mhod. Shah Hussain, Scientist-D level and scientist in charge at Aravalli Biodiversity Park CEMDE, University of Delhi, earned his doctoral degree in wildlife science from Aligarh Muslim University. He has been working as an ecologist (Scientist-D level) since 2005 and is scientist in charge at the Aravalli Biodiversity under the Biodiversity Parks Programme of Centre for Environmental Management of Degraded Ecosystems (CEMDE), Centre of Excellence of MoEF & CC, Govt. of India at University of Delhi. He is Honorary General Secretary of World Pheasant Association, India, an Indian Chapter of the World Pheasant Association UK, handling all administrative work of different projects across India, having association with different agencies for the last nine years.

Mohd. Mukhtyar Hussain, Aligarh Muslim University, India, completed his masters in wildlife sciences at the Department of Wildlife Sciences, Aligarh Muslim University, Aligarh. For his MSc dissertation, he worked on "Evaluating the wetland in Aligarh district using the geospatial technique". At present, he is doing freelance work in the wildlife field.

Ekwal Imam, Aligarh Muslim University, India, received his PhD in wildlife sciences from Aligarh Muslim University, Aligarh, India. His principal research interests are ecology, conservation biology, primatology, Remote Sensing & GIS, Protected Areas management, and biodiversity Assessment. He has implemented several research projects and translocated more than 600 rhesus monkeys, perhaps Asia's largest monkey rehabilitation programme. So far he has published three books and 29 scientific papers in international and national journals having good impact factors. At present Dr. Imam is associated with the Department of Wildlife Sciences.

Shahzada Iqbal, Ph.D. Research Scholar, Aligarh Muslim University, India, is currently enrolled in the Ph.D. programme at Aligarh Muslim University in the Department of Wildlife Sciences. He did a master's dissertation on human-blackbuck interaction. His primary academic interests are wolf ecology and human-wolf interaction. His current project is in the Mahuadanr Wolf Sanctuary in Jharkhand.

Jamal Ahmed Khan, Chairman, Department of Wildlife Sciences, Aligarh Muslim University, India, earned his MPhil and doctoral degree from the Wildlife Institute of India under the able guidance of the late Prof. A.H. Musavi, the late Dr. William Alan Rodgers, Dr. Anna Jawahar Thomas Johnsingh, and Dr. P.K. Mathur. Presently, he is chairman and professor in the Department of Wildlife Sciences, Aligarh Muslim University. He was a member of the National Tiger Conservation Authority, Ministry of Environment, Forests & Climate Change, GOI from 2006 to 2009 and member of the State Wildlife Advisory Board, Uttar Pradesh from 2011 to 2012, IUCN Cat Specialist Group. His current research interests are large carnivore ecology, human wildlife conflict studies, biodiversity conservation, and conservation of endangered taxa such as tiger, lion, swamp deer, etc

Iqbal Ali Khan, Ph.D., Research Scholar, Gurukul Kangri University, India, is a research scholar at Zoological Survey of India, Dehradun, registered with Gurukula Kangri Vishwavidyalaya, Haridwar. Recently, he has been working on the distribution pattern, population status, and behavioral ecology of the Eurasian magpie *Pica pica bactriana* (Family Corvidae) in Ladakh UT, India.

Sharad Kumar, Assistant Professor, Aligarh Muslim University, India, has been working as Assistant Professor in the Department of Wildlife Sciences, Aligarh Muslim University, Aligarh since 2019. He has conducted a long-term study on the ecology of tigers in Corbett Tiger Reserve. He joined The Corbett Foundation in 2008 and worked with the organisation until 2019. He worked as principal investigator in various research and conservation projects. Currently, he is a member of IUCN, Commission on Environment, Economic and Social Policy (CEESP). His research interests lie in developing and working on innovative conservation ideas for large carnivores and birds and mitigation of human-wildlife conflict in and around the Protected Areas.

Anil Kumar, Senior Scientist, Zoological Survey of India, has a working experience of 20 years in the avifauna of Himalayas especially avian species like robins and bulbuls, their behavioural ecology and socio-biology with special reference to their communication systems.

Sumanta Kundu, Senior Programme Officer, The Corbett Foundation, India, holds a master's degree in environmental science and has nearly two decades of experience in wildlife research and conservation. He has been associated with many renowned organisations in India such as the Asian Nature Conservation Foundation (ANCF), Centre for Ecological Sciences (CES, IISc), and Wildlife Trust of India (WTI). His prime interests are human-wildlife conflict mitigation, ecology of elephants, and wildlife corridor connectivity.

Salvadora Lyngdoh, Scientist-D, Wildlife Institute of India, is a Scientist-D in the Wildlife Institute of India with experience of the art techniques in population monitoring, behavioural studies, and conflict studies in western as well as eastern India. His long term interest lies in understanding the movement ecology of mammalian carnivores in the Himalayan ecosystem. His area of specialisation include population estimation, human-wildlife conflicts, animal behaviour, and spatial ecology. His prior work experience includes the investigation of socio-economic drivers of environmental change to animal ecological studies in tropical evergreen forests as well as high altitude Himalayas.

Prashant Mahajan, Project Fellow, Wildlife Institute of India, has completed his master's degree in wildlife sciences from the department of Wildlife Sciences, Aligarh Muslim University. His research focuses are human-wildlife conflicts, prey-predator interactions, and the ecology of large carnivores. He has worked on the ecology of wolves in Rajasthan. Presently, he is working as a research scholar at Dept of Wildlife Sciences, AMU, Aligarh.

Arzoo Malik, Ph.D., Research Scholar, Wildlife and Conservation Biology Research Lab, Hemchandracharya North Gujarat University, Patan (Gujarat) India, is a doctoral student working on the conservation of sloth bears and their habitats. Her research work aimed at alleviating the human-bear conflict in the regions of north Gujarat. For the past four years, she has worked as a Principal Investigator on different research projects at the Wildlife and Conservation Biology Research Lab. Her areas of expertise lie in wildlife biology and GIS, modeling. She was appointed as a member of the IUCN SSC Sloth Bear Expert Team.

Abdul Hamid Malik got his MSc from Baba Ghulam Shah Badshah University, Rajouri. He worked on biodiversity conservation and is looking for the opportunity to pursue his Ph.D. programme.

Zaffar Rais Mir, Assistant Professor, Central University of Kashmir, earned his post-doctoral degree on the human-leopard conflict in Kashmir Himalaya using Remote Sensing and GIS technology. His research interests include carnivore ecology and conservation, predator-prey relationship, resource selection, and animal space use. While at the Wildlife Institute of India, he worked on several other research projects like the Leopard Ecology Project in Kalesar National Park, Haryana, and the blackbuck study in Fatehabad, Haryana. After completion of his PhD, he worked as a field officer with the Wildlife Trust of India for about a year to study the ecology of markhor (*Capra falconeri*) in the Pir Panjal range of the Himalayas. After that he pursued two years in a National Postdoctoral Fellowship Program (funded by SERB-DST) at the Division of Wildlife Sciences, Sher-e-Kashmir University of Agricultural Sciences and Technology of Kashmir (2018–2019).

Krishnendu Mondal, Project Associate, Wildlife Institute of India, obtained his doctoral degree on the ecology of leopards and post-doctoral under National Mission for Himalayan Studies. He was also associated with a tiger reintroduction programme. He was trained by INTERPOL, International Fund for Animal Welfare and Wildlife Trust of India for conducting trainings on Wildlife Crime Prevention for frontline forest staff. He trained more than 3000 forest staff from different states of India and around 900 staff from Bhutan. He has published 25 scientific research articles, four books, ten book chapters, and four popular articles.

Poornima Nailwal, Ph.D. Research Scholar, Zoological Survey of India, is a doctoral student, pursuing her degree focused on a zoological survey of India. Her major research work involves field recordings and acoustical analysis of songs of the Indian robin. Her interests involve birding, reading, and photography.

Anukul Nath, Technical Officer, Wildlife Institute of India, earned his master's degree in wildlife biology from AVC College, Tamil Nadu and PhD in ecology and environment science from Assam University. His research focuses on population and landscape ecology (spatial statistics and species distribution modeling). His interests are in mainstreaming scientific and modelled explorations to advance conservation planning and implementation.

Athar Noor, Principal Project Associate, Wildlife Institute of India, is an avid wildlife ecologist with a doctoral degree. He has been active in the field of wildlife research for over 15 years and has published several research articles in reputed national and international journals. He has worked across a range of species and ecosystems and landscapes. His research interests include carnivore ecology and conservation, predator-prey relationship, resource selection, animal space use, movement ecology, and use of GIS in ecological studies.

Naveen Kumar Pandey, Deputy Director and Veterinary Advisor, The Corbett Foundation, India, is a competent conservation professional having 18 years of experience in India's rural provinces and Protected Areas, focusing on animal welfare, nature conservation, climate change awareness, sustainable livelihood, animal production and food security, and community-based conservation initiatives. He is presently designing and implementing projects favouring rural entrepreneurship, ecosystem protection, wildlife conservation, and disease control in northeast India, central India, and north India. Connecting rural dwellers and policymakers with sustainable agriculture, animal husbandry, and ecosystem services are key work areas. Ongoing projects are a human-elephant conflict mitigation project through the International Elephant Foundation, US, and human-large cat conflict mitigation through the Van Tienhoven Foundation, Netherlands, in a team leader capacity. He is also an Invited Guest Lecturer for the Species Conservation and Ecosystem Health course at the University of Edinburgh and serves The Corbett Foundation as its Deputy Director and Veterinary Advisor in northeast India. He is also affiliated to Centre for Indian Knowledge Systems, IIT Guwahati

Mordecai Panmei, Senior Programme Officer, The Corbett Foundation, India, a member of the indigenous Naga society in Tamenglong, Manipur, holds a master in social work, and serves The Corbett Foundation as Senior Programme Officer. He undertakes conservation-oriented ecological research to assess the health of habitats and wildlife. At the grassroots level, he organises events and awareness programmes in schools, villages, and community gatherings. He is a Green Hub Fellow and has documented the diversity of Garo Hills in Meghalaya, Ziro Valley in Arunachal Pradesh, and Keibul Lamjao National Park in Manipur.

Talat Parveen, Ph.D., Research Scholar, Aligarh Muslim University, India, is working as a Senior Research fellow in CSIR funded research project titled "Conservation status, ecology, and co-existence of sympatric ungulates in Panna Tiger Reserve, M.P., CSIR-Delhi, and Government of India". She is also enrolled for her Ph.D. Programme on "Status, distribution and ecological co-existence of antelopes in Panna Tiger Reserve, Central India"

Kruti Patel, MSc, Wildlife and Conservation Biology Research Lab, Hemchandracharya North Gujarat University, Patan (Gujarat) India, completed her master degree in environmental sciences and is aspiring for her Ph.D. in wildlife sciences. She has her master's dissertation onfood resource sharing regarding human-sloth bear interaction in the Jessore Wildlife Sanctuary and has one year of experience as a volunteer in a research project, the habitat of sloth bears in North Gujrat.

Ayesha Qamar, Professor, Aligarh Muslim University, India, is working as a professor in the Department of Zoology.Her focus areas include physiology, toxicology, pest management, pesticide exposure, and investigating adverse effects on health and the environment using Drosophila and Bombyx as models.

Mohd Qasim did his MSc from Baba Ghulam Shah Badshah University, Rajouri. He worked on the bird community structure of Baba Guam Shah Badshah University Rajouri. At present, he is workfreelanceance and looking for the opportunity to pursue his Ph.D. programme

Qamar Qureshi, Faculty, Wildlife Institute of India, obtained a Master in Philosophy in Wildlife Sciences from Aligarh Muslim University. He has been a faculty member at the Wildlife Institute of India for the last three decades and has undertaken training in geographical information systems (GIS) and remote sensing at Colorado State University, Fort Collins (US). He was a visiting fellow of University College, London (UK) and Natural Resource Ecology Unit, CSU (US). He deals with projects on biodiversity assessment at the landscape and species levels and studying habitat use patterns using remote sensing and GIS. His areas of specialisation are conservation biology, quantitative ecology, and landscape ecology.

Rita Rathi, Professor, Department of Zoology, Dyal Singh College, University of Delhi, India, is a professor of biology with 30 years of combined teaching and research experience in entomology and insect biochemistry at the University of Delhi, New Delhi, India. She has done research as a post-doctoral fellow with the Center of Excellence in Applied Research and Training (CERT), Abu Dhabi, United Arab Emirates, along with the Environmental Agency Abu Dhabi (EAD) under the able guidance of Dr. Tayeb Kamali, Vice Chancellor of Higher College of Technology (HCT) and CEO of CERT.

Dibyjyoti Saikia, Mobilization Officer, The Corbett Foundation, India, hails from Kaziranga, a UNESCO World Heritage site. He has been working with The Corbett Foundation since 2013 for securing alternative livelihood in fringe villages of Kaziranga National Park and other forests in northeast India. He has published papers in reputed international journals. He has a strong cultural understanding of Assamese society.

Kalyanasundaran Sankar, Director, Salim Ali Centre of Ornithology and Natural History, India, completed his doctoral degree in the ecology of large herbivores. Based on 29 years of research experience, he prepared the Species Recovery Programme for tigers in Rajasthan; gaur reintroduction programme in central India; and coordinated tiger census in southern Indian landscape. He is also a member of the IUCN Deer Specialist Group, Asian Wild Cattle Specialist Group, and is a Fulbright Scholar. He has published over 100 scientific publications, 3 books, 13 book chapters, 2 manuals, and 15 popular articles to his credit and has supervised 20 Ph.D. scholars and 20 MSc students for thesis and dissertations respectively in the discipline of Wildlife Science.

Ali Asghar Shah, Associate Professor, Baba Ghulam Shah Badshah University, Rajouri, is an Associate Professor and Head at the Department of Zoology, Baba Ghulam Shah Badshah University, Rajouri. He has done pioneering research work on free-living nematodes of India for which the DST, Ministry of Science and Technology awarded him Young Scientist award and FAST Track fellowship in 2007. He has reported 61 species under 21 genera, falling under three superfamilies, four families, and eight subfamilies. Of these, 26 species and seven genera are new to science. Giving recognition to his work, a UK-based scientist, Dr. M.R. Siddiqi, named a genus "Shahnema" after his name. He has published several research papers in journals of international repute and supervised 11 research degrees.

Junaid Nazeer Shah, Natural Resource Conservation Section, Environmental Department, Dubai Municipality, Dubai, United Arab Emirates, is a wildlife Biologist with specialisation in Avian Ecology. He has am MSc and a PhD obtained in Wildlife Sciences and has worked for several governments and for non-governmental organisations (NGOs) in the United Arab Emirates and in India. He has experience of over 20 years of ecological conservation and environmental issues including 14+ years of expertise in the Arabian Peninsula for terrestrial and marine biodiversity in the areas of environmental compliances and audits, Environmental Impact Assessments (EIA), and Coastal Zone Management. His work includes long-term assessment and monitoring of breeding bird species of global and regional importance; understanding wild birds population dynamics in Arabian Gulf; Satellite Telemetry studies (> 120 PTTs on 14 bird species); population and conservation biology; - GIS, Remote Sensing, and species habitat modeling; alien invasive species; and urban biodiversity

Anam Shakil, MSc, Aligarh Muslim University, India, completed her master's degree in biodiversity and management from Aligarh Muslim University. Her primary areas of interest are in exploring insect diversity and the ecology of birds.

Swaima Sharif, Ph.D. Research Scholar, Aligarh Muslim University, in the Department of Zoology, Aligarh Muslim University. Her research area lies in the field of forensic entomology. Her research experience includes insects associated with animal (stray) and human carcass decomposition, developmental stages of the blow, and sarcophagid flies. She is currently working on the potential of natural weathering of cuticular hydrocarbons of empty puparial cases to estimate the Post Mortem Interval (PMI) in later stages of human and animal decomposition.

Sneha Singh, Ph.D., Research Scholar, Aligarh Muslim University, India in the department of wildlife sciences. working on "Comparative eco-toxicological study on the two sympatric harrier species in its Indian wintering ground" for her thesis. Her area of interest includes ecotoxicology, human-wildlife conflicts, and spatial landscape ecology.

Hilloljyoti Singha, Professor, Centre for Biodiversity & Natural Resources Conservation, Assam University, India, has been a professor and head of the department of Zoology, Bodoland University, Kokrajhar, Assam since 2019. He did his MSc in Zoology (specialised in animal ecology and wildlife biology) at Gauhati University, followed by a PhD in Wildlife Sciences at Aligarh Muslim University. He also worked as a research scientist in the Bombay Natural History Society. He has published about 73 research articles and ten book chapters. He has so far produced 13 PhD scholars and one MPhil student.

Intesar Suhail, Wildlife Manager, Department of Wildlife Protection, Jammu & Kashmir, India has been working as a wildlife manager since 1999 in important protected areas of Jammu and Kashmir and Ladakh, including Dachigam National Park, Hemis National Park, Overa-Aru Wildlife Sanctuary, Changthang Wildlife Sanctuary, Overa-Aru Wildlife Sanctuary, and Hirpora Wildlife Sanctuary.

xxiv   *Contributors*

Aisha Sultana, Wildlife Ecologist, Center of Degraded Ecosystem and Environmental Management, Delhi, has her doctoral degree in wildlife sciences from Aligarh Muslim University. Her deep interest lie in the conservation andecosystem restoration of degraded habitats of wild fauna. She has more than 15 years of professional experience in the restoration and management of ecosystems. She is presently working as Wildlife Ecologist under Biodiversity Parks Programmes. She has published more than 30 research articles in leading journals, conference proceedings, and books.

Subhechha Tapawini, Research Intern, Wildlife Institute of India, holds her master's degree in wildlife sciences from Amity Institute of Forestry and Wildlife Sciences. Her area of interest and focus lies in the sustainable development of flora and fauna.

Huda Tarannum, Aligarh Muslim University, India, did her MSc in wildlife sciences from the Department of Wildlife Sciences, Aligarh Muslim University, Aligarh. For her MSc dissertation, she worked on "Bird community structure in the restored and degraded area at Aravalli Biodiversity Park". At present, she is involved in school teaching.

Gopi Govindan Veeraswami, Scientist-E, Wildlife Institute of India, works as Scientist-E and is head of the department Endangered Species Management at the Wildlife Institute of India, Dehradun. Apart from his teaching responsibilities at M.Sc. level, he is also involved in research and focuses primarily on the ecology and dynamics of wildlife populations, especially in the context of management and conservation problems. His research since the year 2001 in the coastal and associated wetlands has involved studies on marine turtles, saltwater crocodiles, Indian skimmers, and colonial nesting water birds.

Preface

A big journey starts with small steps...and powerful experiences. Wildlife conservation has become subject to numerous national and international academic debates and bio-political regulations in the last few decades. In 2019, the Department of Wildlife Sciences, Aligarh Muslim University organised an International Conservation Conference (ICC-2019), which was attended by several researchers from various reputed government and non-government organisations in India, including ZSI (Zoological Survey of India), WII (Wildlife Institute of India), WTI (Wildlife Trust of India), Global Tiger Forum (GTF), and academic institutions, such as Aligarh Muslim University, The University of Delhi, University of Silchar Assam University, Hemchandracharya North Gujrat University, Sher-e-Kashmir University of Agricultural Sciences and Technology of Kashmir, and Baba Ghulam Shah Badshah University. Some of the research studies presented at the conference inspired and sparked the idea of developing a book on an international forum to enlighten the world about the diverse species of India and their ecology through real-world instances, covering various aspects of the wildlife of India. A book that will serve as a single window of information for its readers, shedding light on the challenges that emerge while conserving wildlife species that need to flourish.

This book, *Case Studies on Wildlife Ecology and Conservation in India*, differ from the other books available on the market as it encompasses various aspects of wildlife ecology and conservation efforts in India as part of sustainable development. Each section of the book is written thoroughly to pique the interest of its readers and urge them to learn more about the topics. The book's primary goal is to introduce its readers to the wildlife of India and gratify their curiosity with real-world examples. Since wildlife ecology is a voluminous subject, we might not have covered the entire gamut. However, we tried to put together as much information as possible. The book, therefore, comprises a unique combination of different but informative studies from eminent scholars in a very comprehensive manner. The first part of the book discusses resource partitioning among species during the unavailability or shortage of natural resources when two species depend on the same resources and distribute them among themselves for their growth and survival. The book's midsection will try to make its readers aware of the repercussions of anthropogenic interventions, leading to habitat fragmentation, habitat loss, and change in land use for developmental activities like urbanisation. This creates an imbalance in the ecosystem and leads to human-wildlife conflicts. The last part of the book describes landscape ecology as a management tool in the long-term conservation of species at a large spatial scale. Thus, each chapter of the book is significantly distinct and suggests reliable scientific solutions to existing challenges that arise during the conservation of the same species in a similar type of habitat.

In developing the book, we, as editors encountered several challenges in avoiding confusion, and discontent among readers. The idea of writing a book had come, still the problem was how to divide all 21 research studies thematically and what themes should be taken into account to cover as many aspects of wildlife in India as possible to make the book worthwhile. It was a challenging and massive task to pick some studies from the collection of the best research, with case studies that are unique and informative so that readers do not get bored even after reading them repetitively. As all chapters of the book are part of research studies conducted by various researchers of different institutes and organisations, coordinating with them together, from seeking authors to collecting precise information from them, was another challenge for us while collating the information to complete the book within the stipulated time. Since wildlife encompasses a wide range of topics and is interdisciplinary, compiling 21 distinct case studies under a single umbrella without compromising the book's integrity, authenticity, or relevance was one of our biggest challenges while designing the book. The book's sole aim is to gain popularity among readers, mostly wildlife enthusiasts who appreciate the aesthetic beauty of nature and feel indebted to mothernature. This book will significantly benefit scholars, researchers, and scientists. Apart from the scientific community and academicians, this book will be helpful for lawmakers, economists, administrators, wildlife managers, policymakers, and even sociologists to develop more effective conservation plans. We hope it will serve as a desktop reference book for years.

It gives us immense pleasure to place this book in the hands of scientific communities working in the area of wildlife. The book could not have been published without the substantial contribution of innovative ideas from the enthusiastic researchers and the support of the publishers. Last but not least, we express our gratitude towards our publisher, Taylor & Francis, for showing confidence in our book.

Acknowledgments

In the name of supreme power, the most gracious and merciful, we bow our heads in reverence to him, whose benign benediction gave us enough zeal to complete this endeavour.

Case Studies on Wildlife Ecology and Conservation in India is the culmination of more than three years of hard work by many individuals and organisations. We are overwhelmed in humbleness and gratefulness when acknowledging our deep appreciation to all those who have supported us in putting this idea well above the level of simplicity and into something concrete. We sincerely appreciate and acknowledge the contribution of Prof. Tariq Mansoor, Vice-chancellor, Aligarh Muslim University as the patron of the International Conservation Conference. We could organise the ICC-2019 because of his generous support and patronage. In particular, we are grateful to Prof. M. Parvez and Prof. M. Sajjad, Center for Advanced Study, Department of History, Aligarh Muslim University, Aligarh, who shared the initial idea with us that inspired this book initiative just after the conclusion of the ICC-2019 conference held at the Department of Wildlife Sciences, Aligarh Muslim University, and encouraged us to take the first step to make our dream of coordinating something as big as a book into reality.

We highly appreciate and acknowledge the foresight, diligent support, and understanding of the publisher, Taylor & Francis Group, and Routledge staff for their enthusiastic response to our book proposal and efficient handling of publishing the book with their international hallmark of excellence. We extend special thanks to Hanna Ferguson, the subject editor, for her painstaking efforts in reviewing the draft proposal of this book and for providing us with invaluable assistance, constant guidance, encouragement, and scientific acumen at all stages, leading to the successful publication of this book. Many thanks to Katie Stokes, editorial assistant, for providing all the technical assistance during the compilation of the book. We thank the Routledge team for completing this project in such a short period.

We, as the editors, would like to acknowledge the solicitous contributions of all the anonymous reviewers who spent their valuable time providing constructive criticism to improve the quality of the chapters. We wish to thank the reviewers of the different manuscripts for their unceasing dedication and significant input during the peer-review process.

This book is a collaborative effort. The editors would like to express their gratitude to all the distinguished authors and co-authors for condensing information and analysis on the huge range of wildlife ecologies and conservation in their specialised fields. We appreciate their patience in revising the first draft of the chapters after assimilating the suggestions and working diligently to bring the book to fruition. Above all, for providing us with the essential text, figures, and graphs as per the requirements of the book to be published before the deadline.

We would also like to acknowledge the teaching and non-teaching staff of the Department of Wildlife Science, Aligarh Muslim University, India, for being our academic partner. We are grateful to the chairperson of the department and other faculty members for their advice, practical support, and friendliness that have sustained and fed our work. We also wish to thank the support staff at the department who do a great job of keeping everything clean, functioning, safe, and organised.

We are thankful to our conference sponsors, Birla Sanskriti Trust, Treo Trend, and Karm Chand Thaper, for providing us with funds under their CSR (Corporate Social Responsibility) to organise the International Conservation Conference 2019, from which we got the inspiration for this book. We hope to continue our relationship and trust in the near future.

We want to acknowledge all our beta readers for taking the time and making an effort not just to read our draft book but also to send detailed suggestions and feedback. Our account has benefited hugely from the thoughts, insights, comments, and corrections of the people who read drafts with generosity and care. We particularly acknowledge Dr. Sharad Kumar, Dr. Ekwal Imam, and Mr. Tariq Aziz, and express our heartfelt appreciation for their invaluable insights and feedback on this project.

Research scholars are the backbone of the department. We are equally indebted to the research scholars Talat Parveen, Sneha Singh, Shahzada Iqbal, Farah Akram, and Mohsin Javed for their exceptional research assistance and help, meticulously copy-editing and checking through the manuscript prior to submission and gathering various information and compiling it from time to time to make this book unique.

The book has served as a point of continuity in the lives of its editors throughout its long gestation, despite significant changes and several personal circumstances. As always, such work is carried out borrowing family times. OI acknowledges the patience and the maturity shown by her teenage daughter Injila Parwez who has at times single-handedly taken up the household responsibilities while OI is away from home, spending long hours in her office editing the manuscripts. OI is also grateful to her siblings Dr Farah Ashher, Dr Hamrah Siddiqui, Mobussera Ilyas and Iram Raina for their constant moral support. AUK acknowledges his spouse Mumtaz Fatima and children Saadullah Khan, Asadullah Khan and Amaanullah Khan. OI thanks to Prof. M. Parwez her spouse for the lengthy discussions about this book's conceptual, methodological, theoretical, and empirical framework to proofreading this entire book, there is nothing he did not do to help her get over the finish line.

Dr. Orus and Prof. Afifullah Khan would like to thank Mr. Sheikh Zafarul Haque Tanveer, additional director of Income Tax (Investigation), Kolkata, for being a silent supporter in the process, without whom the work was impossible to be accomplished. This work is dedicated to their smiles.

That said, we apologise in advance, first for any errors that may have crept into the work, and second, to anyone who assisted with the project but whose contributions we have overlooked above. For reasons of anonymity and space, we are very grateful to them all, though it is, unfortunately, impossible to name them all here for reasons of anonymity and space. Compiling and developing this book has been a rich and rewarding process and a lot of fun! We thank each other for making it so.

1 Introduction

Orus Ilyas

Introduction

Wildlife ecology and conservation are the bedrock of environmental science and are inextricably tied to human existence. The layer of living organisms that occupy the surface of the earth and seas is the most intricate, dynamic, and diverse character of our planet. This amazing, rather unique feature is undergoing dramatic transformation due to pervasive human interventions resulting in climate change. It is one of the major concerns today.

Through the collective metabolic activities of Earth's living organisms, the biosphere, atmosphere, geosphere, and hydrosphere combine to form one environmental system that sustains millions of species, including humans. Obtaining wildlife conservation status is one of the major factors that can have a significant positive and enormous impact on our climate.

Wildlife ecology and conservation are part of biodiversity conservation that are often confused with unmanaged ecosystems, such as wildlands or national parks. However, managed systems, such as plantations, farms, croplands, aquaculture sites, rangelands, or even urban ecosystems, have their biodiversity. Given that cultivated systems account for more than 24% of the earth's terrestrial habitat any decisions concerning biodiversity ecosystem services must address the maintenance of biodiversity in these largely anthropogenic systems.

There are 17 mega-biodiversity countries identified by the World Conservation Monitoring Centre (WCMC) of the United Nation Environment Program. These mega-biodiversity countries house the most significant indices of biodiversity, including several endemic species. This concept was first proposed in 1988 by Russell Mittermeier. Although this work only accounts for about 10% of the earth's surface, the mega-diverse countries house at least 70% of the planet's terrestrial biological diversity, including more than two-thirds of all non-fish vertebrae species and three-quarters of all the higher plant species. The 17 mega-diverse countries are Australia, Brazil, China, Colombia, Ecuador, the United States, Philippines, India, Indonesia, Madagascar, Malaysia, Mexico, Papua New Guinea, Peru, Democratic Republic of Congo, South Africa, and Venezuela.

India is the home of 7.6% of all mammalian species, 12.6% of all avians, 6.2% of all reptilians, 4.4% of all amphibians, 11.7% of all fishes, and 6.% of all

DOI: 10.4324/9781003321422-1

flowering plant species. With only 2.4% of the earth's land area, it accounts for 7–8% of the world's recorded species (17,30,725 species).

India is among the signatory countries to the Convention on Biological Diversity (CBD), 1992. In 2010, the regulatory parties of the CBD adopted the Strategic Plans for Biodiversity 2011–2020, with 20 Achi Biodiversity Targets. Target 11 aims, by 2020, to conserve at least 17% of terrestrial and inland water areas and 10% of coastal and marine areas through effectively and equitably managed, ecologically representative, and well-connected systems of protected areas and other effective area-based conservation measures. The ethical implications of conserving biodiversity are based on what we owe to the millions of plants, animals, and microbial species with whom we share the planet. We must recognise that every species has its intrinsic value, even if it is not presently economically valuable to us. It's a moral obligation to look out for their well-being and to ensure their sustainability. With this view, the proposed title of this book, *Case Studies of Wildlife Ecology and Conservation in India*, is an attempt to compile a total of 21 insightful studies to enlighten different aspects of biodiversity and its application in conservation.

It will benefit environmental scientists to understand the monitoring and conservation of wildlife better. The book has been thematically defined into three parts, starting from Part I, "Resource Selection in Protected and Non-Protected Areas." The distribution of wildlife (animals and plants) is governed by the availability of resources (food, water, and shelter). When the resources are in short supply due to loss of habitat, man-animal conflict arises, as discussed in Part II. Part II, "Human-Wildlife Interaction," describes conflict due to urbanisation and encroachment of forest cover, the home of wild species. The growing needs of an increasing human population have shrunk the forest and left the forest cover in a fragmented state. The fragmentation of the forest is making animal movement very difficult. Keeping these issues in mind, management has to be done at a landscape level. Landscape ecology has been discussed in the last part of the book, Part III, "Landscape Ecology and Conservation." All three parts are the thematic classifications of the process of wildlife conservation.

All the chapters of Part I mainly focus on the resource selection by various species in different regions. Whether carnivores prefer certain prey species or herbivores show habitat preferences they are both based on the availability of food resources to sustain the expanding population of the species. A few chapters in this part also deal with the coexistence of closely related sympatric species through niche partitioning for the same resources. In addition, this part also comprises studies dealing with species diversity of the Suru Valley, Ladakh, and its management implications. The differences in the vocalisation characteristics of the Indian robin and its role in marking territories are discussed in detail.

Part I: Resource Selection in Protected and Non-Protected Areas

After the introductory chapter, the second chapter of this part, entitled "Prey Preferences of the Large Predators of Asia," investigates resource selection in several Asian large predator species. Asia supports the world's second most diverse large predator guilds, with iconic species such as tigers, *Panthera tigris*; leopard, *Panthera pardus*; snow leopard, *Panthera uncia*; dhole, *Cuon alpinus*; wolve, *Canis lupus*; bear, *Melursus ursinus*; and jackal, *Canis aureus*. The authors have reviewed the prey preferences of several species in these large predator guilds to emphasise the critical dietary needs for their conservation. Except for dholes, the researchers concluded that there is limited preferred prey overlap, with all predators exhibiting individual prey preferences. The majority of the species were stable in numbers, with the exception of the sambar deer. They are the sole preferred prey of dholes. Thus, this suggests that dhole conservation could be jeopardised or may be threatened by a declining prey base.

The next chapter in this part, Chapter 3, entitled "Habitat Use Pattern of Indian Leopard in Western Himalaya," aims to understand the ecology of leopards in mountainous habitats by radio collaring three different leopards. The home range of leopards was estimated, and it was discovered that the male leopard had a larger home range than the female. Leopards also prefer the densely forested broad-leaved forest of the Dachigam terrain.

The fourth chapter, "Insights on the Diet of the Golden Jackal," discusses the diet composition of the golden jackal. This study was carried out in the Sariska Tiger Reserve, Rajasthan. The different prey species preferred by the golden jackal include a wide range of vertebrates. It includes large mammals, reptiles, insects, crustaceans, and some fruits. Also, the diet preferences of golden jackals were discovered to be influenced by several factors.

In Chapter 5, "Blackbuck in Agricultural Landscape of Aligarh," the authors evaluated the status of blackbucks. This is a baseline study to estimate the population of blackbuck in the *Prosopis juliflora-* and *Prosopis cineraria*-dominated regions of the study area. It was revealed that a small area (approx. 63 hectares) of Palasallu had a mean group size of 108.9 ±2.76. The dispersion of species can be influenced by the availability of food resources in a particular area, and it varies with the changing environment and the region. The assessment reveals a substantial percentage of blackbucks in the area, suggesting a challenge for wildlife managers to ensure the long-term conservation of blackbucks in a human-dominated area due to the unavailability of food resources to meet the demand of an expanding population.

In the sixth chapter, entitled "Seasonal Variations in the Diet of Four-Horned Antelope," the authors describe the seasonal variation in feeding habits of four-horned antelope. The study was carried out using micro-histology technique based on pellet groups to assess the feeding habits of species. The results suggest that four-horned antelope prefer dicots over monocots in general. The most significant and preferred plant species were identified, and management recommendations were made in accordance.

The next chapter of this part deals with the niche partitioning for space between two sympatric species of sparrows. The seventh chapter, "Do the Niches of Sympatric Sparrows Differ?," explains the differences in one or more niche dimensions in closely related, ecologically similar species. For example, niche differences occur between house sparrows and Eurasian sparrows, which also share a common ancestor. Similarly, the authors discovered that house sparrows occupied a wider range of areas than tree sparrows in Guwahati, Assam. At a macro-scale, both species of sparrows differ in spatial distribution. And within that, they differ as per the urbanisation gradients at a micro-scale level.

In Chapter 8, "Avifauna of Suru Valley: Species Composition and Habitats," the authors document the new records of bird species in the Suru Valley of Kargil district, Ladakh. Of the 140 listed species, a total of 97 species have been recorded, including 12 new records from the Suru Valley. Also recorded the habitat preferences of the bird species on a macro-scale. It was discovered that uncontrolled tourist pressure might change the present scenarios of species composition. The authors also suggest mitigation measures to conserve the biodiversity value of the Suru Valley.

The ninth and last chapter of this part deals with the vocalisation of the Indian robin. In this chapter, "Indian Robin: Song Characteristics and Functions," ten different individual Indian robins were studied in the Chandrapur district of Maharashtra. It was revealed that the song repertoire was comprised of 2 to 11 types of strophes per individual. It was also discovered that the Indian robin delivers two modes of song, namely, stereotyped and eventual ways. Songs were mainly sung by males for territorial advertising.

The availability of substantial resources can help to avoid conflicts between humans and wildlife. However, due to the increase in the population, forests are being cleared to meet this demand. This has increased the rate of human-wildlife conflicts in the last few decades. To avoid conflicts, a sustainable approach is recommended to sustain the present population without depriving future generations of the use of forests. The emphasis in Part II of the book will be on the human-wildlife conflicts that arise due to the unavailability of sufficient resources and their mitigation measures.

Part II: Human-Wildlife Interaction

The first two chapters of this part discuss the existence of human-wildlife conflicts and the factors influencing the incidences of livestock depredation by tigers and leopards in and around the Corbett Tiger Reserve, Uttarakhand. Human-wildlife conflicts usually jeopardise wildlife conservation and sustainable use goals. To a great extent, wildlife conservation depends on local people' perception, which is directly related to the intensity of human-wildlife conflicts. In Chapter 10, "Human-Tiger Conflict in Corbett Tiger Reserve," the authors investigate the various factors influencing human-tiger conflicts in and around the Corbett Tiger Reserve and discover that the incidence of tiger attacks on livestock was higher during the monsoon, perhaps due to less visibility. As the incidences of human-tiger conflict have increased in and around the Corbett Tiger Reserve, mitigating measures have been proposed by the authors to ensure the tiger's long-term conservation.

In Chapter 11, "Ecological Drivers of Livestock Depredation by Large Predators," the authors evaluate the factors influencing livestock depredation by tigers and leopards in the study area. They reveal that both tigers and leopards are ambush predators, and vegetation composition predominantly affects their hunting success. In order to visualise the spatial hotspots of livestock kills across the study area, priority conflict sites were also identified. Simultaneously, besides the ecological perspective on human-carnivore interaction, recommendations are made by the authors to reduce livestock vulnerability to carnivores.

The next chapter of Part II, Chapter 12, "Is Resource Sharing Leading to Human-Bear Interaction? ," discusses food resource sharing between humans and sloth bears and the conflicts that arise from this interaction. The authors evaluated food sharing through food consumption, availability, and resource consumption by humans and sloth bears. The primary diet composition of sloth bears includes fruits in a dry deciduous climate during the cold season compared to food items like ants, termites, and honey. Fruits of the *Ziziphus* species form more than 50% of the sloth bear's diet, and local people collect these fruits and other forest products for their livelihood. However, the study also revealed that food resources are not the sole reason behind the human-sloth bear conflicts. The authors have raised their concerns about the overlapping of resource use by sloth bears and humans and suggested conservation measures to reduce the incidence of conflicts across the study area. Chapter 13, entitled "Bird Community Structure in Restored and Unrestored Areas," is a unique study discussing the conversion of degraded mining sites into restored biodiversity parks. The authors assessed bird community structure in both degraded and restored areas of Aravalli Biodiversity Park and recorded a total number of 103 bird species, 84 species using the species richness method, and 89 species through point count methods.

The majority of protected areas in India are subjected to varying degrees of livestock grazing pressure, having a range of detrimental consequences for protected areas and their biodiversity values. The next chapter, Chapter 14, "Impact of Grazing on Bird Community Structure in Binsar Wildlife Sanctuary, " discusses grazing as one of the factors responsible for habitat degradation and a decline in the populations of wildlife species. The authors tried to understand the impact of grazing on the bird community at Binsar Wildlife Sanctuary in the Almora district of Uttarakhand. The authors suggest several management recommendations to reduce the impacts of grazing on wildlife.

In nature, small organisms play an equally important role as larger ones in maintaining the ecosystem through pollination, seed dispersal, nutrient cycling, and so on, but they are often overlooked. In addition, small organisms play a significant role in wildlife conservation, especially in bird conservation. They provide food for many of the world's most beloved birds and keep other small organisms in check by feeding or parasitising them. The last two chapters of Part II deal with assessing of insect diversity and dipteran flies' role as forensic indicators. In Chapter 15, "Insect Diversity in a Semi-Natural Environment," the authors made efforts to assess the diversity of insects in this human-dominated landscape, and they recorded 80 species of insects representing ten orders and 47 families.

In Chapter 16, "Dipteran Flies Associated with Abuse and Neglect: A Forensic Indicator," the authors discuss the identification, characterisation, and colonisation through forensic analysis of some crucial insects that invade aged or older animals that are incapable of taking care of their health and hygiene. To determine the negligence, efforts have been made by the authors to develop a technical study through forensic entomology.

Part III: Landscape Ecology

The last part of the book is about landscape ecology and conservation. To avoid the human-wildlife conflicts that arise due to niche overlapping or resource sharing, it is essential to conserve species at the landscape level. Rather than having fragmented patches of forest, a prominent landscape is recommended that can be established through the interlinking of these trim patches, thus facilitating gene flow across the area and avoiding inbreeding.

The first chapter of this part, Chapter 17, "Conservation of Pheasants in Jammu and Kashmir: A Review," discusses the most colourful ground-dwelling bird, the pheasant. This review is based on the conservation status of pheasants in Jammu and Kashmir. The author reviewed a total of 50 species of pheasants found in Asia, with 17 species occurring in India and seven species found to have happened in Jammu and Kashmir. These seven species include three threatened species: the western tragopan, the rarest of the Himalayan pheasants; the Himalayan monal, *Lophophorus impejanus*; and the cheer pheasant. A detailed study on the ecology and behaviour of Galliformes, especially pheasants, is recommended by the author for the effective management and long-term conservation of these species in Jammu and Kashmir.

The next chapter is about identifying green space on university campuses to ecological balance and sustainable urban development. A total of 66 species of birds were recorded in "Avifaunal Diversity at Baba Ghulam Shah Badshah University Campus." The omnivorous feeding guild of birds had the most significant overall percentage, while the insectivorous feeding guild had the lowest. The two most dominant bird species in the study area are omnivores and insectivores.

The nineteenth chapter, "Distribution and Conservation of Kashmir Gray Langur," discusses the conservation status of grey langur at the landscape level in the Kashmir Valley. The authors have proposed the International Union for the Conservation of Nature (IUCN) extend the the species' ranges for better conservation and management.

Land use and land cover (LULC) maps are widely used in managing natural terrestrial ecosystems for various purposes, especially in conserving biodiversity. The following two chapters are about LULC classification and mapping of wetlands using remote sensing and GIS as a tool. In Chapter 20, "Forest Classification of Panna Tiger Reserve," the authors classify Panna Tiger Reserve in Madhya

Pradesh to illustrate the change in the cover and density of forest due to increases in the rate of deforestation for agricultural expansion to meet the demand of expanding populations.

Wetlands are essential refuges for birds, including migratory bird species, and need special attention for their conservation. However, they are mostly ignored in the Aligarh district of Uttar Pradesh. In Chapter 21, entitled "Wetland Inventory of Aligarh District," the authors identified and recognised 926 wetlands from the study area, and of the total wetlands, studied five in detail. The authors discovered 85 avian species belonging to 16 orders and 31 families in these five selected wetlands.

The last chapter of the book, Chapter 22, "Web of Socio-Economic Considerations for Wildlife Conservation in Manipur," suggests conservation measures at a landscape level. The authors discuss human-wildlife conflict and socio-economic impact due to the establishment of villages in the fragmented landscapes of Manipur. Through participatory risk mapping meetings, 11 risks were identified influencing the villagers' financial returns. In addition, wild boars were the most sighted among wild animals for crop damage, and to save the crops, villagers trapped the animals or used gunshots. Thus, the availability of wild animals was perceived to be declining. As per the villagers, training for alternative livelihoods was recommended for sustainable development.

This book, *Case Studies of Wildlife Ecology and Conservation in India*, deals with various challenges in wildlife conservation in India, making it different from other books that fall under the heading "Wildlife and biodiversity." The book has been categorised into three significant parts: resource selection in protected and unprotected areas, human-wildlife conflicts, and landscape ecology and conservation. Most of the chapters are field-based studies using various scientific methodologies. Therefore, each chapter of the book under each part is equally relevant.

The availability of resources plays a significant role in governing the dispersion of a species. Some important factors, such as food, water, shelter (escape from prey), space, and mates, determine the distribution of a species in a given space and time frame. Both carnivores and herbivores prefer habitats that contain preferred food items at a safer place. Resource selection is an important process to allow escape from predators with an abundant supply of food. However, due to an increase in urbanisation and industrialisation to meet the demands of an expanding human population, the rate of deforestation for various purposes has also increased, resulting in the reduction of forest cover and density. Because resources are limited in nature and the population is growing at faster rate, competition for the same resources has arisen between species or even among individuals of the same species for sustenance. Hence, it is important to understand the basic requirements of a species, such as adiet preference, habitat preference, and ecological interactions, for mitigating conflicts and for the conservation of the species in the long term through effective conservation measures.

To formulate effective conservation strategies, it is important to assess the environmental factors or threats associated with the habitat at a landscape level.

Some of the chapters of this book are also dedicated to such factors, which play a crucial role in determining the suitable environment for a species. For example, restored degraded biodiversity parks and small unidentified wetlands provide potential habitats for aquatic flora and fauna. The transformation of a degraded area provides a habitat to numerous species and acts as the lungs of a polluted city. Similarly, wetlands recharge the water table, making it suitable for various species as well.

The island biogeography theory, proposed by McArthur and Wilson in 1977, states that small, connected habitats sustain more diverse species than larger fragmented habitats because of the increased gene flow, avoiding inbreeding in the species. However, due to encroachment into forest areas and over-exploitation of natural resources, forests are being fragmented, resulting in no connectivity between the habitats. This habitat-fragmentation has led to the extinction of several species, such as the pink-headed duck (*Rhodonessa caryophyllacea*). Therefore, to develop conservation measures for any species (such as endangered, threatened, endemic, and vulnerable), it is necessary to conserve species at a broader scale to avoid the risk of becoming extinct.

Part I

Resource selection in Protected and Non-Protected areas

2 Prey Preferences of the Large Predators of Asia

Matt W. Hayward, Cassandra Bugir, Salvador Lyngdoh, and Bilal Habib

Introduction

Predator-prey interactions are an arms race between individuals seeking food and others seeking to avoid becoming food. The long evolutionary history of these interactions suggests that predators have evolved strategies to optimally forage on their prey (Charnov, 1976; Pyke, 1984).

An extensive body of literature has explored the optimal foraging of large carnivores to determine what they preferentially prey upon. These studies began in Africa on species such as lion *Panthera leo*, spotted hyena *Crocuta crocuta*, cheetah *Acinonyx jubatus*, and African wild dog *Lycaon pictus* (Hayward, 2006; Hayward et al., 2006a; Hayward & Kerley, 2005; Hayward, 2006b). The studies used Jacobs' electivity index (1974) to determine if a prey species was preferred or avoided by a predator. Africa's large predators exhibit unique prey preferences based on the body mass of the prey, where larger predators and group-hunters prefer larger prey than smaller or solitary hunters. Each predator preferentially preys upon at least one unique prey species, suggesting there is limited scope for redundancy within the African predator guild (Woodroffe & Ginsberg, 2005). This information has been successfully used to predict the diet, home range, and carrying capacity of Africa's large predator guild. (Hayward, 2009; Hayward et al., 2008, 2007a).

Reviews of the prey preferences of Africa's large predator guild have been conducted (Hayward & Kerley, 2008). These concluded that smaller carnivores exhibited the greatest overlap in actual and preferred diet and niche breadth (Hayward & Kerley, 2008), reflecting the higher degree of predation pressure smaller prey face (Sinclair et al., 2003). Species with a higher extinction risk, based on the IUCN Red List categories, had lower preferred and actual dietary niche breadths, which may explain why some predators are inherently scarcer than others (Hayward & Kerley, 2008).

In this review, we focus on Asia's large predator guild—tiger *Panthera tigris*, leopard, dhole, *Cuon alpinus*, and golden jackals—to investigate their commonalities and uniqueness, as well as comparing these with those of Africa's large predator guild. We first summarise what is known about the prey preferences of

DOI: 10.4324/9781003321422-3

Asia's predators and then explore the degree of competition within the guild, and similarities with the African guild.

Prey Preferences of Asian Carnivores

The tiger is the largest predator in Asia with a diet that reflects its size (Sunquist et al., 1987; Sunquist & Sunquist, 1997). Using 3,187 kill records of 32 prey species from all four extant tiger subspecies across seven range states, we found that tigers significantly prefer to kill wild boar (*Sus scrofa*) and sambar deer (*Rusa unicolor*), and are likely to significantly prefer red deer (*Cervus elaphus*) and bar-asingha (*R. duvaucelii*) with a larger sample size (Hayward et al., 2012). These preferences support the hypothesis of Sunquist et al. (1999) that tigers evolved in the Pleistocene following the radiation of cervids in south-east Asia. Tiger's prey preferences are driven solely by the body mass of prey—tigers prefer prey weighing between 60 and 250 kg, and there is a linear relationship between prey body mass (excluding mega-herbivores) and preference that illustrates tigers are Asia's apex predators (Hayward et al., 2012). This information has been used by India's National Tiger Conservation Authority (NTCA) to recommend increasing the abundance of tigers throughout the country by improving the available prey base (Jhala et al., 2011; Jhala et al., 2019).

Leopards have a diverse diet that enables them to live alongside apex predators such as tigers (McDougal, 1988) and even humans (Athreya et al., 2013). We used 33 studies from 41 spatiotemporally unique sites in 13 countries (including four sites in India) to document the prey preferences of leopards based on 8,643 kills of 111 species (Hayward et al., 2006). The most common prey of leopards in Asia was chital deer (*Axis axis*), accounting for 64% of kills where they occur, while langur monkeys (*Presbytis sps*) accounted for 12% (Hayward et al., 2006). Felids and canids are also frequently killed. A larger sample size is likely to see chital as significantly preferred by leopards (Hayward et al., 2006). The preferred weight of the prey for leopards' ranges from 15 to 45 kilograms (Clements et al., 2014), and, as a solitary predator, they have an ideal prey body mass slightly less than their own (Hayward et al., 2006).

Snow leopards occupy a range of 1.83 million km² and, using data from 25 studies involving 1696 scats, we showed that they are a specialised predator with a Levin's niche breadth of 0.29–0.54 (Lyngdoh et al., 2014). Siberian ibex (*Capra sibrica*), Himalayan tahr (*Hemitragus jemlahicus*), goats (*Capra hircus*), sheep (*Ovis aries*), rodents (*Rattus*), and horses (*Equus ferrus*) were killed by snow leopards wherever those prey species were found (Lyngdoh et al., 2014). Siberian ibex and blue sheep were the preferred prey of snow leopards, while Asiatic wild ass (*Equus hemionus*), Tibetan gazelle (*Procapra picti caudata*), and argali (*Ovis ammon*) were significantly avoided (Lyngdoh et al., 2014). The preferred prey weight range of snow leopards was 36–76 kg (Lyngdoh et al., 2014).

Dholes are a group hunting predator that use stamina to chase down prey. We used 24 studies from 16 sites throughout the distribution of dholes to determine the prey preferences of the species based on 8,816 kill records (Hayward et al., 2014). Sambar, chital, and wild boar (*Sus scrofa*) contribute almost two-thirds of the food biomass of the dhole, with sambar alone contributing 38% of the diet and are being significantly preferred (Hayward et al., 2014). It was discovered that dholes avoid wild boar, langur, Asian elephants (*Elephas maximus*), and peafowl (*Pavo cristatus*) (Hayward et al., 2014). Wild prey are preferred significantly more than livestock (Hayward et al., 2014). Sambar are at the upper end of the accessible prey spectrum (30–235 kg) of the dhole, and are marginally above the dhole's preferred weight range of 130–190 kg. Dholes exhibit a negative relationship between prey body mass and preference, signifying that they are not apex predators within Asia (Hayward et al., 2014). Although prey preferences are not affected by pack size, larger packs ultimately take larger prey (Hayward et al., 2014).

Eurasian golden jackals are much smaller than other members of Asia's predator guild. We reviewed 13 studies of golden jackal diets from 23 different spatiotemporal perspectives, reporting on 2,440 kill records to discover that they prefer brown hare (*Lepus europaeus*) (Hayward et al., 2017). Golden jackals significantly avoided cervids in general, plus langur monkeys, pheasants, and small mammals (Hayward et al., 2017). Golden jackals have a preferred weight range of 0–4 kg and a ratio of predator to preferred prey mass of 0.6, reflecting their solitary hunting strategy (Hayward et al., 2017). Body mass is the primary driver of golden jackal prey preferences, with some evidence suggesting that larger predators compete with golden jackals for access to wild boar and pheasants, which occur more frequently in the diets of jackals in the absence of apex predators (Hayward et al., 2017).

Humans are also predators of wildlife around the world. Hominids, including the ancestors of the modern human, are likely to have been important predators since our arrival in Asia 100,000 YBP. Using 19 studies from 9 countries reporting on 235,516 modern human hunter-gatherer kills of 40 different species, we found that hunter-gatherer communities in Asia still practising traditional livelihoods significantly avoid prey weighing less than 6.4 kg and kill prey above this weight in accordance with its availability (Bugir et al., 2021). The Indian crested porcupine (*Hystrix indica*), wild boar (*Sus scrofa*), and barking deer (*Muntiacus muntjak*) were strongly preferred, while Asian elephant (*Elephas maximus*), Asian golden cat (*Catopuma temminckii*), and emerald doves (*Chalcophaps indica*) were avoided (Bugir et al., 2021). The ratio of significantly preferred prey to human body mass ratio is 0.92, which suggests modern human hunter-gatherers are acting as solitary predators in their hunting strategies (Bugir et al., 2021). Extinct groups of anatomically modern humans preferred much larger prey, reflecting the impact of defaunation within Asia since the Pleistocene (Bugir, 2022).

The largest members of Asia's large predator guild exhibit extensive overlap in their preferred weight range (Table 2.1). Except for sambar, which is a preferred

Table 2.1 Summary data of the prey preferences of Asia's predator guild. Statuses of preferred prey are from the IUCN Red List (IUCN, 2018) with LC referring to species of least concern for extinction, and VU referring to vulnerable species. There is no preferred prey weight range for modern human hunter-gatherers in Asia as everything above 6.4 kg is killed in accordance with its availability within the prey community

Species	Ideal	Preferred (low)	Preferred (high)	Ratio	Preferred prey
Tiger	130	60	250	1.1	Wild boar (LC), sambar (VU), red deer (LC), barasingha (LC)
Human	40	—	—	0.92	Indian crested porcupine (LC), wild boar (LC), barking deer (LC)
Leopard	23	15	45	0.79	Chital (LC)
Snow leopard	55	36	76	1.29	Siberian ibex (LC), blue sheep (LC)
Dhole	200	130	190	15	Sambar (VU)
Golden jackal	4	0	4	0.6	Brown hare (LC)

prey of dholes and tigers, there is a little overlap in the prey preferences of the two species (Table 2.1).

Discussion

The accessible prey spectrum of Asia's predators exhibits extensive overlap between dholes, leopards, and tigers, and reflects extensive dietary competition within Asia's large predator guild (Hayward et al., 2014). According to Hayward et al. (2014), dholes and tigers preferentially prey on sambar, and leopards completely overlap in accessible prey with dholes. This is probably driven by the lower diversity of available prey in Asia compared to Africa, particularly in light of the "empty forests" that are beginning to dominate there (Redford, 1992). Variation in the prey preferences of hominids over thousands of years illustrates the impact of defaunation. The preferred prey of Asia's predators can be considered the key dietary resources required for their conservation. Fortunately, the status of these preferred prey species is generally secure (Table 2.1; IUCN, 2018); the suggestion of natural system modifications via a declining prey base generally does not threaten them. However, sambar are listed as vulnerable to extinction and are continuing to decline (IUCN, 2018) and are the key dietary resource for dholes and a preferred prey species of tigers (Table 2.1). This suggests tigers, and dholes in particular, may soon be further threatened by a declining prey base, in addition to habitat loss and persecution. Tigers appear more resilient to such threats than dholes because they also prefer wild boar (IUCN

Status = Least Concern), despite both dholes and tigers being considered as endangered (IUCN, 2018).

The tight relationship between preferred food resources and predator density in Africa (Hayward et al., 2007b) illustrates the importance of bottom-up factors in governing predator densities. The same relationships have been shown for tigers, albeit with large confidence intervals (Karanth et al., 2004), potentially due to the inclusion of non-preferred prey. Hence, we must ensure that focusing on expanding the area available for large predators does not come at the expense of ensuring there is a concomitant increase in available prey, particularly of the key prey resources required by Asia's large predators.

Acknowledgments: Matt Hayward thanks the organizing committee of the International Conservation Congress in Aligarh for inviting him to participate in the 2019 conference and to contribute to this publication.

References

Athreya, V., Odden, M., Linnell, J. D. C., Krishnaswamy, J., & Karanth, U. (2013). Big cats in our backyards: Persistence of large carnivores in a human dominated landscape in India. *PLOS ONE*, 8(3), E57872. https://doi.org/10.1371/Journal.Pone.0057872

Bugir, C. (2022). Prey preferences of hominids. Unpublished PhD thesis, In School of Environmental and Life Sciences. pp 140. University of Newcastle.

Bugir, C. K., Peres, C. A., White, K. S., Montgomery, R. A., Griffin, A. S., Rippon, P., Clulow, J., & Hayward, M. W. (2021). Prey preferences of modern human hunter-gatherers. *Food Webs*, 26, E00183. https://doi.org/10.1016/j.fooweb.2020.e00183

Charnov, E. L. (1976). Optimal foraging, the marginal value theorem. *Theoretical Population biology*, 9, 129.

Clements, H. S., Tambling, C. J., Hayward, M. W., & Kerley, G. I. H. (2014). An objective approach to determining the weight ranges of prey preferred by and accessible to the five large African carnivores. *PLOS ONE*, 9(7), E101054. https://doi.org/10.1371/journal.pone.0101054

Hayward, M. W. (2009). Moving beyond the descriptive: Predicting the responses of top-order predators to reintroduction. In M. W. Hayward & M. J. Somers (Eds.), *Reintroduction of top-order predators* (p. 345). Blackwell Publishing.

Hayward, M. W., Hayward, G. J., Druce, D. J., & Kerley, G. I. H. (2008). Do fences constrain predator movements on an evolutionary scale? Home range, food intake and movement patterns of large predators reintroduced to Addo elephant national park, South Africa. *Biodiversity and Conservation*, 18(4), 887–904. https://doi.org/10.1007/s10531-008-9452-y

Hayward, M. W., Henschel, P., O'brien, J., Hofmeyr, M., Balme, G., & Kerley, G. I. H. (2006). Prey preferences of the leopard (*Panthera pardus*). *Journal of Zoology*, 270(2), 298–313. https://doi.org/10.1111/J.1469-7998.2006.00139.X

Hayward, M. W., Hofmeyr, M., O'brien, J., & Kerley, G. I. H. (2006a). Prey preferences of the cheetah (*Acinonyx jubatus*) (Felidae: Carnivora): Morphological limitations or the need to capture rapidly consumable prey before kleptoparasites arrive? *Journal of Zoology*, 270(4), 615–627. https://doi.org/10.1111/J.1469-7998.2006.00184.X

Hayward, M. W., Jędrzejewski, W., & Jêdrzejewska, B. (2012). Prey preferences of the tiger *Panthera tigris*. *Journal of Zoology*, *286*(3), 221–231. https://doi.org/10.1111/J.1469-7998.2011.00871.X

Hayward, M. W., & Kerley, G. I. H. (2005). Prey preferences of the lion (*Panthera leo*). *Journal of Zoology*, *267*(3), 309–322. https://doi.org/10.1017/S0952836905007508

Hayward, M. W., & Kerley, G. I. H. (2008). Prey preferences and dietary overlap amongst Africa's large predators. *South African Journal of Wildlife Research*, *38*(2), 93–108. https://doi.org/10.3957/0379-4369-38.2.93

Hayward, M. W., Lyngdoh, S., & Habib, B. (2014). Diet and prey preferences of dholes (*Cuon alpinus*): Dietary competition within a Sia's apex predator guild. *Journal of Zoology*, *294*(4), 255–266. https://doi.org/10.1111/Jzo.12171

Hayward, M. W., O'brien, J., Hofmeyr, M., & Kerley, G. I. H. (2006b). Prey preferences of the African wild dog *Lycaon pictus* (Canidae: Carnivora): Ecological requirements for conservation. *Journal of Mammalogy*, *87*(6), 1122–1131. https://doi.org/10.1644/05-Mamm-A-304r2.1

Hayward, M. W., O'brien, J., Hofmeyr, M., & Kerley, G. I. H. (2007b). Testing predictions of the prey of lion derived from modeled prey preferences. *Journal of Wildlife Management*, *71*(5), 1567–1575. https://doi.org/10.2193/2006-264

Hayward, M. W., O'brien, J., & Kerley, G. I. (2007a). Carrying capacity of large African predators: Predictions and tests. *Biological Conservation*, *139*(1–2), 219–229. https://doi.org/10.1016/J.Biocon.2007.06.018

Hayward, M. W., Porter, L., Lanszki, J., Kamler, J. F., Beck, J. M., Kerley, G. I., Macdonald, D. W., Montgomery, R. A., Parker, D. M., Scott, D. M., O'brien, J. & Yarnell, R. W. (2017). Factors affecting the prey preferences of jackals (Canidae). *Mammalian Biology*, *85*, 70–82. https://doi.org/10.1016/J.Mambio.2017.02.005

IUCN. (2018). *2018 Red list of threatened species*. International Union for The Conservation of Nature and Natural Resources (IUCN). Retrieved November 6, 2019, from www.Redlist.org

Jacobs, J. (1974). Quantitative measurement of food selection. *Oecologia*, *14*(4), 413–417. https://doi.org/10.1007/Bf00384581

Jhala, Y. V., Qureshi, Q., & Gopal, R. (2011). Can the abundance of tigers be assessed from their signs? *Journal of Applied Ecology*, *48*(1), 14–24. https://doi.org/10.1111/j.1365-2664.2010.01901.x

Jhala, Y. V., Qureshi, Q., & Nayak, A. K. (Eds.). (2019). *Status of tigers, co-predators and prey in India 2018*. Summary Report. National Tiger Conservation Authority & Wildlife Institute of India.

Karanth, K. U., Nichols, J. D., Kumar, N. S., Link, W. A., & Hines, J. E. (2004). Tigers and their prey: Predicting carnivore densities from prey abundance. *Proceedings of the National Academy of Sciences*, *101*(14), 4854–4858. https://doi.org/10.1073/Pnas.0306210101

Lyngdoh, S., Shrotriya, S., Goyal, S. P., Clements, H., Hayward, M. W., & Habib, B. (2014). Prey preferences of the snow leopard (*Panthera uncia*): Regional diet specificity holds global significance for conservation. *PLOS ONE*, *9*(2), E88349. https://doi.org/10.1371/Journal.Pone.0088349

McDougal, C. (1988). Leopard and tiger interactions at Royal Chitwan national park, Nepal. *Journal of the Bombay Natural History Society*, *85*, 609–610.

Pyke, G. H. (1984). Optimal foraging theory: A critical review. *Annual Review of Ecology and Systematics*, *15*(1), 523–575. https://doi.org/10.1146/Annurev.Es.15.110184.002515

Redford, K. H. (1992). The empty forest. *BioScience*, *42*(6), 412–422. https://doi.org/10.2307/1311860

Sinclair, A. R. E., Mduma, S., & Brashares, J. S. (2003). Patterns of predation in a diverse predator–prey system. *Nature, 425*(6955), 288–290. https://doi.org/10.1038/Nature01934

Sunquist, M. E., Karanth, K. U., & Sunquist, F. (1987). Ecology, behavior and resilience of the tiger and its conservation needs. In R. L. Tilson & U. S. Seal (Eds.), *Tigers of the world: The biology, bio-politics, management and conservation of an endangered species* (pp. 5–18). Noyes Publishing.

Sunquist, M. E., Karanth, K. U., & Sunquist, F. (1999). Ecology, behavior and resilience of the tiger and its conservation needs. In R. L. Tilson & P. J. Nyhus (Eds.), *Tigers of the world: The science, politics and conservation of Panthera tigris* (pp. 5–18). Academic Press.

Sunquist, M. E., & Sunquist, F. (1997). Ecological constraints on predation by large felids. In J. Seidensticker, S. Christie, & P. Jackson (Eds.), *Riding the tiger: Tiger conservation in human-dominated landscapes* (pp. 283–301). The Zoological Society of London and Cambridge University Press.

Woodroffe, R., & Ginsberg, J. R. (2005). King of the beasts? Evidence for guild redundancy among large mammalian carnivores. In J. C. Ray, K. H. Redford, R. S. Steneck, & J. Berger (Eds.), *Large carnivores and the conservation of biodiversity* (pp. 154–176). Island Press.

3   Habitat Use Pattern of Indian Leopard in Western Himalaya

Athar Noor, Zaffar Rais Mir,
Gopi Govindan Veeraswami, and Bilal Habib

Introduction

The leopard (*Panthera pardus*) is the most successful cat species among all the four large felids in terms of their geographic distribution and abundance. It occupies variable environments as different as open and semi-arid deserts, savannahs, and tropical forests, reflecting the broadness of its ecological niche (Nowell & Jackson, 1996). In Africa and India, the leopard is sympatric with other large felids like lion (*Panthera leo*), tiger (*Panthera. tigris*), and canids like dhole (*Cuon alpinus*), by tolerating and adjusting itself spatial-temporally with large carnivores and synchronising its activity with its prey (Seidensticker, 1976; Bertram, 1982; Ramesh et al., 2012; Chaudhary et al., 2020). In other places, such as the mountainous regions of the South Africa and the Himalayas, where other such large carnivores are absent, the leopard becomes the apex predator, regulating the prey populations and, thus, maintaining the ecological balance (Noor et al., 2020). Leopards are the most adaptable large carnivore that can survive in a wide variety of habitat types, recorded even up to elevations of 5,200 m above mean sea level in the Himalayas (Jackson, 1984; Stein et al., 2016) but mostly below the tree line (Roberts, 1977; Green, 1987). The leopard's flexibility is due, in part, to its universal choice of diverse prey items and sustenance on medium to large ungulates, supplemented by a selection of small prey animals such as primates, rodents, reptiles, and birds (Ilani, 1981; Odden & Wegge, 2005; Hayward et al., 2006; Lovari et al., 2013a, 2013b).

However there are few studies on the leopard's home ranges in the Indian subcontinent except for the moist-temperate region of the western Himalaya, particularly the Kashmir region of India. The density estimates of leopards in this study area are lower than in other parts of the country because of the low prey abundance, even though there is quality habitat available (Noor et al., 2020). The main reason could be the region's more than two decades of armed insurgency, conflict, and political instability (Akbar, 2002; Bhatnagar et al., 2009) that led to forests to becoming empty of sufficient prey base (Noor et al., 2020). These factors, together with human-leopard conflict in the region (Singh et al., 2007; Mir, 2016), could have significant implications for the setting of reliable management and conservation strategies. Thus, estimation of robust home range

DOI: 10.4324/9781003321422-4

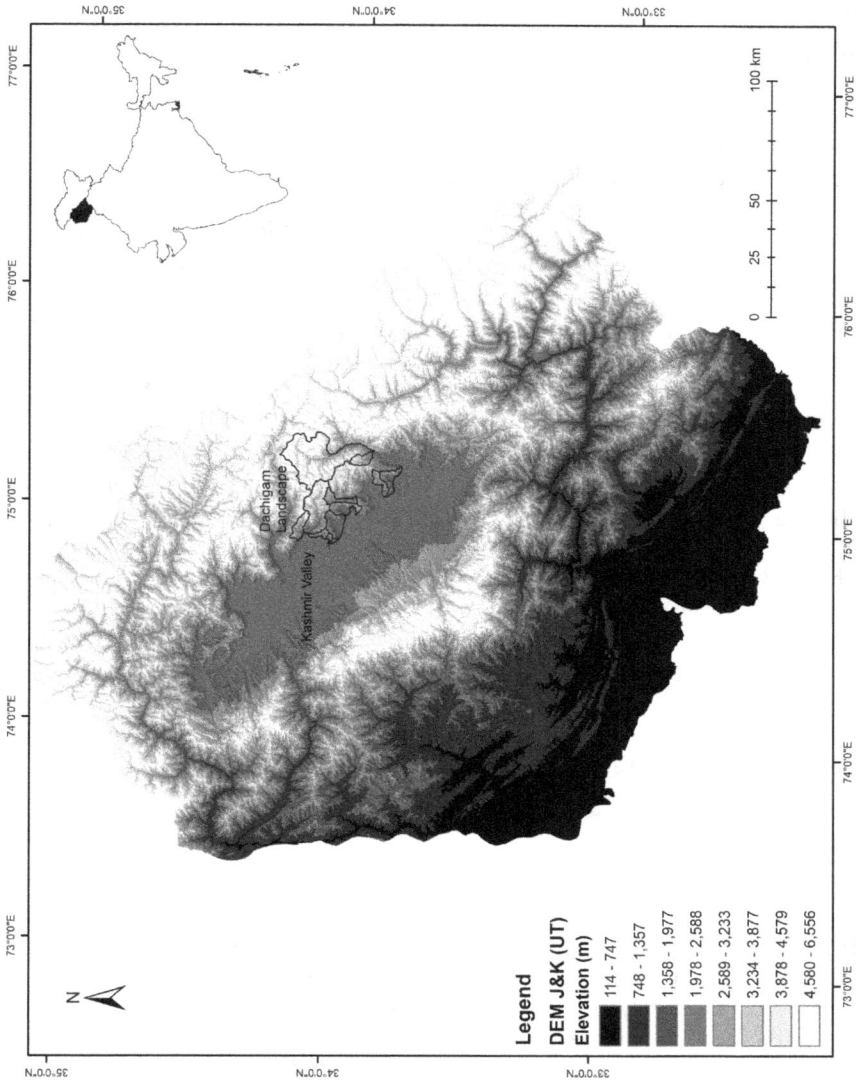

Figure 3.1 Dachigam landscape comprising several protected areas in Jammu and Kashmir.

estimates and a sound understanding of leopard spatial and habitat requirements are required, not only within the protected area (PA) system but also in areas outside of the PA system.

We used GPS location data from three collared leopards to evaluate their habitat preferences. We used this information to test if leopards prefer dense cover habitats as has been reported earlier e.g., Simcharoen et al. (2008), Karanth et al. (2009), and Kshettry et al. (2017).

Study Area

The Dachigam National Park (henceforth Dachigam NP; approximately 141 km²; 34°05′–34°12′ N and 74°54′–75°09′ E; 1,650–4,400 m above sea level) (Figure 3.1) is one of India's most important protected areas falling under the Northwest Himalayan biotic province (2A) of India (Rodgers et al., 2000). The climate in Dachigam is of the sub-Mediterranean type, with a bi-xeric regime having two spells of dryness between April and June, and September and November (Singh & Kachroo 1978). The area experiences irregular weather conditions with a considerable variation in the amount of precipitation. Snow is the main source of precipitation and, in some higher parts, melts by June. The park also serves as the catchment area of the famous Dal Lake and has received (between 2006 and 2013) an average annual rainfall of 734.9 mm (range = 577.2 mm–1,020.4 mm). There is no specific rainy season but four distinct seasons.

Administratively, the Dachigam NP has been divided into the Lower Dachigam in the west and the Upper Dachigam in the east. It is surrounded by several conservation reserves and wildlife sanctuaries that are contiguous and collectively form the Dachigam Landscape (Naqash & Sharma, 2011) (Figure 3.1). The vegetation is broadly classified as Himalayan moist-temperate (Champion & Seth 1968). The mammalian elements species recorded in Dachigam NP include the endemic and critically endangered Kashmir stag or Hangul (*Cervus hanglu ssp. hanglu*) and other endangered species such as Kashmir gray langur (*Semnopithecus ajax*), leopard (*Panthera pardus*), Kashmir musk deer (*Moschus cupreus*), Himalayan serow (*Capricornis thar*), Himalayan brown bear (*Ursus arctos*), Asiatic black bear (*Ursus thibetanus*), red fox (*Vulpes vulpes*), wild pig (*Sus scrofa*), yellow-throated marten (*Martes flavigula*), jungle cat (*Felis chaus*), leopard cat (*Prionailurus bengalensis*), and Indian porcupine (*Hystrix indica*).

Methodology

Data Collection

To evaluate the habitat use, and preference by of the leopard, three leopard individuals (two adult females and one adult male) were fixed with GPS collars during May 2011 and July 2013. The GPS location data generated from the collars were then used to estimate the habitat use and preferences. The GPS Plus collars (Vectronic Aerospace GmbH) were used to collect location data from the collared leopards. Each collar offered the ability to measure and store the position of the animal, the ambient temperature profile of the collar, and the activity of the collared animal. All the data was stored in the on-board non-volatile memory and was downloaded via cable, VHF/UHF link, or satellite.

The number of fixes that one sets the collars to record and the upload interval scheduled determine the battery life of the collars, therefore a trade-off was needed to find a suitable number of fixes and a good battery life. The GPS collars were scheduled to log GPS fixes for a maximum of five times a day at variable intervals set using three schedule rules (cyclic, discrete, and rollover) provided by Vectronic Aerospace. A minimum time interval of two hours was kept between two consecutive GPS positions recorded by the collars. Details of the live capture and home range analysis are available in Habib et al. (2014).

Home Range Analysis

Although there are several analytical methods to estimate the home ranges of collared individuals, the Minimum Convex Polygon (MCP) (Mohr, 1947; Nilsen et al., 2008) was used for the home range estimation. The simplest and also the most widely used home range estimator is 100% MCP, however it is mainly used for comparative purposes only. During habitat use, a 100% MCP polygon boundary encompasses all location fixes. This 100% MCP polygon boundary encompassing all location fixes was used in the habitat use evaluation.

Habitat Classification

The Global Landsat 8 data with a resolution of 30 m was used to generate a false-colour composite classification of the habitat types identified. A habitat map for the Dachigam landscape was generated by adopting a hybrid method of classification in the GIS domain using ArcGIS 10.2.2 and ERDAS 10.1, with an accuracy of 82%. Habitat mapping was done following Sharma et al. (2010) and Naqash and Sharma (2011), with reclassification of habitat types relevant in terms of leopard habitat use. These are:

Broadleaved Forest

This type of habitat is mainly a moist-temperate deciduous forest which is found at the bottom of the valley between 1,600 m and 1,900 m, with dominating tree species like *Parrotiopsis jacquemontiana*, *Morus alba*, *Salix sp.*, *Populus alba*, *Acer caesium*, *Juglans regia*, etc. This habitat also comprises riverine vegetation found along the vicinity of Dachigam nallah and the seasonal drainages where other broadleaved associates species such as *Quercus robur*, *Aesculus indica*, *Prunus cerasifera*, *Prunus tomentosa*, etc. are found. The shrub associates of this habitat type include *Rosa brunonii*, *Robinia sp.*, *Indigofera heterantha*, *Vibernum sp.*, and *Berberis sp.* The undergrowth comprises mainly *Alliaria sp.*, *Viola odoratae*, *Geranium sp.*,

Solenanthus circinatus, *Climatis grate*, *Vitis vinifera*, etc. This habitat type is the densest and has the highest canopy cover.

Conifer Mixed Forest

This habitat category represents temperate forest, comprising conifers and several other non-coniferous species of at lower elevations (starting at approx. 1,900 m up to 2,400 m). The non-conifer tree species include *Juglans regia*, *Asculuas indica*, *Rus sp.*, and *Populus ciliate*, whereas *Prunus sp.*, *Berberis sp.*, *Vibernum sp.*, *Rosa webbiana*, etc. constitute dominant shrub species. The higher elevations of this habitat category (up to 2,700 m) are represented by conifer species like *Pinus wallichiana*, *Taxus wallichiana*, and *Cedrus deodara* in association with *Abies pindrow* as tree storey. The shrub layer is represented by species such as *Isodon plectranthoides*, *Indigofera heterantha*, *Rosa webbiana*, *R. macrophylla*, and *Viburnum sp*. This habitat type is dense with good canopy cover.

Temperate Grassland

This habitat mainly comprises *Parrotiopsis jacquemontiana*, *Prunus armeniaca*, *Celtis australis*, and *Ulmus wallichiana* as its main tree species, which are scattered throughout the habitat. This habitat forms scrublands in the area where the undergrowth includes several species of xerophytic plants such as *Indigofera heterantha*, *Rosa brunonii*, *Rosa webbiana*, *Lonicera quinqelocularis*, *Geranium nepalensis*, *Dipsacus mitis*, *Dactylis gloerate*, *Colochicum luteum*, *Stipa sibirica*, *Zizypus species*, *Fragaria verca*, and *Kocleuria cristata*. This habitat usually spreads from 1,900 m to 2,700 m elevation. This habitat type is quite open with sparse canopy cover.

Alpine

This habitat category comprises the sub-alpine Birch-Rhododendron forest (2,700 m to 3,500 m) and alpine pastures (above 3,500 m). The dominant species in this sub-alpine birch-rhododendron forest type include *Betula utilis*, *Rhododendron campanulatum*, *Syringa emodi*, *Lonicera discolor*, *R. anthopogon*, *Juniperus recurva*, etc., and the ground cover is dominated by herbs such as *Anemone obtusiloba*, *Sieversia elata*, *Aster thomsonii*, *Iris hookeriana*, *Fragaria vesca*, and *Stachys sericea*. The alpine pastures are composed of perennial mesophytic herbs with very little grasses, conspicuous amongst which are *Primula spp.*, *Anemone spp.*, *Fritillaria imperialis*, *Iris spp.*, and *Gentiana spp.*, with several members from the *Ranunculaceae*, *Cruciferae*, *Compositae*, and *Caryophyllaceae* plant families. This habitat type is moderately dense in the sub-alpine region with less ground cover and very poor cover in the alpine zone.

Orchard or Plantation

There are several plantations or orchards which that are near the periphery of the National Park. Mostly, cherries (*Prunus sp.*) and apples (*Malus sp.*) are cultivated in these orchards. The low-growing height of trees in these plantations provides fair cover and often attracts wildlife during the fruiting season.

Agriculture

This includes the open agricultural fields and wasteland patches in and around the Dachigam landscape. It has very poor or no cover at all.

Habitat Use by Leopard

Habitat use by leopards was estimated as the percent number of locations in each habitat type (White & Garrot, 1990). The 100% MCP home range represents the total area from within which an animal has the opportunity to choose different habitat types. Therefore, the availability of different habitat types to a leopard was computed as the area of a habitat within its 100% MCP home range in a GIS domain (Hooge & Eicnehlaub, 2000). We used the chi-square test to determine if a leopard used the available habitat types non-randomly. Leopard habitat selection was estimated by Jacobs' preference index (Jacobs, 1974) and calculated as:

$$D = (R - P)/(R + P - 2RP)$$

where R is the ratio of the locations found in a specific habitat type to the total number of locations, and P is the ratio of the area of a specific habitat to the total size of the home range (Jacobs, 1974). The value of selectivity (D), varies from -1 (avoided) through 0 (indifference) to +1 (preferred habitat).

Results

The collars deployed on the leopards worked well with a variable number of locations (Table 3.1). Maximum number of locations was obtained from the collar of female leopard, F74, because it was tracked for the longest period (438 days), whereas the other collared female leopard, F71, lost contact after 20 days, and

Table 3.1 Number of days GPS collared leopard individuals tracked, number of locations collected, and 100% MCP home range estimates of collared leopards in Dachigam National Park

Collared animal ID	No. of days	No. of locations	100% MCP home range (km²)
Female—F74	438	2312	94
Female—F71	20	160	-
Male—M73	78	543	145

only 160 locations could be obtained through this individual. Only one male leopard could be captured and collared, which was tracked for 78 days, yielding 543 locations, after which contact was lost with this leopard. In the incremental analysis of 100% MCP range sizes versus sample sizes, asymptotes were reached for female F74 and the male M73, suggesting adequacy of sampling only for F74 and M73. Therefore, the home range was not estimated for F71 and the location data were excluded from further analysis. The largest home range was that of M73 (145 km^2), followed by F74 (94 km^2) (Table 3.1). Since the amount of time F74 was monitored spanned more than a year, therefore seasonal habitat use was also computed apart from habitat use during the day and night.

A total of seven habitat types were identified as relevant for assessing leopard habitat use. Of these, six were related to vegetation and agriculture categories, and one was the human habitation category. The final output map of habitats classified in the Dachigam landscape is provided in Figure 3.2. Both collared leopards' home ranges (100% MCP) were overlaid on the habitat map for habitat use analysis.

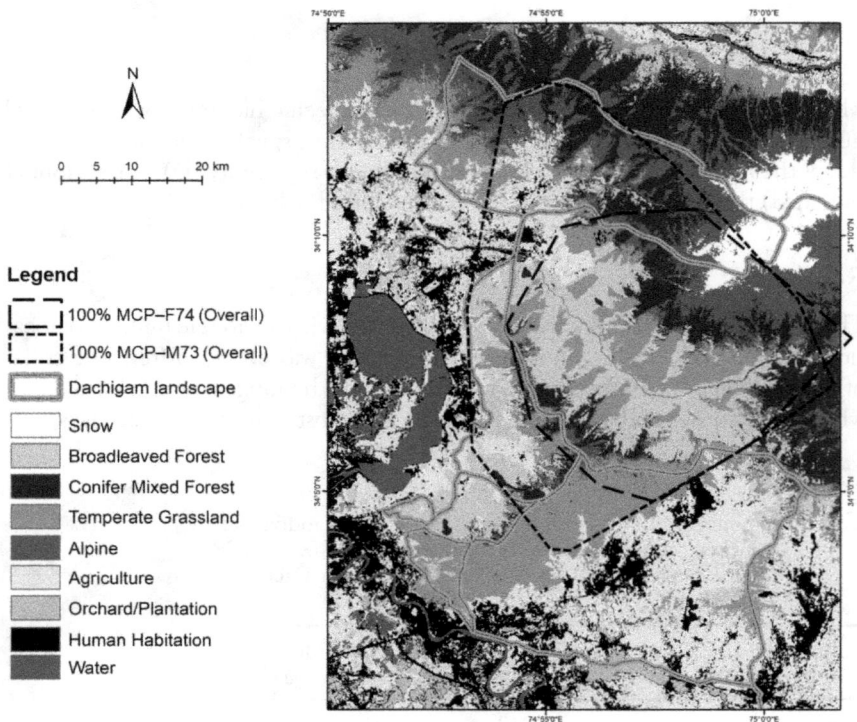

N

0   5   10   20 km

Legend

- ⌐‐‐¬ 100% MCP–F74 (Overall)
- ⌐‐‐‐¬ 100% MCP–M73 (Overall)
- ☐ Dachigam landscape
- ☐ Snow
- ▨ Broadleaved Forest
- ▪ Conifer Mixed Forest
- ▥ Temperate Grassland
- ▨ Alpine
- ☐ Agriculture
- ▨ Orchard/Plantation
- ▪ Human Habitation
- ▪ Water

Figure 3.2 Different habitat types of Dachigam landscape with the 100% MCPs of the two GPS collared leopards.

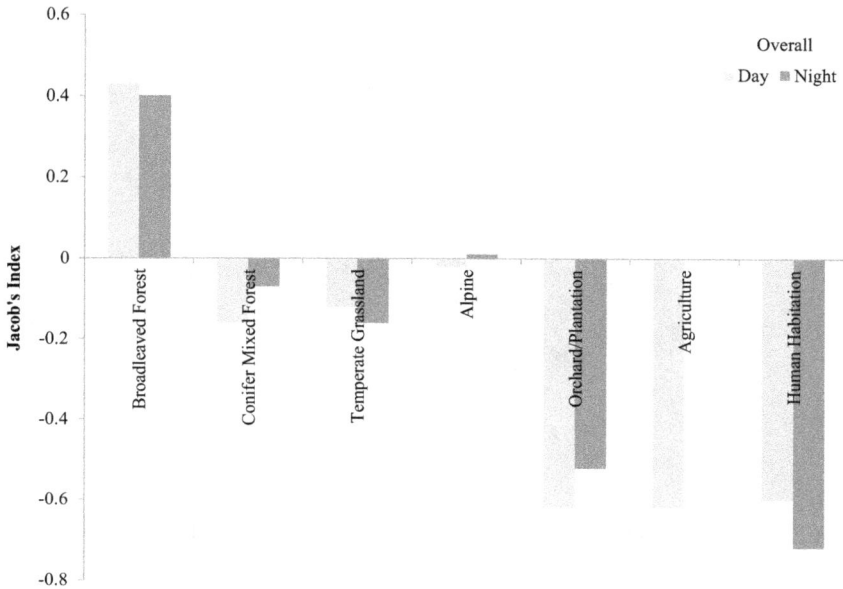

Figure 3.3 Overall day and night habitat preference of two GPS collared leopards (M73 and F74) in Dachigam landscape.

Overall, available habitat types were used non-randomly during the day ($\chi^2 = 1234.714$, df = 5, P = 0.000) and night ($\chi^2 = 765.809$, df = 5, P = 0.000) by leopards, and preference was shown for broadleaved forests in the Dachigam landscape (Figure 3.3). The male leopard (M73) used the available habitat selectively during the day ($\chi^2 = 84.394$, df = 5, P = 0.000) and night ($\chi^2 = 67.857$, df = 5, P = 0.000) showing preference for alpine habitats, including sub-alpine meadows the most, followed by broadleaved forests (Figure 3.4). Similarly, the female leopard F74 also used habitats selectively during the day ($\chi^2 = 1260.241$, df = 5, P = 0.000) and at night ($\chi^2 = 494.009$, df = 4, P = 0.000), showing preference for broadleaved forest (Figure 3.5). In terms of seasonal habitat use by F74, broadleaved forest was the preferred habitat during the three seasons, viz. winter, spring, and autumn, whereas this preference changed to sub-alpine habitats during the summer season (Figure 3.6). Both the leopards showed non-uniform usage (P = 0.000) of elevation ranges (Figure 3.7). Similarly, female leopard F74 also showed non-uniform usage (P = 0.000) of elevation range across the four seasons (Figure 3.8).

Discussion

Studies on the home ranges and habitat use of leopards in the mountainous ecosystems of the Himalayas are few, resulting in a very limited understanding of the aspects of leopard ecology. This study presents 100% MCP home ranges and

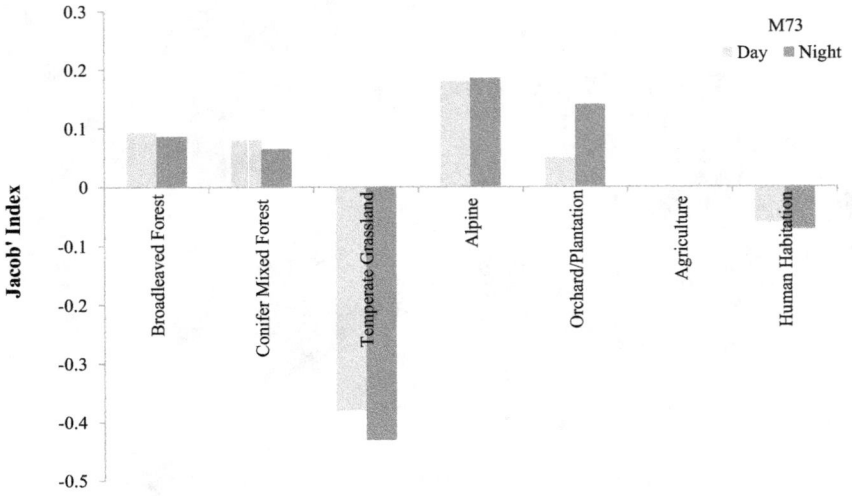

Figure 3.4 Day and night habitat preference of male GPS collared leopard (M73) in Dachigam land.

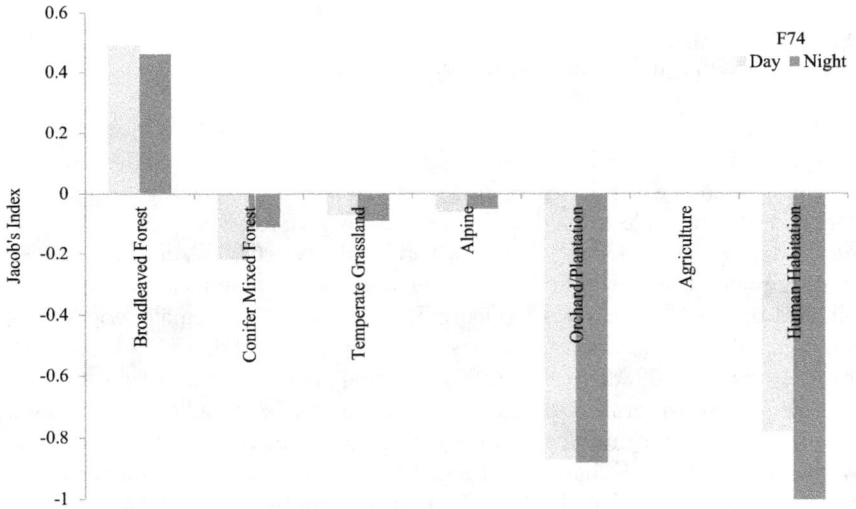

Figure 3.5 Day and night habitat preference by female GPS collared leopard (F74) in Dachigam land.

habitat use by leopards by using GPS locations from collared leopards in the moist-temperate forest of the Kashmir valley, western Himalaya. The home range estimates were larger compared to estimates reported from similar ecosystems in the Indian subcontinent (e.g., Seidensticker et al., 1990; Odden & Wegge, 2005;

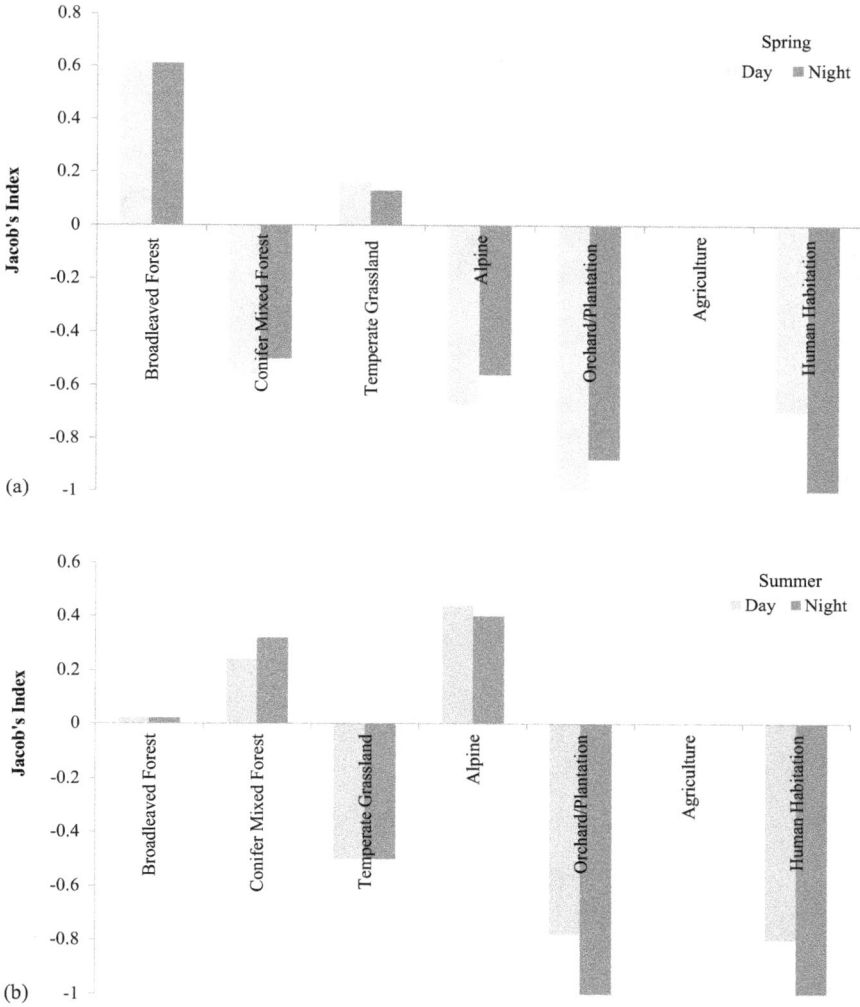

Figure 3.6 Seasonal habitat preferences of female GPS collared leopard (F74) in Dachigam land.

Khan et al., 2018). This large size of the home range estimate can be attributed to low prey availability in this study area, which is considered to influence the home range sizes of carnivores (Gittleman & Harvey, 1982; Saitoh, 1991). The collared leopards selected the broadleaved forest, which consisted of dense cover and high plant diversity, which is similar to that reported in other studies (e.g., Karanth et al., 2009; Athreya et al., 2015; Kshettry et al., 2017). This forest, in general, is the most diverse in terms of plant species and also very dense in canopy and ground cover. Considering the male leopard, it preferred the alpine habitat followed by

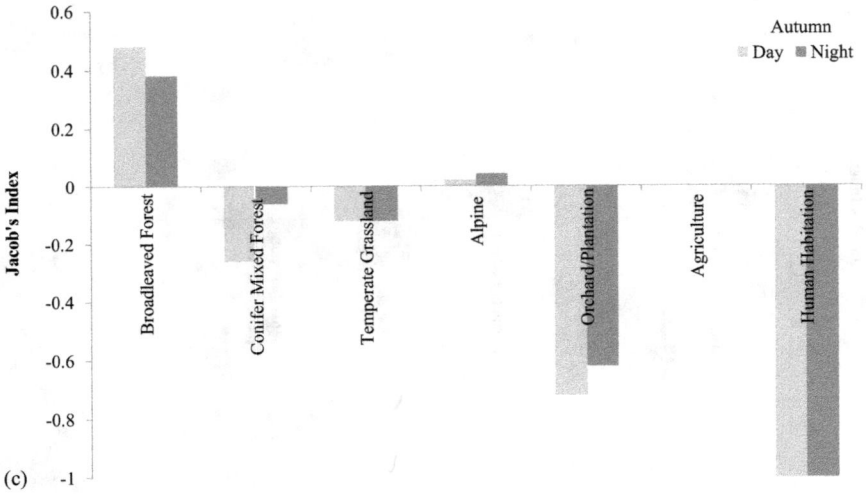

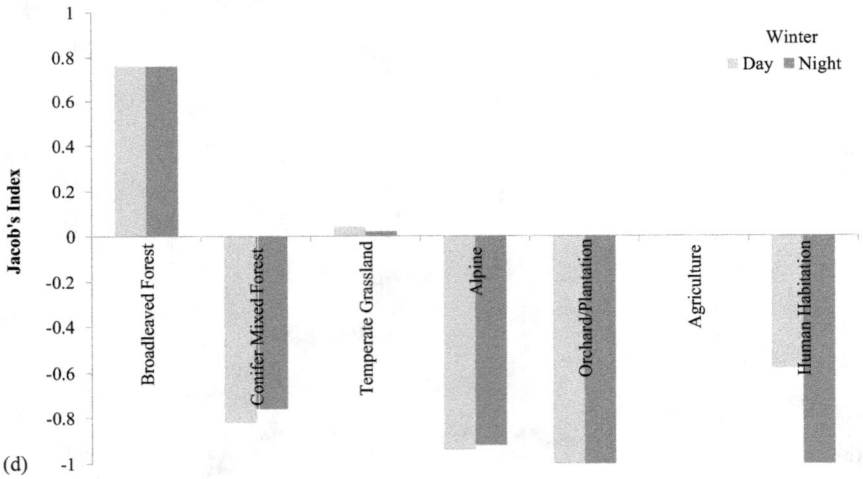

Figure 3.6 (Continued)

broadleaved and conifer mixed forests in Dachigam NP (Figure 3.4). As far as the preference of alpine habitat is concerned, it was mainly the sub-alpine areas within the alpine category that were used mostly, as evident from the elevation use presented in Figure 3.7. This habitat in the sub-alpine zone has dense to moderately dense cover and was used by this individual during the summer season. There were no differences in the preferences of habitat categories during the day and night by the male leopard, except for the orchard or plantation category, where M73 showed preference during the night time (Figure 3.4). There

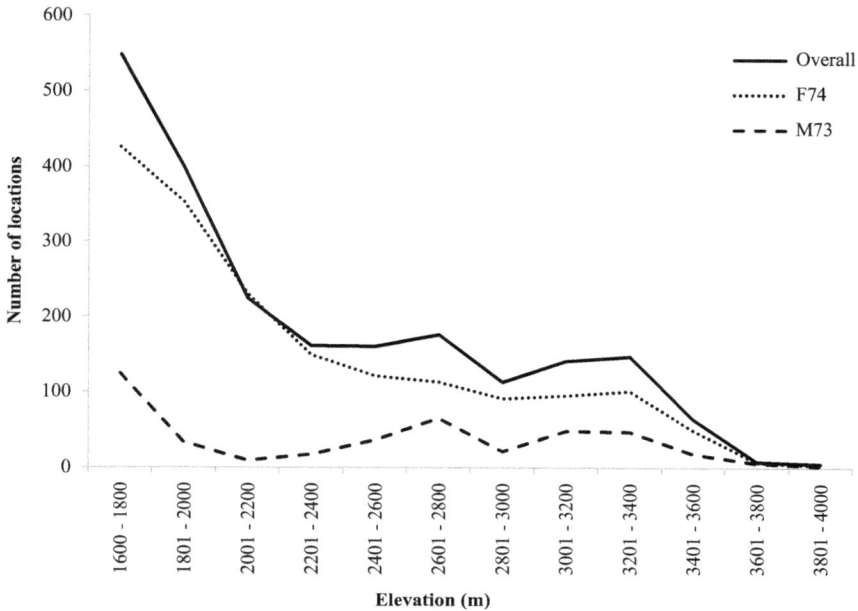

Figure 3.7 Overall and individual elevation range use by GPS collared leopards (M73 and F74) in Dachigam landscape.

are several orchards and croplands interspersed in the Dachigam landscape which provide cover and are frequented by wildlife (Charoo et al., 2011). As a result, therefore, use of these orchards by M73 at night could be a strategy to minimise human-leopard interaction. The male leopard was captured and tracked during the summer season, when the abundance of most of wild prey species, such as Kashmir gray langur (*Semnopithecus ajax*) and Kashmir stag (*Cervus hanglu ssp. hanglu*), was lowest (Mir et al., 2015; Mir, 2016) due to seasonal dispersal to higher elevations to forage (Sharma et al., 2010). In the case of the female leopard (F74), this individual mainly preferred the densest broadleaved forest available on the floor of the Dachigam valley, as clearly evident from Figure 3.5, even though almost all elevation ranges up to sub-alpine habitats were used (Figure 3.7). It is thought that the vegetation on the valley floor provides good cover and stealth for ambushes or stalking prey for leopards, allowing them to obtain prey (Gittleman & Harvey, 1982; Karanth & Sunquist, 1995, 2000), as the prey mainly inhabits forested habitats (Karanth & Sunquist 1995, 2000). Since this individual was tracked for more than a year (Table 3.1), seasonal variation in habitat preference was also studied. The dense broadleaved forest was the main habitat preferred across the four seasons, except during the summer (Figure 3.6), when other moderately dense areas of conifer mixed and alpine habitats were used, similar to the preference shown by M73. During the summer and autumn

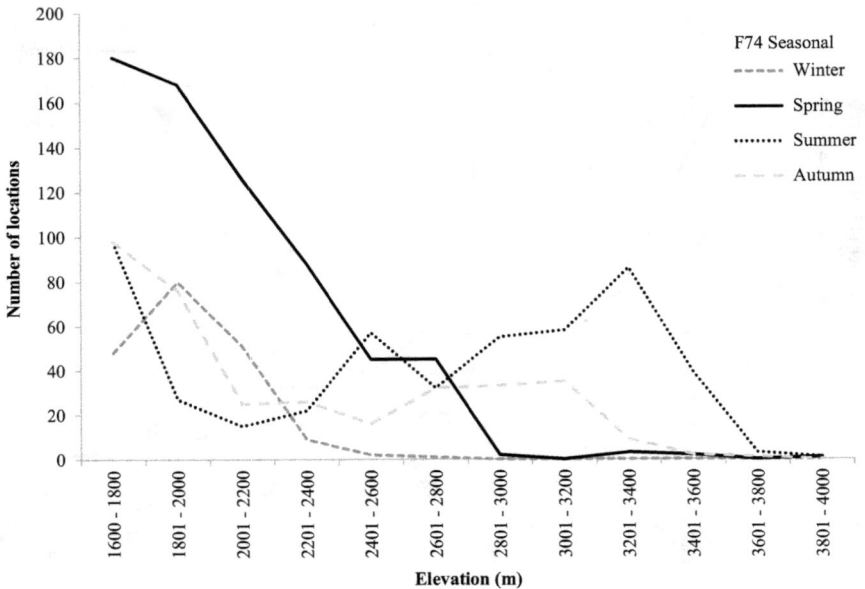

Figure 3.8 Seasonal elevation range usage by GPS collared female leopard (F74) in Dachigam landscape.

seasons, only the elevation used by F74 ranged from 3,000 m to 3,400 m, which corresponds to temperate and sub-alpine forests. Interestingly, leopards did not show any preference for areas of human habitation, although they have been recently reported to do so in other areas of India (Athreya et al., 2013; Odden et al., 2014). In the study area, leopards appear to cope with temporal and spatial variation in resources as their diets fluctuate with the seasons (Mir, 2016).

Conservation and Management Implications

This is the first time a study on home range size and habitat use has been conducted on the leopard population of the Dachigam National Park with its high altitude and rugged mountainous terrain, however, with sample size limitations. This study has also established an important baseline and benchmark which can serve as a foundation for a long-term monitoring programme for wildlife as a whole and for leopards in particular in the Kashmir region. These baseline estimates for the only large cat in the area are essential for monitoring the effectiveness of conservation activities. While planning for the leopard conservation and management, the strategies should take into consideration the approaches to improving the natural prey base and suitable habitat conditions.

Acknowledgments: This work was part of a larger study on the ecological aspects of the leopard in the Kashmir Valley and was supported and funded by the Department of Science and Technology, Govt. of India. We are thankful to the Department of Wildlife Protection, Govt. of Jammu and Kashmir for granting us the necessary permissions to conduct this research in the protected area. We also thank the Director and Dean of the Wildlife Institute of India for motivation and support. Thanks are due to the field assistants who assisted us in data collection.

References

Akbar, M. J. (2002). *Kashmir: Behind the vale*. Roli Books.

Athreya, V., Odden, M., Linnell, J. D. C., Krishnaswamy, J., & Karanth, U. (2013). Big cats in our backyards: Persistence of large carnivores in a human dominated landscape in India. *PLOS ONE, 8*(3), e57872. https://doi.org/10.1371/journal.pone.0057872

Athreya, V., Srivathsa, A., Puri, M., Karanth, K. K., Kumar, N. S., & Karanth, K. U. (2015). Spotted in the news: Using media reports to examine leopard distribution, depredation, and management practices outside protected areas in Southern India. *PLOS ONE, 10*(11), e0142647. https://doi.org/10.1371/journal.pone.0142647

Bertram, B. C. R. (1982). Leopard ecology as studied by radio tracking. *Symposium Zoological Society London, 49,* 341–352.

Bhatnagar, Y. V., Ahmad, R., Kyarong, S. S., Ranjitsinh, M., Seth, C., Lone, I. A., Easa, P., Kaul, R., & Raghunath, R. (2009). Endangered markhor Capra falconeri in India: Through war and insurgency. *Oryx, 43*(3), 407. https://doi.org/10.1017/s0030605309001288

Champion, H. G., & Seth, S. K. (1968). *A revised survey of the forest types of India*. Government of India Publication.

Charoo, S. A., Sharma, L. K., & Sathyakumar, S. (2011). Asiatic black bear–human interactions around Dachigam National Park, Kashmir, India. *Ursus, 22*(2), 106–113. https://doi.org/10.2192/ursus-d-10-00021.1

Chaudhary, R., Zehra, N., Musavi, A., & Khan, J. A. (2020). Spatio-temporal partitioning and coexistence between leopard (*Panthera pardus fusca*) and Asiatic lion (*Panthera leo persica*) in Gir protected area, Gujarat, India. *PLOS ONE, 15*(3), e0229045. https://doi.org/10.1371/journal.pone.0229045

Gittleman, J. L., & Harvey, P. H. (1982). Carnivore home-range size, metabolic needs and ecology. *Behavioral Ecology and Sociobiology, 10*(1), 57–63. https://doi.org/10.1007/bf00296396

Green, M. J. B. (1987). *The conservation status of the leopard, goral and serow in Bangladesh, Bhutan, northern India and southern Tibet* [Unpublished report]. World Conservation Monitoring Centre.

Habib, B., Gopi, G. V., Noor, A., & Mir, Z. R. (2014). *Ecology of Leopard (Panthera pardus) in relation to prey abundance and land use pattern in Kashmir Valley. Project completion report submitted to department of science and technology, Govt. of India*. Wildlife Institute of India, pp. 72.

Hayward, M. W., Henschel, P., O'Brien, J., Hofmeyr, M., Balme, G., & Kerley, G. I. H. (2006). Prey preferences of the leopard (*Panthera pardus*). *Journal of Zoology, 270*(2), 298–313. https://doi.org/10.1111/j.1469-7998.2006.00139.x

Hooge, P. N., & Eichenlaub, B. (2000). *Animal movement extension to ArcView Version 2.0. Alaska science center – Biological science office*. US Geological Survey.

Ilani, G. (1981). The leopards of the Judean Desert. *Israel Land and Nature*, 6, 59–71.

Jackson, R. (1984). The snow leopard. In *The plight of the cats: Proceedings of the of the meeting and workshop of the IUCN/SSC Cat Specialist Group at Kanha national park, Madhya Pradesh, India, 9–12 April 1984*. [Unpublished report] (pp. 197–198). IUCN/ SSC Cat Specialist Group.

Jacobs, J. (1974). Quantitative measurement of food selection. *Oecologia*, 14(4), 413–417. https://doi.org/10.1007/bf00384581

Karanth, K. K., Nichols, J. D., Hines, J. E., Karanth, K. U., & Christensen, N. L. (2009). Patterns and determinants of mammal species occurrence in India. *Journal of Applied Ecology*. https://doi.org/10.1111/j.1365-2664.2009.01710.x

Karanth, K. U., & Sunquist, M. E. (1995). Prey selection by tiger, leopard and dhole in tropical forests. *Journal of Animal Ecology*, 64(4), 439. https://doi.org/10.2307/5647

Karanth, K. U., & Sunquist, M. E. (2000). Behavioural correlates of predation by tiger (*Panthera tigris*), leopard (*Panthera pardus*) and dhole (*Cuon alpinus*) in Nagarahole, India. *Journal of Zoology*, 250(2), 255–265. https://doi.org/10.1111/j.1469-7998.2000 .tb01076.x

Khan, U., Lovari, S., Shah, S. A., & Ferretti, F. (2018). Predator, prey and humans in a mountainous area: Loss of biological diversity leads to trouble. *Biodiversity and Conservation*. https://doi.org/10.1007/s10531-018-1570-6

Kshettry, A., Vaidyanathan, S., & Athreya, V. (2017). Leopard in a tea-cup: A study of leopard habitat-use and human-leopard interactions in North-Eastern India. *PLOS ONE*, 12(5), e0177013. https://doi.org/10.1371/journal.pone.0177013

Lovari, S., Minder, I., Ferretti, F., Mucci, N., Randi, E., & Pellizzi, B. (2013). Common and snow leopards share prey, but not habitats: Competition avoidance by large predators? *Journal of Zoology*, 291(2), 127–135. https://doi.org/10.1111/jzo.12053

Lovari, S., Ventimiglia, M., & Minder, I. (2013). Food habits of two leopard species, competition, climate change and upper tree line: A way to the decrease of an endangered species? Ethology Ecology and amphibian. *Evolution*, 25(4), 305–318. https://doi.org/10.1080/03949370.2013.806362

Mir, Z. R. (2016). Monitoring prey dynamics and diet fluctuation of leopard (*Panthera pardus*) in Dachigam National Park, Srinagar, Jammu and Kashmir. Ph.D. dissertation submitted to Saurashtra University, Rajkot, Gujarat. 253 p.

Mir, Z. R., Noor, A., Habib, B., & Veeraswami, G. G. (2015). Seasonal population density and winter survival strategies of endangered Kashmir gray langur (*Semnopithecus ajax*) in Dachigam National Park, Kashmir, India. *SpringerPlus*, 4, 562–569. https://doi.org /10.1186/s40064-015-1366-z.

Mohr, C. O. (1947). Table of equivalent populations of North American small mammals. *American Midland and Naturalist*, 37(1), 223. https://doi.org/10.2307/2421652.

Naqash, R. Y., & Sharma, L. K. (2011). *Management plan (2011–2016) Dachigam National Park*. Department of Wildlife Protection, Govt. of Jammu and Kashmir, India.

Nilsen, E. B., Pedersen, S., & Linnell, J. D. C. (2008). Can minimum convex polygon home ranges be used to draw biologically meaningful conclusions? *Ecological Research*, 23(3), 635–639.

Noor, A., Mir, Z. R., Veeraswami, G. G., & Habib, B. (2020). Density of leopard in a moist-temperate forest of western Himalaya, India. *Tropical Ecology*, 61(3), 301–310. https://doi.org/10.1007/s42965-020-00090-w

Nowell, K., & Jackson, P. (1996). *Wild cats: Status survey and conservation action plan*. IUCN/SSC Cat Specialist Group, IUCN.

Odden, M., Athreya, V., Rattan, S., & Linnell, J. D. C. (2014). Adaptable neighbours: Movement patterns of GPS-collared leopards in human dominated landscapes in India. *PLOS ONE, 9*(11), e112044. https://doi.org/10.1371/journal.pone.0112044

Odden, M., & Wegge, P. (2005). Spacing and activity patterns of leopards *Panthera pardus* in the Royal Bardia National Park, Nepal. *Wildlife Biology, 11*(2), 145–152.

Ramesh, T., Kalle, R., Sankar, K., & Qureshi, Q. (2012). Spatio-temporal partitioning among large carnivores in relation to major prey species in Western Ghats. *Journal of Zoology, 287*(4), 269–275. https://doi.org/10.1111/j.1469-7998.2012.00908.x

Roberts, T. J. (1977). *The mammals of Pakistan.* Ernest Benn.

Rodgers, W. A., Panwar, H. S., & Mathur, V. B. (2000). Planning a wildlife protected area network in India. *Executive summary.* Wildlife Institute of India.

Saitoh, T. (1991). The effects and limits of territoriality on population regulation in grey red-backed voles, *Clethrionomys rufocanus bedfordiae. Researches on Population Ecology, 33*(2), 367–386. https://doi.org/10.1007/bf02513561

Seidensticker, J. (1976). On the ecological separation between tigers and leopards. *Biotropica, 8*(4), 225. https://doi.org/10.2307/2989714

Seidensticker, J. C., Sunquist, M. E., & McDougal, C. W. (1990). Leopards living at the edge of Royal Chitawan National Park, Nepal. In J. C. Daniel & J. S. Serraro (Eds.), *Conservation in developing countries: Problems and prospects* (pp. 415–423). Bombay Natural History Society, Bombay, India and Oxford University Press.

Sharma, L. K., Charoo, S. A., & Sathyakumar, S. (2010). Habitat use and food habits of Kashmir red deer or hangul (*Cervus elaphus hanglu*) at Dachigam National Park, Kashmir, India. *Galemys, 22,* 309–329.

Simcharoen, S., Barlow, A. C., Simcharoen, A., & Smith, J. L. (2008). Home range size and daytime habitat selection of leopards in Huai Kha Khaeng Wildlife Sanctuary, Thailand. *Biological Conservation, 141*(9), 2242–2250. https://doi.org/10.1016/j.biocon.2008.06.015

Singh, G., & Kachroo, P. (1978). *Plant community characteristics in Dachigam Sanctuary, Kashmir.* Natraj Publishers, Publications Division.

Singh, S. U., Singh, R., Satyanarayan, K., Seshamani, G., Suhail, I., & Parsa, M. A. (2007). *Investigation on human-leopard conflict in Jammu and Kashmir.* Wildlife SOS.

Stein, A., Athreya, V., Balme, G., Henschel, P., Karanth, U., & Miquelle, D. (2016). Panthera pardus. The IUCN red list of threatened species. http://dx.doi.org/10.2305/IUCN.UK.2016-1.RLTS.T15954A50659089.en.

White, G. C., & Garrott, R. A. (1990). *Analysis of wildlife radio-tracking data.* Harcourt Brace Jovanovich.

4   Insight on the Diet of the Golden Jackal

Pooja Chourasia, Krishnendu Mondal,
Kalyanasundaram Sankar, and Qamar Qureshi

Introduction

The golden jackal (*Canis aureus*) is a medium-sized canid, considered the most typical representative of the genus Canis, and is one of the 37 species of Canidae (Moehlem, 1983) that occur in the world. The golden jackal is widely distributed from East Africa through the Middle East and into South Asia. According to the IUCN Red List and the Indian Wildlife (Protection) Act, 1972, they are placed at lower risk and in Schedule II, respectively. Three Indian races of jackal are recognised: Canis aureus *aureus*, Canis aureus *indicus*, and Canis aureus *naria* (Prater, 1980). The Jackal is a generalist predator and occurs in a variety of habitats, from savannah and woodland in protected areas (Moehlman, 1983; Fuller et al., 1989) to farmland around human habitation (Jeager et al., 2001). In areas, particularly where large predators have been eliminated, they are the most abundant carnivores (Yom-Tov et al., 1995; Krystufek et al., 1997). The opportunistic nature of golden jackals is illustrated by their universal diet and their ability to flourish where human garbage is abundant (Yom-Tov et al., 1995). Food utilisation is an important aspect in the study of carnivore ecology, since trophic resources dominate several aspects of their biology (Macdonald, 1983; Bekoff et al., 1984). Scat analyses are widely used for determining the food habits of carnivores. However, the degree to which the relative frequencies of identifiable remains represent the actual proportion of prey types eaten is not clearly known. The golden jackal (*Canis aureus* L.) is a food-opportunistic canid, showing high flexibility in hunting strategies according to the actual food availability (Macdonald, 1983; Chourasia et al., 2012). Its diet patterns vary across its wide distribution range (Demeter & Spassov, 1993, Yom-Tov et al., 1995, Krytufek et al., 1997, Jhala & Moehlman, 2008), where its food spectrum ranges from plants, arthropods, and reptiles to birds, lagomorphs, rodents, and poultry; it preys upon young domestic or wild ungulates but also scavenges on dead animals. Jackal scat analyses present problems because of the great disparity in size and weight of prey eaten by them. Johnson and Aldred (1982) and Weaver and Hoffman (1979) have documented the differential digestibility of smaller mammals and the problems of enumerating small mammals in scats. For instance, insects were frequently eaten by the jackals; their contribution may be over-estimated when using a simple frequency of occurrence measure (Angerbjörn et al., 1999).

DOI: 10.4324/9781003321422-5

For terrestrial carnivores, various methods of interpreting scat-analysis data exist, such as, frequency of occurrence, volume estimation in scats, estimation of the mass of food items in scats and biomass calculation methods. However, the choice of method can have a significant impact on the results of dietary analysis and can lead to very different conclusions about a species' ecology. The frequency of occurrence has the least ecological significance, and results can be misleading (Klare et al., 2011). Whereas biomass calculation probably provides the best approximation to true diets when models were being available for the study species or for a closely related species with a similar food spectrum (Klare et al., 2011). The conversion factor for differential digestibility in coyotes is available (Weaver & Hoffman, 1979). No conversion factor is available for golden jackal to compute prey biomass from scats. Hence, we conducted feeding trials on jackal and intended to compare it with that of coyote.

The purpose of this study was to understand the dietary habits of jackals in a forested area (in the presence of other large carnivores) and to develop an equation predicting the prey consumed per collectable scat, following the procedure used by Jethva & Jhala (2004) for wolves. Jackal scats collected in the field were analysed using frequency of occurrence (expressed as a percentage of the total number of scats), numerical frequency (NF, expressed as a percentage of the total number of occurrences of all food items) (Corbett, 1989; Reynolds & Aebischer, 1991; Paltridge, 2002), "whole scat equivalents" (WSE was derived as a percent of the dry volume of prey in scats through visual estimation) (Angerbjorn et al., 1999, Elmhagen et al., 2002), and biomass calculation to demonstrate the possible magnitude of differences in results between these techniques. The Index of Relative Importance (IRI) (Pinkas et al., 1971; Short et al., 1999; Paltridge, 2002; Medina et al., 2008) was also used to determine the importance of different prey items in the diet of the golden jackal (Home & Jhala, 2009).

Materials and Methods

The study was conducted in the Sariska Tiger Reserve (Sariska TR) from April 2011 to March 2013. Sariska TR is situated in the Aravalli hill range in the semi-arid part of Western India (27°05'–27°33' N; 76°15'–76°35' E). The total area of the Tiger Reserve is 881 km², of which 273.8 km² is notified as the Sariska National Park. The vegetation of Sariska corresponds to northern tropical dry deciduous and northern tropical thorn forests (Champion & Seth, 1968). Plain areas are covered with scrub forests dominated by shrubs such as *Zizyphus nummularia*, *Capparis sepiaria*, *Capparis decidua*, *Adathoda vasica*, *Prosopis juliflora*, and *Acacia sp.* The valleys are dominated by *Zizyphus mauritiana* in mixed forest. The gentle slopes are dominated by *Anogeissus pendula* forest, and the steep slopes are occupied by *Boswellia serrata* forest.

Other than golden jackal, the park supports various carnivores and omnivores such as tiger (*Panthera tigris*), leopard (*Panthera pardus*), striped hyena (*Hyena hyena*), jungle cat (*Felis chaus*), common mongoose (*Herpestes edwardsii*), small

Indian mongoose (*H. auropunctatus*), ruddy mongoose (*H. smithi*), common palm civet (*Paradoxurus hermaphroditus*), small Indian civet (*Viverricula indica*), ratel (*Mellivora capensis*), and prey species like chital (*Axis axis*), sambar (*Rusa unicolor*), nilgai (*Boselaphus tragocamelus*), common langur (*Seminopithecus entellus*), wild pig (*Sus scrofa*), Indian crested porcupine (*Hystrix indica*), rufous-tailed hare (*Lepus nigricollis ruficaudatus*), and Indian peafowl (*Pavo cristatus*) (Sankar et. al., 2009). There are 30 villages within Sariska TR, of which six are located in the intensive study area of 160 km². A large number of cows (*Bos indicus*), buffaloes (*Bubalus bubalis*), goats (*Capra hircus*), and sheep (*Ovis aries*) are kept by people living in villages.

Estimation of Prey Population and Scat Collection

Ungulates and ground bird densities were estimated using line transects (Anderson et al., 1979; Burnham et al., 1980; Buckland et al., 1993). Thirty transects were laid in the intensive study area. The length of transects varied from 2 km to 3 km. All transects (total length = 61.0 km) were walked thrice in the winter and summer seasons during the study period. The total effort was 183 km in each season. Rodent and shrew abundance was estimated using web trapping (Otis et al., 1978; Wilson & Anderson, 1985). Rodent trapping was conducted at 14 locations to assess seasonal changes in relative abundance. Each location was operated by a single trapping web using 41 Sherman traps (H. B. Sherman Traps, Inc., Tallahassee, Florida). Trapping was conducted for two years at each location, once in summer and winter, amounting to a total effort of 5740 trap-nights in each season. The data was analysed using DISTANCE 5.0 software (Buckland et al., 2004). Scats were collected from a total of 18 trails, each 5 km in length, which were systematically surveyed once a month for three years, amounting to a total effort of approximately 720 km, and also along roads and trails opportunistically.

Scat Analysis

Jackal scats were identified by their characteristic shape, size, and nearby tracks, and analysis was done using the method described by Mukherjee et al. (2004). Dietary components were identified at the species/taxa level based on pertinent anatomical elements such as hair, mandibles, scales, feathers, wings, fruit cuticles, and seeds. No attempt was made to identify rodents at species level from the remains of jackal scats. Hair remains were identified with the help of reference slides available at the Research Laboratory of the Wildlife Institute of India. The index of relative importance was calculated as: IRI = (Relative Frequency + Whole Scat Equivalent) frequency of occurrence. The results of the scat analysis were subjected to re-sampling with bootstrapping using SIMSTAT 2.0 (Peladeau, 2000) and randomising the original order of scat samples (1000 iterations) using the software Estimate S (Colwell, 1996).

Prey Biomass Consumption Using Feeding Trials on Jackal

Feeding trials were conducted on jackals at Jaipur Zoo, Rajasthan from January to March 2013 using the procedure followed by Floyd et al. (1978), Weaver (1993), and Jethva and Jhala (2004) for wolves. The whole carcasses of species such as rabbits (*Oryctolagus spp.*), goats (*Capra hircus*), sheep (*Ovis aries*), chickens (*Gallus gallus domesticus*), buffalo (*Bubalus bubalus*), and fruits (*Zizyphus sp.* and dates) were weighed and fed to the group of six jackals in captivity. In most of their range in India, jackals thrive on domestic livestock, and hence the feeding trials covered the entire range of weights for prey normally consumed by jackals, except for small rodents like rats and shrews. Rats and shrews could not be included in the trials because of the quarantine issues in the zoo environment. The digestibility was determined for different prey using the method described by Wachter et al. (2012). The moisture loss was estimated from a separated portion of dressed fresh carcasses of bait provided for 72 hrs. The carcasses were accurately weighed every 12 hours and left exposed to the environment. Percent weight loss from the entire carcass and from a separated portion of bones and hide was regressed against time to esti-mate moisture loss (Jethva & Jhala, 2004). Biomass consumed was calculated as the numerical frequency of prey remains multiplied by the correction factor y, i.e., the number of collectable scats produced per prey. Data on utilisation (from scat analysis) and availability (from distance sampling analysis) were compared using an index of selection for each prey species by applying Ivlev's index (Ivlev, 1961).

Results

Estimation of Prey Abundance

Scat Analysis

A total of 14 prey species were recorded in 737 jackal's scats during 2010–2013. These were ungulate species (chital, sambar, nilgai, and wild pig), small mammals (hare, rodent), livestock (cow, buffalo, and goat), and birds (peafowl, francolin, and quail). 68% of scats contained more than one food type (mean number of food types per scat: 2.28±0.04 SE). The proportion of prey occurrences versus the cumulative number of scats stabilised between 160 and 170 scats, suggesting adequacy of sampling.

In terms of frequency of occurrence, numerical frequency, WES and IRI scores, grass was the most abundant in jackal scat, followed by nilgai, chital and fruit remains (mainly *Ziziphus* sp.) (Table 4.1 and 4.2). The presence of grass/grass seeds in the scat is consistent with other studies mostly concluding a possible role in scouring the intestine during digestion (Kalle, 2013).

Estimation Biomass Consumption Using Feeding Trials on Golden Jackal

Of the scats produced, the percentage of non-collectable scats ranged from 2.6% to 18.4%, with an average of 9.9%. The consumption of carcasses by jackals in 15

Table 4.1 Importance of prey categories in the diet of golden jackal (*Canis aureus*) in Sariska, based on the Index of Relative Importance (IRI)

S. No.	Food item	IRI
1	Vegetation	1114.19
2	Chital	694.47
3	Nilgai	662.96
4	Fruit	510.11
5	Cattle	276.50
6	Rodent	266.36
7	Hare	244.36
8	Bird	195.54
9	Sambar	121.15
10	Insect	86.87
11	Grass seeds	64.77
12	Goat	61.64
13	Reptile scale	25.80
14	Wild pig	1.60

Table 4.2 Food items recorded in scats of the golden jackal (*Canis aureus*) from Sariska, Rajasthan

Food Item	% Frequency of occurrence	% Numerical frequency	% Whole scat equivalence
Vegetation	37.72 (34.2–41.2)	16.56	14.83 (11.7–15.7)
Nilgai	26.73 (23.5–29.9)	11.73	13.07 (11.1–14.9)
Chital	26.32 (23.1–29.5)	11.55	12.98 (12.7–16.9)
Fruit	24.69 (21.5–27.8)	10.84	9.82 (8.4–11.2)
Cattle	17.50 (14.7–20.2)	7.68	8.11 (6.6–9.6)
Rodent	17.10 (14.3–19.8)	7.50	8.08 (6.6–9.6)
Hare	16.15 (13.5–18.8)	7.09	8.05 (6.5–9.6)
Bird	15.20 (12.6–17.8)	6.67	6.20 (6.5–9.6)
Sambar	11.26 (8.9–13.5)	4.94	5.81 (4.5–7.2)
Insect	10.72 (8.5–12.9)	4.71	4.38 (2.7–5.2)
Grass–seeds	9.36 (7.5–11.5)	4.11	3.40 (2.1–3.5)
Goat	7.87 (5.9–9.8)	3.45	2.81 (3.1–5.6)
Reptile scale	5.97 (4.2–8.0)	2.62	1.70 (1.2–2.2)
Wild pig	1.22 (0.4–2.0)	0.54	0.77 (0.2–1.3)

feeding trials ranged from 25.7% to 100%, with an average of 79.4%, and was negatively correlated with the size of prey ($r = 0.72$, $P = 0.001$). Loss of moisture from the prey bait carcass accounted for 0.5% (in chicken) to 50% (in buffalo) weight loss in 72 hrs. Moisture loss was recorded as less for birds and fruits, whereas it was high for rabbits and buffalo. The maximum consumption by jackals over a 24 hr period was 132 g/jackal. The conversion of prey body mass into a number of scats and their weights excreted by an individual jackal after prey consumption was inferred accurately using seven quantities (Q) as proposed by Wachter et al. (2012) (Table 4.3).

Table 4.3 Prey species and prey body mass provided to jackal during 15 feeding trials, prey mass-consumed and scats excreted by jackals, prey mass consumed per collectable scat, and digestibility of prey species

Prey/food items	Prey provided No.	Prey provided Mean kg (Q1)	Group size of jackal (Q2)	Prey consumed Total	Prey consumed mean/animal (Q3)	Scat Collectable	Scat Non collectable	Collectable scat/animal (Q4)	Prey consumed/collectable scat in kg (Q5=Q3/Q4)	Mean wt. of collectable scat/animal (Q6)	Digestibility % (Q7) (Q3-Q6)	(Q3-Q6)/Q3*100
Chicken	20	0.950[a]	6	14.7	0.123	68	6	0.567	0.216	0.073	0.050	40.8*
Chicken	20	0.980[a]	6	18	0.150	77	2	0.642	0.234	0.085	0.065	43.3*
Chicken	19	1.053[a]	6	19	0.167	80	1	0.702	0.238	0.056	0.111	66.6
Rabbit	19	1.105[a]	6	16	0.140	63	2	0.553	0.254	0.085	0.056	39.7*
Rabbit	15	1.200[a]	6	16	0.178	65	7	0.722	0.246	0.041	0.137	77.0
Sheep	1	20.800	6	14.8	2.467	60	5	10.000	0.247	0.044	2.422	98.2
Goat	1	22.000	6	17	2.833	54	3	9.000	0.315	0.045	2.788	98.4
Goat	1	32.000	6	13.5	2.250	38	5	6.333	0.355	0.357	1.893	84.1
Buffalo	1	35.000	6	9	1.500	27	3	4.500	0.333	0.222	1.278	85.2
Buffalo	1	39.000	6	19	3.167	46	4	7.667	0.413	0.124	3.043	96.1
Zizyphus	—	0.002[b]	6	6	0.000	67	—	0.004	0.090	0.036	—	—
Zizyphus	—	0.002[b]	6	5.8	0.000	69	—	0.004	0.084	0.028	—	—
Zizyphus	—	0.002[b]	6	4	0.000	47	—	0.004	0.085	0.013	—	—
Dates	—	0.002[b]	6	2	0.000	30	—	0.004	0.067	0.012	—	—
Dates	—	0.002[b]	6	3	0.000	41	—	0.004	0.073	0.014	—	—

Correction factor as determined by fitting logistic regression to consumed prey mass per collectable scat (Q5) as a function of mean prey body mass provided per feeding experiment Q1 ([a]number of chicken and rabbit with their total weight in kg were eaten entirely). Calculations were based on total weight of n animal/number of animals. Mean weight (kg) of collectable scats per jackal are based on mean weights of collectable scats, multiplied by the number of collectable scats, divided by the number of collectable scats per jackal ([b]two to six kg of fruits were provided in five trials were number of fruits ranged from 1177 to 2830). Since mean weight of each fruit was very less digestibility could not be calculated. *Species with low digestibility.

Digestibility was recorded less for birds (chickens) and more for heavy-weighted prey baits (sheep, goat, and buffalo). The equation fitted to the data is given as follows.

$$y = 0.003 \, x + 0.228$$

(p 0.001) (R2 = 0.825; F = 37.6; df = 1)

Consumption of prey biomass (y) per collectable scat increased with an increase in live body weight of prey (x).

In the above equation, fruits were not included, as their digestibility differed in comparison to other food items in the jackal diet. The equation provides a better prediction in cases where detailed scat analysis permits prey to be assigned to groups of different weight classes. As a result, data from similar weight class prey species were plotted separately, yielding three equations for higher weight class, lower weight class, and fruits (Figure 4.1). Data was corrected for prey species using a "common equation" for all prey classes, three different "weight-wise equations" for each prey class, and coyote's "differential detectability equation" (Weaver & Hoffman, 1979). These three equations were subjected to calculating three respective values of prey consumed per field collectable scat (y). The comparison revealed that the differences in values between the three methods were insignificant (Friedman test; 2 = 1.4, p = 0.49, n = 10). The number of scats produced for each prey species was calculated by dividing the average body weight of each prey species by (y). Also, to obtain utilisation values, prey biomass consumption was calculated using the frequency of occurrence, WSE, and numerical frequency. There was no significant difference observed between WSE and NF values (Wilcoxon test; z = –1.99, p = 0.06, n = 10), whereas a significant difference was observed with frequency of occurrence (Friedman test; 2 = 16.8, p = 0.00, n = 10). Data on utilisation (from scat analysis) and availability (from distance sampling analysis) were compared using an index of selection for each prey species by applying Ivlev's index (Ivlev, 1961). Numerical frequencies of prey items were used to estimate the utilisation values for the electivity index.

In the study period, the remains of *Zizyphus* fruits, rodents, hare, and chital were represented more than their availability in jackal scats, while goats, wild pig, cattle, sambar, and nilgai were represented less than their availability. The values of selection for chital and bird were calculated more for the index of data corrected for "all weight classes" (Chart A), as compared to the other two corrections (Charts B and C). In the index, for the data corrected "weight class wise" (Chart B), the selection value for hares was calculated more than for rodents. Chital and bird values were close to zero for the data corrected for "weight class" and "differential detectability" (Chart B and C), suggesting its representation with respect to their availability in jackal scat. However, the electivity index for prey items using three different methods of data correction was not significantly different from each other (p = 0.74) (Figure 4.2).

Finding food is one of the most challenging tasks for the survival of any species in any environment. This requires well-adapted strategies for food acquisition.

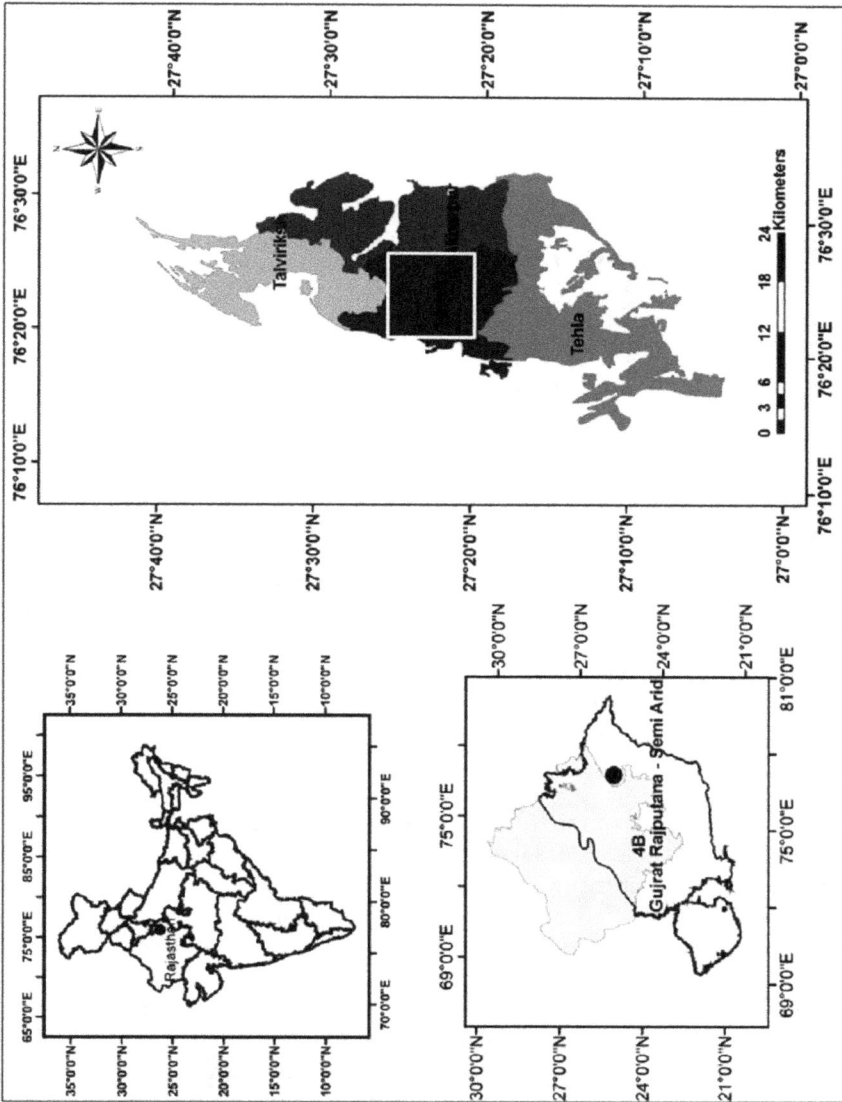

Figure 4.1 Location, administrative boundaries, and intensive study area in Sariska Tiger Reserve, Rajasthan.

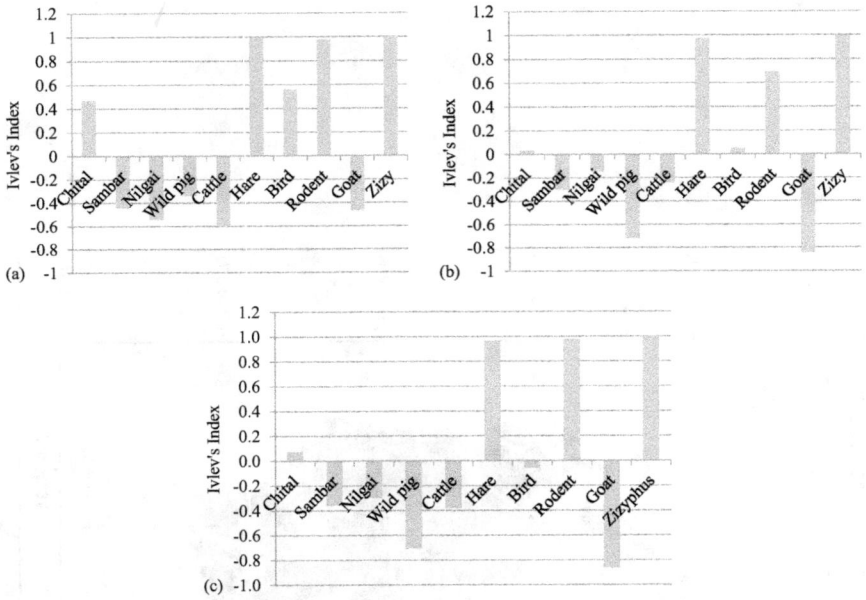

Figure 4.2 Prey selection by jackal in the study area based on availability of food items and utilisation through scat data in 2010–2013.

A. Electivity index corrected for '*all weight classes*':
   y = 0.003x + 0.228... Eq.1

B. Electivity index corrected for '*weight class wise*':
   a. Higher Weight Class: y = 0.006x + 0.136.. Eq.2
   b. Lower Weight Class: y = 0.114x + 0.116... Eq.3
   c. Fruits: y = 38.85x + 0.003... Eq.4

C. Electivity index corrected for '*differential detectability*':
   y = 104.23 log10 x – 111.43 .... Eq.5
   (Weaver and Hoffman, 1979)

   Note: Utilization values obtained using numerical frequency NF.

One strategy is to specialise in one exploitable source, while the alternative is to generalise. The golden jackal generally displays a general carnivore category, and the same is observed in the study area. Sariska Tiger Reserve is a matrix of open scrubs, barren lands, moderate to dense forest, human settlements, and agricultural land in and around it. In the present study, the responses of jackals to fragmented habitat were comparable with the ecological characteristics of the species reported from Greece (Giannatos et al., 2005), Israel (Borkowski et al., 2011), South Africa (Humphries et al., 2016), and Europe (Salek et al., 2014). As meso-predators, the potential food availability increases for jackals as dietary items previously unattainable (large herbivore species, e.g., sambar, cattle) are killed by apex predators, which could be a possible reason for their fair representation in scats (Table 4.2). The remains of livestock were present in the collected scats but less than their availability (Figure 4.1). It can be inferred here that livestock remains in jackal scat samples are factored by anthropogenic influence where the habitats are subject to grazing, and it is perhaps by chance that the samples captured the scavenging events. The jackal is often regarded as benefiting from agriculturally induced habitat fragmentation (e.g., Matthiae & Stearns, 1981) and edge habitats (Saunders et al., 1991; Noss & Csuti, 1994), exploiting a wide range of available food resources from respective regions.

The present study showed the dominance of mammalian prey species in the diet of golden jackals. Reptiles and invertebrates were also found in the diets of jackals, but their identification and quantification were not possible. The important food source for jackals was small mammals that were high in energy (Mukherjee et al., 2004) and also a crucial source of preformed water, which is important for inhabiting arid habitats (Williams et al., 2002). In the present study, rodents constituted 17% of the total mammalian diet of the golden jackal. While golden jackals consumed invertebrates frequently, their remains were difficult to identify at species level. Because large beetles and locusts are protein-rich and easy to catch, a rather large proportion was expected in the jackal scats. Although insect remains were found in many of the samples (represented by 4.4%), they were in a lower proportion than other prey items (representing 10.7%), which is however, comparable to the study conducted in Hungary, Greece, and Israel (Lanszki et al., 2006).

Consumption of alternative food sources such as fruits, which is associated with low foraging costs, is characteristic of generalist predators (Lucherini et al., 1995; Martinoli et al., 2001). *Zizyphus mauritiana* fruits form an important seasonal resource in Sariska, with fruiting occurring between December and February. *Z. mauritiana* comprised the bulk of the fruit diet for golden jackals (24.7% of overall fruits), being represented in 9.8% of the scats. IRI scores of *Zizyphus mauritiana* (510.11) indicate that the golden jackal actively exploits such seasonal resources and therefore could be playing an important role in seed dispersal (Herrera, 1989; Silva et al., 2005) in the semi-arid region of Sariska. This study demonstrates that plant parts were fairly represented in golden jackal scats (37.7%), which demonstrates the opportunistic aspect of their foraging behaviour. The presence of grass and grass seeds in the scat is consistent with other studies, mostly concluding a possible role in scouring the intestine during digestion (Kalle et al., 2013).

The bird remains were found in 15.2% of jackal scats in Sariska, which is lower than other studies on golden jackals reported from Bangladesh (31%), Africa (23.7%) (Mc Shane & Grettenberger, 1984) and other studies in India (Mukherjee, 1998; Gupta, 2011). This may be attributed to the fact that the availability of other easy food resources is more abundant in the study area in terms of energy economy for jackals, compared to chasing and catching bird pray. The general feeding habits of jackals in Sariska appear to confirm the results from available studies (Moehlman 1983, 1986, 1989; Khan & Beg 1986; Sankar, 1988; Mukherjee, 1998; Lanszki & Heltai, 2002).

Research shows that the flexibility of jackals in their food choices and their ability to adopt suitable foraging behaviour enables them to survive in different habitats (Rowe-Rowe, 1976; Stuart, 1976; Lamprecht, 1978; Avery et al., 1987). Jackals are opportunistic in general, but prefer highly nourishing, protein-rich food sources even in periods of low food abundance (MacArthur & Pianka, 1966), supporting optimal foraging theory. Similar results were observed for coyotes in the Chihuahuan Desert (Hernández et al., 2002).

The derivation of an equation of correction factor was attempted to compute biomass consumption using feeding trials on jackals. The relationship obtained using average prey weight consumed per collectable scat with the average weight of prey fed to jackals suggested that prey consumption per collectable scat increased with an increase in prey weight. Assuming random sampling, percent occurrence in scats indicates how pervasive the food item is in the diet of the predator population and is not unduly biased by the food preferences of individual predators (Weaver & Hoffman, 1979). The biomass consumed per collectable scat (y) is calculated from the average weight of the prey species (x) with the use of the derived equations. To control overrepresentation of prey items, prey biomass consumption was estimated by multiplying correction factor "y" with the numerical frequency (NF) of the species in total scats.

According to the current study, jackals used more carcass in smaller prey than in larger prey, which is consistent with the findings of Jethva and Jhala (2004). When feeding on larger prey, jackals are selective in the body parts they feed on, so small-bodied preyers are usually completely consumed. This resulted in higher digestibility for larger prey. Data for prey consumption was corrected for moisture loss as it could account for an average 9.9% overestimation of consumption by jackal. In the case of large prey, where a greater proportion of the carcass is not consumed at once, moisture loss was significant and could account for as much as an 18% overestimation of consumption. The relationship obtained (Figure 4.1) supports Mech's (1970) hypothesis and is in agreement with Floyd et al. (1978) and Weaver (1993) that the percentage frequency of occurrence of prey species tends to over-represent smaller prey in terms of biomass and to under-represent them in terms of numbers when compared to larger prey species.

A major limitation of the present study was the inability to conduct feeding trials using rats and insects because of quarantine issues for rodents in Jaipur Zoo. Also, the insects that were eaten by jackals in natural conditions were not available in the open market, and hence they were not used in feeding trial experiments. The feeding trial on jackals was based on scats containing the remains of one prey species only, whereas golden jackal scats regularly contain more than one prey species. To rectify it, WSE was derived as the percent dry volume of prey in scats through visual estimation and was used to calculate prey biomass consumption for each prey species along with NF and frequency of occurrence. The findings of Ciucci et al. (1996) and Spaulding et al. (1997) state that using the frequency of occurrence directly for biomass estimation without correcting for the proportional contribution of a prey item to a scat for scats with more than one prey item gave results similar to those obtained with the use of proportional contribution data. However, in contrast, a significant difference in the frequency of occurrence was observed (Friedman test; 2 = 16.8, p = 0.00, n = 10) in comparison with the other two methods (i.e., WSE and numerical frequency). Although, with respect to the replicability of the methodology, it is more justifiable to use numerical frequency as observer bias is expected in visual estimation of proportional prey occurrence in scats. The selectivity values for chital and birds were calculated for the index when data was corrected for all weight classes (Figure 4.2, A). Since these species are well observed in scats and differences in light and heavy prey were compromised when corrected for all weight classes, such high selection values resulted. Whereas the index, when data is corrected weight class wise (Figure 4.2, B), more selectivity was observed for hare in comparison to rodents, ecologically justifying the fact that hare, being heavier and comparatively easier to catch, is a selected prey species in the jackal's diet.

The results of the jackal feeding trial confirmed that the accuracy of the correction factor depends on the number of feeding experiments and the range of prey sizes included. The present study, like previous studies, determined a correction factor that fitted a linear regression through the data. Watcher et al. (2012) proposed that consuming prey mass to excrete one scat reaches such an asymptote because there will be a limit to the prey mass consumed by a carnivore which can be digested per excreted scat and there will be a limit to the size of scats that can be excreted. Inclusion of prey sizes that a carnivore may not be able to overcome alone, but only in a group, or not even then, provides information on the overall physiological characteristics of the digestive system. Even if the predator in the wild only feeds on a subset of the presented range of prey sizes, understanding the digestive properties of prey weights spanning several orders of magnitude improves the accuracy of the correction factor for any individual prey size factor derived from an exponential function is likely to be biologically more meaningful and physiologically more realistic than a linear function (Wachter et al., 2012). Because of logistical limitations, the weight of prey provided to jackals during the experiment was restrained to 50 kg. Hence, the asymptote could not be reached. The conversion of prey body mass into a number of scats and their weights excreted by an individual carnivore after prey consumption is a crucial step that diet studies based on scats ought to infer accurately (Wachter et al., 2012). The seven quantities were used to evaluate the steps required to establish the consumed prey mass and the number of consumed prey individuals derived from carnivore scats collected in the field.

Conclusion

No single method of predator scat analysis is ideal; each provides unique information and allows interpretations not possible with others. Adaptations in foraging behaviour in relation to abundance and quality of food sources are expected to be highly pronounced in an extreme habitat and in the presence of larger predators. For carnivores such as golden jackals, that mainly feeding on fruits, invertebrates, birds, and small mammals, the diet can be better determined using the specific con-version factors for these food items. More feeding trials on jackals will be required to establish exponential relationships over a range of prey sizes. The effects of the size of the prey, the presence of non-vertebrate constituents (i.e., insects and rep-tiles), and the metabolic state of the jackals on prey detectability need quantitative evaluation. Considering the high proportion of rodent remains in jackal's scat, it is advisable to promote it as a pest controlling agent in the agro ecosystem to improve its acceptance, which is required for long term conservation of this species.

Acknowledgements: We are thankful to Rajasthan Forest Department and Director, Jaipur Zoo for facilitating this work in Sariska and Jaipur Zoo respec-tively, as a part of research project conducted by the Wildlife Institute of India. We thank Director and Dean, WII, for their encouragement and support pro-vided for the study. We thank Dr. Sutirtha Dutta and Mr. Stotra Chakrabarti for their help in data analysis. We thank our field assistants Mamraj, Jairam, Khagesh, Ratan, and Rajesh for their assistance in field.

References

Anderson, D. R., Laake, J. L., Crain, B. R., & Burnham, K. P. (1979). Guidelines for the transect sampling of biological populations. *Journal of Wildlife Management, 43*(1), 70–78.

Angerbjörn, A., Tannerfeldt, M., & Erlinge, S. (1999). Predator-prey relationships: Arctic foxes and lemmings. *Journal of Animal Ecology, 68*(1), 34–49.

Avery, G., Avery, D. M., Braine, S., & Loutit, R. (1987). Prey of coastal black-backed jackal *Canis mesomelas* (Mammalia: Canidae) in the Skeleton Coast Park, Namibia. *Journal of Zoology, 213*(1), 81–94.

Bekoff, M., Daniels, T. J., & Gittleman, J. L. (1984). Life history patterns and comparative social ecology of carnivores. *Annual Review of Ecology, Evolution, and Systematics, 15*, 191–232.

Borkowski, J., Zalewski, A., & Manor, R. (2011). Diet composition of golden jackals in Israel. *Annales Zoologicifennici, 48*(2), 108–118. https://doi.org/10.5735/086.048.020

Buckland, S. T., Anderson, D. R., Burnham, K. P., & Laake, J. L. (1993). *Distance sampling: Estimating abundance of biological populations*. Chapman and Hall.

Buckland, S. T., Anderson, D. R., Burnham, K. P., Laake, J. L., Borchers, D. L., & Thomas, L. (Eds.). (2004). *Advanced distance sampling*. Oxford University Press.

Burnham, K. P., Anderson, D. R., & Laake, J. L. (1980). Estimation of density from line transect sampling of biological populations. *Wildlife Monographs, 72*, 1–202.

Champion, H. G., & Seth, S. K. (1968). *A revised survey of forest types of India*. Manager of Publications, Government of India, pp. 404.

Chourasia, P., Mondal, K., Sankar, K., & Qureshi, Q. (2012). Food habits of golden jackal (*Canis aureus*) and striped hyena (*Hyaena hyaena*) in Sariska Tiger Reserve, Western India. *World Journal of Zoology, 7*(2), 106–112.

Ciucci, P., Boitani, L., Pelliccioni, E. R., Rocco, M., & Guy, I. (1996). A comparison of scat analysis methods to assess the diet of the Wolf *Canis lupus*. *Wildlife Biology, 2*(1), 37–48.

Colwell, R. K. (1996). *Biota: The biodiversity database manager*, Sinauer Associates, xvi. ISBN 0-87893-128-7.

Corbett, L. K. (1989). Assessing the diet of dingoes from feces: A comparison of 3 methods. *Journal of Wildlife Management, 53*(2), 343–346.

Demeter, A., & Spassov, N. (1993). *Canis aureus* Linnaeus, 1758. In J. Niethammer & F. Krapp (Eds.), *Handbuch Der SäugetiereEuropas* (pp. 107–138). Band 5/I. Aula-Verlag.

Elmhagen, B., Tannerfeldt, M., & Angerbjorn, A. (2002). Food-niche overlap between Arctic and Red Foxes. *Canadian Journal of Zoology, 80*(7), 1274–1285.

Floyd, T. J., Mech, L. D., & Jordan, P. A. (1978). Relating wolf scat content to prey consumed. *Journal of Wildlife Management, 42*(3), 528–532.

Fuller, M. F., McWilliam, R., Wang, T. C., & Giles, L. R. (1989). The optimum dietary amino acid pattern for growing pigs. 2. Requirments for maintenance and for tissue protein accretion. *British Journal of Nutrition, 62*(2), 255–267.

Giannatos, G., Marinos, Y., Maragou, P., & Catsasorakis, G. (2005). The golden jackal (*Canis aureaus*) in Greece. *Belgian Journal of Zoology, 135*, 145–149.

Gupta, S. (2011). *Ecology of medium and small sized carnivores in Sariska tiger Reserve, Rajasthan, India* [PhD Thesis]. Saurashtra University, pp. 156.

Herna´Ndez, L., Parmenter, R. R., Dewitt, J. W., Lightfoot, D. C., & Laundre, J. W. (2002). Coyote diets in the Chihuahuan Desert, more evidence for optimal foraging. *Journal of Arid Environments, 51*(4), 613–624.

Herrera, C. M. (1989). Frugivory and seed dispersal by carnivorous mammals, and associated food characteristics, in undisturbed Mediterranean habitats. *Oikos, 55*(2), 250–262.

Home, C., & Jhala, Y. V. (2009). Food habits of the Indian fox (*Vulpes bengalensis*) in Kutch, Gujarat, India. *Mammalian Biology, 74*(5), 403–411.

Humphries, B. D., Ramesh, T., Hill, T. R., & Downs, C. T. (2016). Habitat use and home range of black-backed jackals (*Canis mesomelas*) on farmlands in the midlands of KwaZulu-Natal, South Africa. *African Zoology, 51*(1), 37–45. https://doi.org/10.1080/15627020.2015.1128356

Ivlev, V. S. (1961). *Experimental ecology of the feeding of fishes*. Yale University.

Jaeger, C., Renn, O., Rosa, E., & Webler, T. (2001). *Risk: Uncertainty and rational action*. Earthscan.

Jethva, B. D., & Jhala, Y. V. (2004). Computing biomass consumption from prey occurrences in Indian wolf scats. *Journal of Zoo Biology, 23*(6), 513–520.

Jhala, Y. V., & Moehlman, P. D. (2008). *Canis aureus*. In IUCN 2010. IUCN Red List of Threatened Species, Version 2010.2.

Johnson, M. K., & Aldred, D. R. (1982). Mammalian prey digestibility by bobcats. *Journal of Wildlife Management, 46*(2), 530.

Kalle, R., Ramesh, T., Qureshi, Q., & Sankar, K. (2013). Predicting the distribution pattern of small carnivores in response to environmental factors in the Western Ghats. *PLOS ONE, 8*(11), E79295. https://doi.org/10.1371/journal.pone.0079295

Khan, A. A., & Beg, M. A. (1986). Food of some mammalian predators in the cultivated areas of Punjab. *Pakistan Journal of Zoology, 18*, 71–79.

Klare, U., Kamler, J. F., & Macdonald, D. W. (2011). A comparison and critique of different scat-analysis methods for determining carnivore diet. *Mammal Review, 41*(4), 294–312. https://doi.org/10.1111/j.1365-2907.2011.00183.X

Krystufek, B., Murariu, D., & Kurtonur, C. (1997). Present distribution of the golden jackal *Canis aureus* in the Balkans and adjacent regions. *Mammal Review, 27*(2), 109–114.

Lamprecht, J. (1978). On diet, foraging behavior and interspecific food competition of jackals in the Serengeti National Park, East Africa. *ZeitschriftFürSäugetierkunde, 43*, 210–223.

Lanszki, J., & Heltai, M. (2002). Feeding habits of golden jackal and red fox in south Western Hungary during winter and spring. *Mammalian Biology, 67*(3), 129–136.

Lanszki, J., Heltai, M., & Szabo, L. (2006). Feeding habits and trophic niche overlap between sympatric golden jackal (*Canis aureus*) and red fox (*Vulpes vulpes*) in the Pannonian ecoregion (Hungary). *Canadian Journal of Zoology, 84*(11), 1647–1656.

Lucherini, M., Lovari, S., & Crema, G. (1995). Habitat use and ranging behavior of the red fox (*Vulpes vulpes*) in a mediterranean rural area: Is shelter availability a key factor? *Journal of Zoology, 237,* 57–591.

Macarthur, R. H., & Pianka, E. R. (1966). On optimal use of a patchy environment. *American Naturalist, 100*(916), 603–609.

Macdonald, D. W. (1983). The ecology of carnivore social behavior. *Nature, 301*(5899), 379–384.

Martinoli, A., Preatoni, D. G., Chiarenzi, B., Wauters, L. A., & Tosi, G. (2001). Diet of stoats (*Mustela erminea*) in an Alpine habitat: The importance of fruit consumption in summer. *Acta Oecologica, 22*(1), 45–53.

Matthiae, P. E., & Stearns, F. (1981). Mammals in forest islands in Southeastern Wisconsin. In R. L. Burgess & D. M. Sharpe (Eds.), *Forest island dynamics in man-dominated landscapes* (pp. 55–66). Springer Verlag.

McShane, T. O., & Grettenberger, J. F. (1984). Food of the golden jackal (*Canis aureus*) in central Niger. *African Journal of Ecology, 22*(1), 49–53.

Mech, L. D. (1970). *The wolf; the ecology and behaviour of an endangered species.* Natural History press, pp. 389.

Medina, F. M., López-Darias, M., Nogales, M., & Garchia, R. (2008). Food habits of feral cats (Felis sylvestris catus L) in insular semiarid environments. *European Journal of Wildlife Research, 35,* 162–169.

Moehlman, P. D. (1983). Socioecology of silverbacked and golden jackals (*Canis mesomelas and Canis aureus*). In J. F. Eisenberg & D. G. Kleiman (Eds.), *Recent advances in the study of mammalian behavior* (pp. 423–453). American Society of Mammalogists.

Moehlman, P. D. (1986). Ecology of co-operation in canids. In D. I. Rubenstein & R. W. Wrangham (Eds.), *Ecological aspects of social evolution: Birds and mammals* (pp. 64–86). Princeton University Press.

Moehlman, P. D. (1989). Intraspecific variation in canid social systems. In J. L. Gittleman (Ed.), *Carnivore behavior, ecology, and evolution* (pp. 164–182). Cornell University Press.

Mukherjee, S. (1998). *Habitat use in sympatric small carnivores in Sariska Tiger Reserve, Rajasthan, Western India* [PhD Thesis]. University of Saurashtra, Biosciences Department, India.

Mukherjee, S., Goyal, S. P., Johnsingh, A. J. T., & Leite Pitman, M. R. P. (2004). The importance of rodents in the diet of Jungle Cat (*Felis chaus*), caracal (*Caracal caracal*) and golden jackal (*Canis aureus*) in Sariska Tiger Reserve, Rajasthan, India. *Journal of Zoology, 262*(4), 405–411.

Noss, R. F., & Csuti, B. (1994). Habitat fragmentarinn. In G. K. Meffe & R. C. Carroll (Eds.), *Principles of conservation biology* (pp. 237–264). Sinauer Associates, Inc.

Otis, D. L., Burnham, K. P., White, G. C., & Anderson, D. R. (1978). Statistical inference from capture data on closed animal populations. *Wildlife Monographs, 62,* 1–135.

Paltridge, R. (2002). The diets of cats, foxes and dingoes in relation to prey availability in the Tanami Desert, Northern Territory. *European Journal of Wildlife Research, 29*(4), 389–403.

Péladeau, N. (2000). *Simstat, V 2.0* [Software]. Provalis Research.

Pinkas, L., Oliphant, M. S., & Iverson, I. L. K. (1971). *Food habits of albacore, bluefin tuna and bonito in California waters.* California Department of Fish and Game, Fish Bulletin 152, 1–83.

Prater, S. H. (1980). *The book of Indian animals* (rev. ed.). Bombay Natural History Society, pp. 316.

Reynolds, J. C., & Aebischer, N. J. (1991). Comparison and quantification of carnivore diet by faecal analysis: A critique, with recommendations, based on a study of the red fox *Vulpes Vulpes*. *Mammal Review*, 21(3), 97–122.

Rowe-Rowe, D. T. (1976). Food of the black-backed jackal in nature conservation and farming areas in natal. *East Africa Journal of Wildlife Management*, 14, 345–348.

Šálek, M., Červinka, J., Banea, O. C., Krofel, M., Ćirović, D., Selanec, I., Penezić, A., Grill, S., & Riegert, J. (2014). Population densities and habitat use of the golden jackal (*Canis aureus*) in farmlands across the Balkan Peninsula. *European Journal of Wildlife Research*, 60(2), 193–200. https://doi.org/10.1007/s10344-013-0765-0

Sankar, K. (1988). Some observations on food habits of jackals (*Canis aureus*) in Keolaeo National Park, Bharatpur, as shown by scat analysis. *Journal of the Bombay Natural History Society*, 85, 185–186.

Sankar, K., Qureshi, Q., Mondal, K., Worah, D., Srivastava, T., Gupta, S., & Basu, S. (2009). Ecological studies in Sariska Tiger Reserve, Final report submitted to National Tiger Conservation Authority, Govt. of India, and Wildlife Institute of India, pp. 145.

Saunders, D. A., Hobbs, R. J., & Margules, C. R. (1991). Biological consequences of ecosystem fragmentation: A review. *Conservation Biology*, 5(1), 18–32.

Short, J., Calver, M. C., & Risbey, D. A. (1999). The impact of cats and foxes on the small vertebrate fauna of Heirisson prong, Western Australia; exploring potential impact using diet analysis. *Wildlife Research*, 26(5), 621–630.

Silva, S. I., Bozinovic, F., & Jaksic, F. M. (2005). Frugivory and seed dispersal by foxes in relation to mammalian prey abundance in a semiarid thorn scrub. *Austral Ecology*, 30(7), 739–746.

Spaulding, R. L., Krausman, P. R., & Ballard, W. B. (1997). Calculation of prey biomass consumed by wolves in Northwest Alaska. *Journal of Wildlife Research*, 2, 128–132.

Stuart, C. T. (1976). Diet of the black-backed jackal, *Canis mesomelas*, in the Central Namib Desert, South West Africa. *Zoologica Africana*, 11(1), 193–205.

Wachter, B., Blanc, A. S., Melzheimer, J., Honer, O. P., & Jago, M. (2012). An advanced method to assess the diet of free-ranging large carnivores based on scats. *PLOS ONE*, 7(6), E38066. https://doi.org/10.1371/journal.Pone.0038066

Weaver, J. L. (1993). Refining the equation for interpretating prey occurrence in gray wolf scats. *Journal of Wildlife Management*, 57(3), 534–538.

Weaver, J. L., & Hoffman, S. W. (1979). Differential detectability of rodents in coyote scats. *Journal of Wildlife Management*, 43(3), 783–786.

Williams, J. B., Lenain, D., Ostrowski, S., Tieleman, B. I., & Seddon, P. J. (2002). Energy expenditure and water flux of Rüppell's foxes in Saudi Arabia. *Physiological and Biochemical Zoology*, 75(5), 479–488.

Wilson, K. R., & Anderson, D. R. (1985). Evaluation of two density estimators of small mammal population size. *Journal of Mammalogy*, 66(1), 13–21.

Yom-Tov, Y., Ashkenazi, S., & Viner, O. (1995). Cattle predation by the golden jackal *Canis aureus* in the Golan Heights, Israel. *Biological Conservation*, 73, 19–22.

5 Blackbuck in Agricultural Landscape of Aligarh

Shahzada Iqbal and Orus Ilyas

Introduction

The blackbuck is one of the most beautiful antelope species. It has been described as a sacred animal in Hindu mythology. It is endemic to the Indian subcontinent and was once a ubiquitous larger wild animals of this region (Ranjitsinh, 1982), and they are truly Indian subcontinental in distribution (Rahmani, 1991). They are primarily grazing animals that avoid forest areas but survive in semi-desert regions, as long as there is sufficient scattered vegetation (Roberts, 1977), and also in cultivated land (Majupuria, 1982). The medium-sized antelope is readily adaptable well to wasteland, marginal agricultural land, and cultivated areas. It lives in open plains with low growing grasses and avoids hilly and dense forested areas. In areas of severe habitat decline, it depends on crops. A preference for early crop growing stages is an important consideration for the management of blackbuck habitat (Jhala, 1991).

There is no scientific information available so far on blackbuck population dynamics and land use patterns by the species in and around Blackbuck Conservation Centre. Considering limited resources, and the increasing demand of a growing human population, it is important to conserve species through a holistic research approach for their long-term conservation.

Study Area

Blackbuck Conservation Centre (here after BCC) was set up between 2013 and 2014 and comes under the jurisdiction of the district forest department, Aligarh (Figure 5.1). It is situated in the village of Palasallu (tehsil of Gabhana), 18 kilometres away from the district headquarter in Aligarh, Uttar Pradesh, India. It is spread over an area of 63 hectares, containing grasses like *Cynodon dactylon* and *Cenchrus ciliaris*, and herbs like *Ocimum sanctum*, *Abrus precatorius*, and *Boerthavia diffusa*, but it is majorly dominated by *Prosopis juliflora*, and *Prosopis cineraria*. The golden jackal (*Canis aureus*), palm civet (*Paradoxurus hermaphrodites*), jungle cat (*Felis chaus*), rufous-tailed hare (*Lepus nigricollis*), nilgai (*Boselaphus tragocamelus*), blackbuck (*Antilope cervicapra*), and wild boar (*Sus scrofa*) are the mammalian fauna which are found inside the BCC.

DOI: 10.4324/9781003321422-6

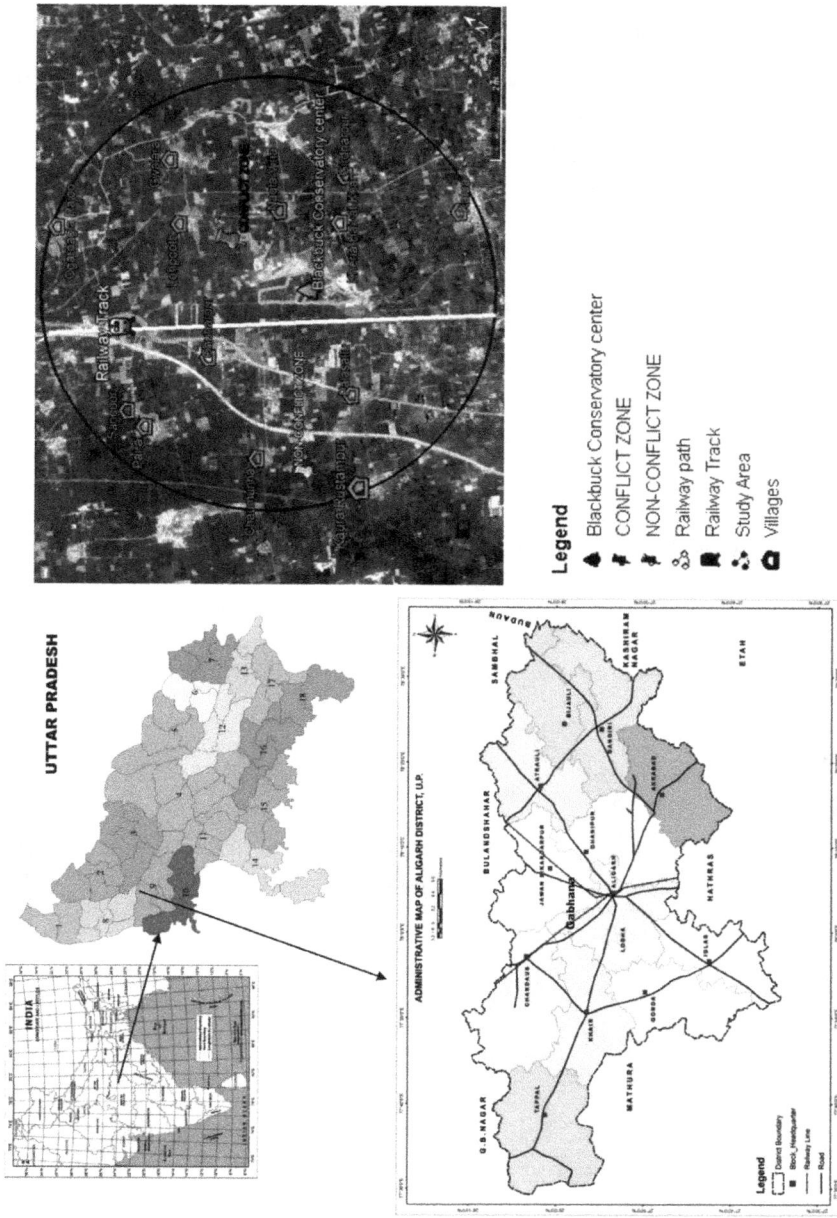

Figure 5.1 Map of the study area.

An area of 3 km radius from BCC was selected for the investigation, from which there are 13 villages comprising 3,460 households (Census 2011, GoI). A railway track (Delhi–Aligarh main route) passes at just 0.5 km from the southern boundary of BCC. There are eight villages near the Blackbuck Conservation Centre and five villages beyond the railway track, altogether within a 3 km radius from the Blackbuck Conservation Centre. Typically, there is no human-blackbuck conflict beyond the railway track, as per the villagers. The complete study area was divided into two zones, (1) the conflict zone which comprises eight villages with 2,319 households and (2) the non-conflict zone which comprises five villages with 1,141 households (Census 2011, GoI).

Methodology

Population estimation was done by total count as used by many researchers (Jhala, 1991; Khan, 2014) during the early morning hours (Davis & Winstead, 1980). A census was conducted at the interval of every sixth day which amounted to ten times during the two months of fieldwork (Madin, 1989). Counting was done usually from 6:30 a.m. to 10:30 a.m. by foot and also with the help of a vehicle. To avoid double counting, different groups were counted simultaneously with the help of forest ground staff from the Centre. Blackbucks are easy to spot in open grassland habitats when they are on their feet and moving (Jhala, 1991), but in this case, croplands act as a grassland. During early morning, blackbucks have long and high-intensity grazing periods (Mungall, 1978), during which they actively feed and slowly move to form a large aggregation. These two behavioural traits were used as an advantage in conducting a census. The blackbucks were classified into different categories based on group size, age, and sex during the census. Age classes were limited to four groups discernible in the field by mere observation.

Results

Abundance Estimation

Ten total counts were conducted, and a total of 1,089 individuals were recorded with the mean of 108.9±2.76; the minimum individuals were counted to be 95, and the maximum individuals were counted to be 119, (SD=8.73, CV=8.02).

Population Structure

1. *Population Composition*
    The population was divided into adult male (> 2 years), adult female (>2 years), sub-adult male (1–2 years), sub-adult females (1–2 years), yearling (6–12 months), and fawn (< 6 months). During the study, we have recorded 13.5±0.34 adult male, 21.9±0.99 adult female, 19.6±1.05 sub-adult male, 28±1.39 sub-adult female, and 25.9±1.12 yearling. No fawn was recorded during the study.

2. *Herd Size*

During the study period, 121 blackbuck groups were recorded. Solitary animals composed 37.2% of the total observed groups, 47.1% were observed in the group ranging from 2–15 individuals, and 15.7% were observed in a group ranging beyond 15 individuals. The largest observed herd was of 68 individuals. The mean group size recorded was 68 (n=121) individuals if solitary animals were excluded.

3. *Age-Sex Ratio*

Among the identified individuals of blackbuck, males were 330 (30.30%), females were 499 (45.8%), and yearlings were 259 (23.78%). Fawns were not recorded during the study period.

The male to female ratio was 1:1.66. The adult individuals numbered were 354 (32.50%), the sub-adults were 475 (43.6%), and the remaining 23.78% were yearlings. Out of the adult population, 135 (12.39%) were adult males and 219 (20.11%) were adult females. So, the adult sex ratio was 1 adult male to 1.61 adult females.

4. *Herd Composition*

76 groups with 1,044 individual blackbucks were analysed. Among those, there were 135 adult males, 219 adult females, 195 sub-adult males, 280 sub-adult females, and 259 yearlings (Table 5.1). The ratio of adults to sub-adults to yearling was 1.36:1.83:1. The average number of adult females in groups varied from 1 to 13 individuals in each group (n=75). Twelve groups (9.12%) did not include any adult females, and 31.92% of all groups included more than one adult female. There was no adult male in 19.76% of the groups and no yearling in 28.12% of the groups. Sub-adult males and sub-adult females were found in 28.12% and 38.76% of the groups, respectively. 34.96% of the individuals were recorded in a mixed herd and 3.8% in bachelor herd, while

Table 5.1 Population structure of blackbuck in and around Blackbuck Conservation Centre

S. No	Category	No./Percentage
1	Total blackbuck observed	1089
2	Total no. of adult individuals observed	354
3	Total no. of sub-adult individuals in the observed population	475
4	% of sub-adult individuals in the observed population	43.6%
5	Total no. of male individuals observed	330
6	% of males in the observed population	30.30%
7	Total no. of female individuals in the observed population	499
8	% of female in the observed population	45.8
9	Total no. of yearlings observed	259
10	% of yearlings in the observed population	23.78%
11	Total no. of adult males in the observed population	135
12	% of adult males in the observed population	12.39%
13	Total no. of adult females in the observed population	219
14	% of adult females in the observed population	25.71%

the rest were in the female and territorial male herd. In the mixed herd, 67.16% were female, 32.9% male, and rest were yearlings.

Discussion

Large mammalian herbivores, such as ruminants, make up groups that are easily recognisable in the field and consist of individuals located at short distances from one another, often engaged in a common activity, like feeding, travelling, or resting (Estes, 1991). The size of the group is often considered a fundamental attribute of the social organisation of such species (Jarman, 1974; Wilson, 1975). The variation of group size with habitat openness was assumed to be a biological adaptation, encoded in the individual (Caughley, 1981). As a consequence, natural selection retained individuals preferring to be within small groups when in a closed habitat, and within large groups when in an open landscape (Jarman, 1974). Herd size of group-living species increases with habitat openness: whereas groups are small in forested habitats, they are much larger in grassland and other open landscapes (Leuthold, 1970; LaGory, 1986). This was supported by the observation of Ranjitsinh (1982), who recorded large groups of blackbuck composed of 430 individuals in Velavadar National Park, characterised by large open habitat. In the present study, the largest herd recorded had 68 individuals, hypothetically due to the scattered open space in between the crop field and much lower predation pressure. As Jarman (1974) advocates, in a closed habitat, herbivores can earlier reduce the probability of being detected by predation while living in a small group. In species that exhibit social systems, it is suggested that they will form large groups when high-quality forage is abundant but will be forced into smaller groups when food supply is less abundant and dispersed in distribution. This was supported by the result of the present study, as the group size of blackbuck varied in relation to the abundance of food availability according to the habitat. Group size showed a decline with the advancement of habitat change (less in Prosopis patch). It was highest in the crop field due to abundant availability of food and lowest in the Prosopis patch due to less food availability. Similarly, Bharucha and Asher (1993) recorded herd of blackbuck with a size varying from 2–200 in Rahekuri Blackbuck Sanctuary and found that changes in group size and composition coincide with a change in vegetation type. Thus, from the present study, it is concluded that the group size of blackbuck in and around Blackbuck Conservation Centre was found to be influenced by habitat type, including a direct relation with food availability. The sex ratio (male to female) of blackbuck recorded by various authors on different study sites were 1:2 in Point Calimere by Danial (1968) and 1:11 (rutting season) by Sharma (1980), and 1:37 in Rehekuri by Bharucha and Asher (1993). In the present study area, it is 1:1.66 which is very close to the sex ratio observed by Bharucha and Asher (1993) in Rehekuri Blackbuck Sanctuary. The data recorded in the present study indicate that blackbuck is gregarious. Although blackbuck mainly lived in a group, a few solitary individuals were also observed.

The study indicates that strong males of solitary nature occurred throughout the study period; the frequency of finding them was higher in the crop field rather than in other habitats. The departure of sub-adults from their mother's group also appears to be a factor responsible for group size variation. Furthermore, human disturbance may be a factor for the group size variation in the study area.

References

Bharucha, E., & Asher, K. (1993). Behaviour patterns of the blackbuck, Antilope cervicapra, under sub-optimal habitat condition. *Journal of Bombay Natural History Society*, 90(3), 371–393.

Caughley, G. (1981). Overpopulation. In P. A. Jewell & S. Holt (Eds.), *Problems in management of locally abundant wild mammals* (pp. 7–19). Academic Press.

Census, NSSO, Government of India Census. (2011). Primary census abstract, registrar general of India, Ministry of Home Affairs. Government of India.

Daniel, J. C. (1968). The Point Calimere Sanctuary, Madras state—May 1967. *Journal of Bombay Natural History Society*, 64(3), 512–523.

Davis, D. E., & Winstead, R. L. (1980). Estimating the numbers of the wildlife population: 221–246. In S. D. Schemnitz (Ed.), *The wildlife management techniques manual* (pp. 686). Wildlife Society.

Estes, R. D. (1991). *The behaviour guide to African mammals*. University of California Press.

Jarman, P. T. (1974). The social organization of antelope in relation to the ecology. *Behaviour*, 46, 215–266.

Jhala, Y. V. (1991). *Habitat and population dynamics of wolves and blackbuck in Velavadar National Park, Gujrat, India* [PhD. Dissertation]. Virginia Polytechnic Institute and State University.

Khan, K. A. (2014). *Current status and distribution of nilgai in Aligarh District* [M. Phil. Dissertation]. Dept. of Wildlife Sciences. A.M.U.

Lagory, K. E. (1986). Habitat, group size, and the behaviour of white-tailed deer. *Behaviour*, 98(1–4), 168–179.

Leuthold, W. (1970). Observation on the social organization of impala. *Zeitschrift für Tierpsychologie*, 27, 693–721.

Madin, C. D. (1989). Variability in total ground counts on a Kenyan game ranch. *African Journal of Ecology*, 27(4), 297–303.

Majpuria, T. C. (1982). *Wild is beautiful: Introduction to the magnificent rich and varied fauna and wildlife of Nepal*. Thacker Spink and Co.

Mungall, E. C. (1978). The Indian blackbuck (*Antilope cervicapra*): A Texas view. *Kleberg Studies in Natural Resources*, 184.

Rahmani, A. R. (1991). Present distribution of blackbuck in India with special reference to the lesser known population. *Journal of Bombay Natural History Society*, 1, 35–46.

Ranjitsinh, M. K. (1982). *Ecology and behaviour of the Indian blackbuck with special reference to the Velavadar National Park* [PhD. Thesis]. Saurashtra University India.

Robert, T. J. (1977). *The mammals of Pakistan*. Ernest Benn Limited, pp. 361.

Sharma, I. K. (1980). Habitat preferences, feeding and survival of the blackbuck. *Tiger Paper*, 7(8), 4–6.

Wilson, E. O. (1975). *Socio-biology: The new synthesis*. Harvard University Press.

6 Seasonal Variation in the Diet of Four-Horned Antelope

Abdul Haleem and Orus Ilyas

Introduction

The four-horned antelope (*Tetracerus quadricornis*), also known as the chousin-gha, is one of the smallest Asian bovids and is the only antelope with two pairs of horns of which the front pair is always shorter than the back (Prater, 1971). Males have horns, whereas in females the horns are absent. It is endemic to the Indian subcontinent. Approximately 95% of its current global population occurs in India (Rahmani, 2001), with the remaining 5% found in Nepal (Krishna et al., 2009). It is distributed in peninsular India south to the Himalayas where the country is wooded and hilly but not too dense with forested areas (Prater, 1971). The four-horned antelope prefers dry deciduous forests in hilly areas with open canopy (Baskaran et al., 2009) and secondly tree-savanna deciduous habitats with a high degree of deciduousness (Krishna et al., 2008, 2009). It has been reported that the four-horned antelopes are distributed in all of the Indian states south from Uttar Pradesh, except Kerala (Rice, 1990).

Despite being widely distributed in India, this species has received little scientific attention. The review of literature shows that this species is gaining attention in very few protected areas, especially with regard to status and distribution (Krishnan, 1972; Sharatchandra & Gadgil, 1975; Karanth & Sunquist, 1992; Berwick, 1974; Rice, 1990; Krishna, 2006; Sharma et al., 2005; Baskaran & Desai, 1999; Singh & Swain, 2003).

The most important and consistent activity determining the survival, health, and power of movement in the life of animals is feeding, which has not yet been studied in the case of chousingha. Feeding activities dominate the lives of most animals, resulting in their ever-lasting quest for food (Shukla, 1990). The nature of foods and the ways in which these are obtained play a very important role in shaping the structure and behaviour of animals. Almost all important features and behavioural mechanisms are also linked to the need to locate and obtain proper nourishment. The problem of food selection is common to all species and plays a pivotal role in the evolutionary divergence of many groups of animals (Mc Farland, 1981). The energy requirements of ruminants depend majorly on their body size (Bell, 1970; Jarman, 1974; Green, 1985). Energy requirements are proportional to the body weight of the species raised to the power of 0.75 (Kleiber, 1961; Green, 1985).

DOI: 10.4324/9781003321422-7

For the conservation and management of the species, it is important to know ecological aspects such as dietary preference. Since four-horned antelope is a shy species and data collection through direct observation of feeding is very difficult, we rely on micro-histological studies. Examining faecal samples by micro-histological technique is the most commonly used method for determining the vegetation composition of a range of herbivore diets (Holechek, 1982; Alipayo, 1992). Moreover, the advantage of faecal analysis over intestinal or ruminal content analysis is that it is a non-invasive method. Consequently, in the present study, this technique was used to understand the feeding habits and the preferred food material of chousingha along with seasonal variation in the preference of their diet.

Study Area

Pench Tiger Reserve (PTR) (77° 55' to 79° 35' E and 21° 08' to 22° 00' N), lies in the south-western region of the state of Madhya Pradesh, India (Figure 6.1). The Tiger Reserve comprises of the Sanctuary and the National Park of the same name, covering an area of 757.85 sq km.

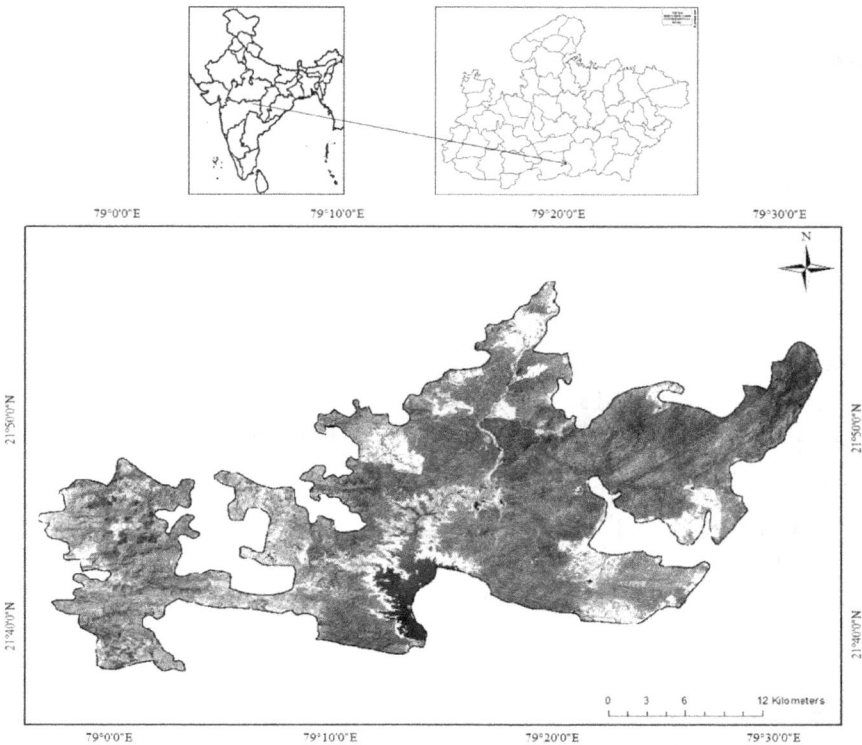

Figure 6.1 Map of the study area- the Pench Tiger Reserve.

PTR lies in the southern lower reaches of Satpura Hill ranges. The terrain is gently undulating and crisscrossed by small streams and nullahs; most of these are seasonal. The Pench river runs through the Park covering a length of 24 km and bisecting it into almost equal halves.

The area has four seasons: summer, monsoon, post-monsoon, and winter, with temperatures ranging from 0 °C in peak winter to 45 °C in the peak summer; it receives an average annual rainfall of 1400 mm (Sankar et al., 2000). The PTR is a dry deciduous forest dominated by teak (*Tectona grandis*), *Boswellia serrata*, *Anogeissus latifolia*, *Sterculia urens*, and *Gardenia latifolia*.

The Pench has a good predator and prey population such as the tiger (*Panthera tigris*), leopard (*Panthera pardus*), dhole (*Cuon alpinus*), jungle cat (*Felis chaus*), small Indian civet (*Viverricula indica*), striped hyena (*Hyaena hyaena*), sloth bear (*Melursus ursinus*), jackal (*Canis aureus*), and common palm civet (*Paradoxurus hermaphroditus*), and among the herbivores are the chital (*Axis axis*), sambar (*Rusa unicolor*), gaur (*Bos frontalis*), nilgai (*Boselaphus tragocamelus*), barking deer (*Muntiacus muntjac*), and chousingha (*Tetraceros quadricornis*).

Methodology

Data Collection

For detailed analysis of feeding ecology of chousingha, pellets were collected for diet analysis. Permanent slides of all available potential food plants were prepared to use as reference slides. Reference slides for food plants of the study area and pellet samples were prepared according to Sparks and Malechek (1968) and another modified method (Holechek, 1982; Ilyas et al., 1998; Syed & Ilyas, 2012; Haleem et al., 2014). For preparation of reference slides, a few bits of leaves were taken from available plants. These were shredded coarsely and placed in a test tube. Concentrated nitric acid (HNO_3) and distilled water were added to the test tube in the ratio of 1:3. The test tube was allowed to heat in a water bath for a minute or two until the material of the test tube became transparent (Chlorophyll free). The highly coloured and tough material was boiled for a second time with fresh quantities of HNO_3 and distilled water. The tube was allowed to cool down and the liquid drained off. The sample was repeatedly washed in distilled water. The sample was stained in safranin for two minutes to stain properly. After staining, the sample was washed repeatedly in distilled water and dehydrated by passing it through a mixture of alcohol and distilled water in the ratio of 1:3, 1:1, and 3:1 respectively. The sample was finally placed into absolute alcohol. The mounting was done in Canada balsam, and a photograph of each sample was captured using digital microscope (Barska, AY11374).

Fifteen permanent transects (three in each habitat), each of two km were established, and permanent plots of 10 m radius were marked at every 200 m interval. These permanent plots were cleared at the start of different seasons. Pellets of four-horned antelopes were identified on the basis of shape and size and collected separately from each plot in different identified habitat for three different seasons vis-à-vis summer, post-monsoon, and winter season. Random searches were also conducted in the study area to collect the pellets. These pellets were used for micro-histological studies. Pellets of species from different plots in different habitat were mixed thoroughly for each season. A total of three major samples were prepared each representing summer, post-monsoon, and winter season.

The major samples were put in a tray, and they were sampled following Ilyas and Khan (2004); the final samples of pellets were oven-dried and then ground into coarse powder form. The grinding did not fractionate the particles but merely separated the agglomerates into single particles. Two successive sieves of 30 mm and 60 mm were placed one above the other, and the ground sample was sieved. The portion of the sample left on the top sieve was rejected. The portion remaining on the second sieve was retained for analysis. The entire sample was first halved and then quartered. One quarter was selected as final sample, and three quarters were kept as reserve in case of loss of sample in subsequent process-ing. A small portion of the final sample was then transferred into test tube with a ratio of 1:3 of concentrated nitric acid (HNO_3) and water (H_2O) that was heated in a water bath for few minutes. After settling down, fresh nitric acid and water were added to the precipitate, and the sample was boiled again in order to obtain a fairly transparent or clear powder. The material was then washed with water repeatedly until the nitric acid was washed off completely to preparing the slides for analysis. The washed samples were used for preparing slides for analysis.

Out of starting samples, 32 random samples were taken, and from these ran-dom samples, four slides from each sample were prepared for the species, and a total of 128 (32 samples x 4 slides) slides were prepared.

While identifying the plant fragments from a slide, five fields of view (FOV) were considered. Plant fragments from pellet groups were identified with the help of reference slides of the plant on the basis of micro-anatomical characters such as cell wall, cell shapes, silica bodies, trichrome, stomata, and companion cells, following Satkopan (1972). Frequency of occurrence was based on the presence or absence of a species in each of 100 fields for each pellet sample. Because pellets are relatively uniform in size, the average relative frequency of occurrence repre-sents the relative abundance of different species in the pellet sample.

Data Analysis

The information on identified browse and grass particles was analysed to calcu-late the proportion of different tree, shrub, herb, and grass at the species level in the diet of chousingha. The proportion of each species in the diet composition of chousingha was compared with the food plants available in Pench Tiger Reserve using Bonferroni 95% confidence intervals following Neu et al. (1974). The for-mula for calculating 95% Bonferroni confidence intervals is as follows:

$$P_{ie} - Z_{\alpha/2k}\sqrt{P_{ie}(1-P_{ie})}/n \le P_{ie} \le P_{ie} + Z_{\alpha/2k}\sqrt{P_{ie}(1-P_{ie})}/n$$

where P_{ie} is the observed proportional utilisation of each species, $Z_{\alpha/2k}$ is the upper standard normal table value corresponding to a probability tail area of $_{\alpha/2k}$, and n is the total number of species recorded in sampling. For each species, its proportional availability (Pio) in the sample was calculated by dividing its total number (n) from that of total number of all individuals (N) sampled for all species (Pio= n/ΣN). Utilisation was significantly greater or less than expected when Pio lay outside the 95% confidence limits constructed for proportional utilisation (Pie) which was calculated by dividing number of plant fragments identified for each species (b) by the total number of identified browse fragment (B) of all species (Pie=b/ΣB). The statistical tests were conducted following Zar (1999) and Fowler & Cohen (1986).

Results

During summer season, a total of 1048 plant fragment were evaluated, out of which 24 species (18 browse and 6 grass species) of plant were identified from pellets of chousingha. Out of all 18 browse species identified from the pellets of chousingha, proportion of *Albizia odoratissima* was found maximum (9.16%), and minimum for *Terminalia tomentosa* (0. 57%). Similarly, for grasses it was found that the proportion of *Eragrostis tenella* was maximum (24.52%), and minimum (0.19%) was recorded for *Imperata cylindrica* (Table 6.1). In the diet of chousingha the percentage of grass was found maximum (45.42%), followed by tree (33.78%), shrub (9.92%, herbs (7.540%), and minimum for climber (3.34). During summers, the diet of chousingha constitutes 54.58% browse food items and 45.41% grass food items. Table 6.2 provides the values of 95% Bonferroni confidence intervals values for different plant species utilised by chousingha during the summer season. In summer, 84 species of plants were recorded in the diet of chousingha. The results reveal that chousingha preferred ten trees, one shrub, two climbers, two herbs, and three grass species. *Buchanania cochinchinensis*, *Cassia fistula*, *Bauhinia variegata*, *Gardenia latifolia*, *Albizia odoratissima*, *Grewia latifolia*, *Milosa tomentosa*, *Ixora arborea*, *Madhuca indica*, *Lannea coromandalica*, *Hilectro ixora*, *Bauhinia vahlii*, *Ziziphus oenoplia*, *Guizotia abyssinica*, *Ventilago maderaspatana*, *Eragrostis tenella*, *Heteropogon contortus*, and *Dicanthium spp.* are the plants which were utilised more than their availability. The species which were utilised less than their availability were *Lagerstroemia parviflora* and *Imperata cylindrica*. *Ziziphus xylopyrus*, *Terminalia tomentosa*, and *Eulaliopsis binate* were utilised in proportion to their availability. Except these, 65 species were sampled but not recorded from the diet, which suggests total avoidance by the chousingha.

A total of 2,635 plant fragments were examined to determine feeding habit of chousingha during the post-monsoon season. During post-monsoon, 43 plant species (37 browse and six grass species) were identified from the pellet group of chousingha. Out of 37 identified browse species, the proportion of *Guizotia abyssinic* was found maximum (10.967%), and minimum (0.0759%) for *Gardenia latifolia* and *Gymnosporia spinosa*. For grasses during post-monsoon season, the

Table 6.1 Chousingha diet in different seasons (N = number of individuals per sampled species, n = number of plant fragment recorded in diet, p = percent of occurrence)

Plant Species	Family	Seasons								
		Summer								
		N	n	P	N	n	P	n	N	P
Trees										
Buchanania lanzan	Anacardiaceae	82	15	1.43	99	29	1.1	96	15	0.41
Alangium salviifolium	Alangiaceae	8	0	0	4	0	0	4	0	0
Cassia fistula	Caesalpinaceae	130	74	7.06	152	89	3.38	184	126	3.44
Terminalia arjuna	Combretaceae	4	0	0	1	0	0	1	0	0
Bauhinia barigata	Caesalpinaceae	86	9	0.86	48	12	0.46	77	49	1.34
Emblica officinalis	Euphorbiaceae	52	9	0	77	29	1.1	63	55	1.5
Gardinia latifolia	Rubiaceae	17	9	0.86	12	2	0.08	13	11	0.3
Ficus benghalensis	Moraceae	1	0	0	1	0	0	1	0	0
Pterocarpus marsupium	Fabaceae	1	0	0	1	0	0	1	0	0
Gymnosporia spinosa	Celastraceae	12	0	0	18	2	0.08	13	0	0
Aegle marmelos	Rutaceae	37	0	0	16	6	0.23	34	46	1.26
Kydia calycina	Malvaceae	34	0	0	27	0	0	26	18	0.49
Semecarpus anacardium	Anacardiaceae	27	0	0	27	0	0	27	0	0
Chloroxylon swietenia	Rutaceae	122	0	0	111	39	1.48	185	32	0.87
Hymenodiction excelsum	Rubiaceae	2	0	0	5	0	0	2	0	0
Delbergia sissoo	Fabaceae	6	0	0	0	0	0	0	0	0
Albizia odoratissima	Mimosaceae	5	96	9.16	5	73	2.77	5	99	2.71
Grewia latifolia	Tiliaceae	70	10	0.95	72	16	0.61	62	36	0.98
Anogeissus latifolia	Combretaceae	63	0	0	74	24	0.91	77	0	0
Delbergia paniculata	Fabaceae	9	0	0	9	0	0	9	0	0
Gardinia gummifera	Rubiaceae	0	0	0	0	0	0	4	0	0
Cleistanthus Collinus	Euphorbiaceae	2	0	0	4	0	0	2	48	1.31
Ziziphus xylopyra	Rhamnaceae	31	22	2.1	42	39	1.48	42	36	0.98

Table 6.1 (Continued)

Plant Species	Family	Seasons								
		Summer								
		N	n	P	N	n	P	n	N	P
Adina cordifolia	Rubiaceae	3	0	0	3	0	0	3	0	0
Acacia leucophloea	Mimosaceae	5	0	0	5	0	0	5	0	0
Syzygiumcumini	Myrtaceae	271	0	0	167	69	2.62	193	185	5.058
Flacourtiaindica	Flacourtiaceae	45	0	0	10	35	1.33	20	0	0
Milosatomentosa	Annonaceae	192	19	1.81	240	41	1.56	208	74	2.02
Brideliaretusa	Euphorbiaceae	26	0	0	29	0	0	29	21	0.57
Ficushispida	Moraceae	7	0	0	4	0	0	25	0	0
Garugapinnata	Burseraceae	1	0	0	1	0	0	1	0	0
Acacia catechu	Mimosaceae	23	0	0	24	0	0	24	18	0.49
Schleicheraoleosa	Santalaceae	20	0	0	16	0	0	17	0	0
Careyaarborea	Lecythidaceae	2	0	0	1	0	0	1	0	0
Sterculiaurens	Sterculiaceae	1	0	0	1	0	0	3	0	0
Lagerstroemia parviflora	Lythraceae	650	16	1.53	609	87	3.3	727	42	1.15
Ixoraarborea	Rubiaceae	78	26	2.48	121	79	2.99	78	93	2.54
Madhucaindica	Sapotaceae	25	33	3.15	36	24	0.91	30	54	1.48
Randiadumetorum	Rubiaceae	71	0	0	2	4	0.158	22	0	0
Schreberaswietenioides	Oleaceae	1	0	0	1	0	0	1	0	0
Leniacoromandalica	Anacardiaceae	117	19	1.81	148	21	0.8	132	29	0.79
Mitragynaparvifolia	Rubiaceae	40	0	0	24	9	0.34	44	41	1.12
Ficusinfectoria	Moraceae	2	0	0	2	0	0	2	0	0
Buteamonosperma	Fabaceae	27	0	0	47	34	1.29	41	59	1.61
Soymidafebrifuga	Meliaceae	21	0	0	21	0	0	21	43	1.18
Terminaliatomentosa	Combretaceae	104	6	0.57	107	19	0.72	117	94	2.57
Boswelliaserrata	Burseraceae	5	0	0	5	0	0	5	0	0
Bombaxceiba	Bombacaceae	2	0	0	2	0	0	2	0	0

Species	Family									
Annona squamosa	Annonaceae	1	0	0	0	0	0	5	7	0.19
Tectona grandis	Verbenaceae	1035	0	0	890	16	0.61	1091	67	1.83
Diospyros melanoxylon	Ebenaceae	872	0	0	428	0	0	657	51	1.39
Ougeinia oojeinensis	Fabaceae	44	0	0	60	0	0	51	0	0
Casearia tomentosa	Flacourtiaceae	59	0	0	25	0	0	26	0	0
Ficus glomerata	Moraceae	1	0	0	1	0	0	1	0	0
Diospyros montana	Ebenaceae	28	0	0	10	0	0	10	0	0
Bauhinia vahlii	Caesalpinaceae	5	35	3.34	5	48	1.82	9	87	2.38
Ziziphus oenoplia	Rhamnaceae	120	53	5.06	147	81	3.07	112	94	2.57
Butea parviflora	Fabaceae	1	0	0	7	5	0.19	27	17	0.46
Asparagus racemosum	Liliaceae	0	0	0	0	0	0	21	0	0
Shrub										
Hilectruxizora	Sterculiaceae	140	51	4.87	175	23	0.87	115	39	1.07
Bamboo spp.	Poaceae	353	0	0	508	7	0.27	419	28	0.77
Pheonix aquilis	Arecaceae	129	0	0	262	0	0	170	21	0.57
Grewia hirsute	Tiliaceae	81	0	0	259	0	0	234	0	0
Grewia spp	Tiliaceae	58	0	0	20	0	0	3	0	0
Holarrhena antidysenterica	Apocynaceae	10	0	0	0	0	0	0	0	0
Lantana camara	Verbenaceae	1155	0	0	1160	29	1.1	999	51	1.39
Vitex negundo	Verbenaceae	0	0	0	58	0	0	0	0	0
Herb										
Phyllanthus amarus	Euphorbiaceae	127	0	0	787	0	0	208	45	1.23
Marsilea quadrifolia	Marsileaceae	0	0	0	1588	209	7.93	2768	89	2.43
Casiatora	Caesalpiniaceae	1441	0	0	1463	0	0	1165	0	0
Vallaris solanacea	Apocynaceae	158	0	0	314	0	0	219	32	0.87
Tribulus terrestris	Zygophylaceae	258	0	0	59	0	0	28	0	0
Elephantopus scaber	Asteraceae	402	0	0	635	8	0.3	370	128	3.5
Guizotia abyssinica	Asteraceae	949	0	0	1040	289	10.97	1178	289	7.9
Sida acuta	Malvaceae	367	0	0	1121	0	0	980	0	0
Desmodium spp.	Fabaceae	3	0	0	49	0	0	0	0	0
Xanthium strumaxium	Asteraceae	57	0	0	0	0	0	0	0	0
Sida spp.	Malvaceae	9	0	0	324	59	2.24	63	71	1.94

(Continued)

Table 6.1 (Continued)

Plant Species	Family	Seasons								
		Summer								
		N	n	P	N	n	P	N	n	P
Ventilago maderaspatana	Rhamnaceae	8	9	0.86	0	0	0	0	0	0
Spirodela polyrhiza	Lemnaceae	7	0	0	148	0	0	0	0	0
Sida spp.	Malvaceae	777	0	0	865	0	0	1079	0	0
Ocimum canum	Lamiaceae	158	0	0	453	99	3.76	386	326	8.91
Parthenium hysterophorus	Asteraceae	0	0	0	372	0	0	36	0	0
Ocimum basilicum	Lamiaceae	0	0	0	156	0	0	0	0	0
Desmodium triflorum	Fabaceae	0	0	0	842	255	9.68	0	0	0
Grasses										
Eragrostis tenella	Poaceae	3798	257	24.52	5477	338	12.83	4871	247	6.75
Imperata cylindrical	Poaceae	51	2	0.19	184	0	0	103	0	0
Pennisetum pedicellatum	Poaceae	3	0	0	94	0	0	46	23	0.63
Cynodon dactylon	Poaceae	18	0	0	504	87	3.3	55	89	2.43
Themeda quadrivalvis	Poaceae	218	0	0	334	0	0	98	39	1.07
Chloris barbata	Poaceae	208	0	0	277	0	0	182	51	1.39
Cyperus Scariosus	Cyperaceae	74	0	0	12	0	0	20	0	0
Dicanthium spp.	Poaceae	512	64	6.11	1989	86	3.26	1311	254	6.94
Eulaliopsis binata	Poaceae	254	12	1.15	387	41	1.56	55	0	0
Heteropogon contortus	Poaceae	1306	141	13.45	1811	112	4.25	1842	189	5.17
Apluda mutica	Poaceae	0	0	0	395	48	1.82	0	0	0
Sub-Total		17795	1048	100	28126	2635	100	23692	3658	100
Browse Unidentified		0	1065		0	816		0	355	
Grass Unidentified		0	459		0	327		0	112	

Table 6.2 Bonferroni confidence limits for available (P_{io}) and utilised proportion of different plants species (Pie), 95% bonferroni confidence limits for Pie, and rating of preferences or avoidances of the chousingha during summer, post-monsoon, and winter seasons. a = utilised more than availability, b = utilised less than availability, c = utilisation in proportion to availability, d =species available but not recorded in diet composition (totally avoidance)

Plant species	Summer			Post-monsoon				Winter				
	Pio	Pie	95% Bonferroni	R	Pio	Pie	95% bonferroni	R	Pio	Pie	95% Bonferroni	R
Trees												
Acacia catechu	0.001	0	0.00≤Pi≤0.00	d	0.001	0	0.00≤Pi≤0.00	d	0.001	0.005	0.003≤Pi≤0.006	a
Acacia leucophloca	0.000	0	0.00≤Pi≤0.00	d	0.000	0	0.00≤Pi≤0.00	d	0.0002	0	0.00≤Pi≤0.00	d
Adina cordifolia	0.000	0	0.00≤Pi≤0.00	d	0.000	0	0.00≤Pi≤0.00	d	0.0001	0	0.00≤Pi≤0.00	d
Aeglemarmelos	0.002	0	0.00≤Pi≤0.00	d	0.001	0.002	0.001≤Pi≤0.003	a	0.0014	0.013	0.010≤Pi≤0.015	a
Alangiumsalviifolium	0.000	0	0.00≤Pi≤0.00	d	0.000	0	0.00≤Pi≤0.00	d	0.0002	0	0.00≤Pi≤0.00	d
Albiziaodoratissima	0.001	0.092	0.084≤Pi≤0.098	a	0.000	0.028	0.024≤Pi≤0.031	a	0.0002	0.027	0.023≤Pi≤0.031	a
Annonasquamosa	5.62E-05	0	0.00≤Pi≤0.00	d	0	0	0		0.0002	0.002	0.000≤Pi≤0.003	a
Anogeissuslatifolia	0.003	0	0.00≤Pi≤0.00	a	0.003	0.009	0.007≤Pi≤0.011	a	0.003	0	0.00≤Pi≤0.00	d
Bahuniabarigata	0.005	0.009	0.006≤Pi≤0.011	a	0.002	0.005	0.003≤Pi≤0.006	a	0.003	0.013	0.011≤Pi≤0.0159	a
Bombaxceiba	0.000	0	0.00≤Pi≤0.00	d	7.11E-05	0	0.00≤Pi≤0.00	d	1.00E-04	0	0.00≤Pi≤0.00	d
Boswelliaserrata	0.000	0	0.00≤Pi≤0.00	d	0.0002	0	0.00≤Pi≤0.00	d	0.0002	0	0.00≤Pi≤0.00	d
Brideliaretusa	0.002	0	0.00≤Pi≤0.00	d	0.001	0	0.00≤Pi≤0.00	d	0.001	0.006	0.004≤Pi≤0.007	a
Buchananialanzan	0.005	0.014	0.011≤Pi≤0.017	a	0.004	0.011	0.009≤Pi≤0.013	a	0.004	0.004	0.00≤Pi≤0.00	d
Buteamonosperma	0.002	0	0.00≤Pi≤0.00	d	0.002	0.013	0.011≤Pi≤0.015	a	0.002	0.016	0.013≤Pi≤0.019	a
Careyaarborea	0.000	0	0.00≤Pi≤0.00	d	3.56E-05	0	0.00≤Pi≤0.00	d	1.00E-04	0	0.00≤Pi≤0.00	d
Caseariatomentosa	0.003	0	0.00≤Pi≤0.00	d	0.001	0	0.00≤Pi≤0.00	d	0.001	0	0.00≤Pi≤0.00	d
Cassia fistula	0.007	0.071	0.06≤Pi≤0.07	a	0.005	0.034	0.030≤Pi≤0.037	a	0.008	0.035	0.030≤Pi≤0.039	a

(Continued)

Table 6.2 (Continued)

Plant species	Summer			Post-monsoon				Winter				
	Pio	Pie	95% Bonferroni	R	Pio	Pie	95% bonferroni	R	Pio	Pie	95% Bonferroni	R
Chloroxylon swietenia	0.007	0	0.00≤Pi≤0.00	d	0.004	0.015	0.012≤Pi≤0.017	a	0.008	0.009	0.007≤Pi≤0.011	c
Cleistanthuscollinus	0.000	0	0.00≤Pi≤0.00	c	0.0001	0	0.00≤Pi≤0.00	d	1.00E-04	0.013	0.011≤Pi≤0.016	a
Delbergiapeniculata	0.000	0	0.00≤Pi≤0.00	d	0.0003	0	0.00≤Pi≤0.00	d	0.0004	0	0.00≤Pi≤0.00	d
Delbergiasissoo	0.000	0	0.00≤Pi≤0.00	a	0	0	0		0	0	0	d
Diospyrosmelanoxylon	0.049	0	0.00≤Pi≤0.00	d	0.015	0.006	0.004≤Pi≤0.007	b	0.028	0.014	0.011≤Pi≤0.017	b
Diospyrosmontana	0.002	0	0.00≤Pi≤0.00	d	0.0004	0	0.00≤Pi≤0.00	d	0.0004	0	0.00≤Pi≤0.00	d
Emblicaofficinalis	0.003	0	0.00≤Pi≤0.00	d	0.003	0.011	0.008≤Pi≤0.013	a	0.003	0.015	0.012≤Pi≤0.018	a
Ficusbenghalensis	5.62E-05	0	0.00≤Pi≤0.00	d	3.56E-05	0	0.00≤Pi≤0.00	d	1.00E-04	0	0.00≤Pi≤0.00	d
Ficusglomerata	5.62E-05	0	0.00≤Pi≤0.00	d	3.56E-05	0	0.00≤Pi≤0.00	d	1.00E-04	0	0.00≤Pi≤0.00	d
Ficushispida	0.000	0	0.00≤Pi≤0.00	d	0.0001	0	0.00≤Pi≤0.00	d	0.001	0	0.00≤Pi≤0.00	d
Ficusinfectoria	0.000	0	0.00≤Pi≤0.00	d	7.11E-05	0	0.00≤Pi≤0.00	d	1.00E-04	0	0.00≤Pi≤0.00	d
Flacourtiaindica	0.003	0	0.00≤Pi≤0.00	a	0.0004	0.013	0.011≤Pi≤0.016	a	0.001	0	0.00≤Pi≤0.00	d
Gardiniagummifera	0	0	0		0.00E+00	0	0		0.0001	0	0.00≤Pi≤0.00	d
Gardinialatifolia	0.001	0.009	0.006≤Pi≤0.011	a	0.0004	0.001	0.000≤Pi≤0.001	c	0.001	0.003	0.002≤Pi≤0.004	a
Garugapinnata	5.62E-05	0	0.00≤Pi≤0.00	d	3.56E-05	0	0.00≤Pi≤0.00	d	4.22E-05	0	0.00≤Pi≤0.00	d
Grewialatifolia	0.004	0.009	0.007≤Pi≤0.011	d	0.003	0.006	0.004≤Pi≤0.007	a	0.003	0.009	0.008≤Pi≤0.012	a
Gymnosporiaspinosa	0.001	0	0.00≤Pi≤0.00	d	0.001	0.001	0.000≤Pi≤0.001	c	0.001	0	0.00≤Pi≤0.00	d
Hymenodictyonexcelsum	0.000	0	0.00≤Pi≤0.00	d	0.0002	0	0.00≤Pi≤0.00	d	1.00E-04	0	0.00≤Pi≤0.00	d
Ixorraarborea	0.004	0.025	0.020≤Pi≤0.028	a	0.004	0.029	0.026≤Pi≤0.033	a	0.003	0.025	0.022≤Pi≤0.029	a
Kydiacalycina	0.002	0	0.00≤Pi≤0.00	d	0.001	0	0.00≤Pi≤0.00	d	0.001	0.005	0.003≤Pi≤0.006	a

Species											
Lagerstroemia parviflora	0.037	$0.012 \leq P_i \leq 0.018$	a	0.022	0.033	$0.029 \leq P_i \leq 0.036$	c	0.031	0.012	$0.009 \leq P_i \leq 0.014$	b
Leniacoromandalica	0.007	$0.014 \leq P_i \leq 0.021$	d	0.005	0.008	$0.006 \leq P_i \leq 0.009$	a	0.006	0.008	$0.006 \leq P_i \leq 0.010$	c
Madhucaindica	0.001	$0.027 \leq P_i \leq 0.035$	d	0.001	0.009	$0.007 \leq P_i \leq 0.011$	a	0.001	0.015	$0.012 \leq P_i \leq 0.017$	a
Milosatomentosa	0.011	$0.014 \leq P_i \leq 0.021$	d	0.009	0.016	$0.013 \leq P_i \leq 0.018$	a	0.009	0.02	$0.017 \leq P_i \leq 0.023$	a
Mitragynaparvifolia	0.002	$0.00 \leq P_i \leq 0.00$	d	0.001	0.003	$0.002 \leq P_i \leq 0.004$	a	0.002	0.011	$0.009 \leq P_i \leq 0.014$	a
Ougeiniaoojeinensis	0.002	$0.00 \leq P_i \leq 0.00$	d	0.002	0	$0.00 \leq P_i \leq 0.00$	d	0.002	0	$0.00 \leq P_i \leq 0.00$	d
Pterocarpusmarsupium	5.62E-05	$0.00 \leq P_i \leq 0.00$	d	3.56E-05	0	$0.00 \leq P_i \leq 0.002$	d	1.00E-04	0	$0.00 \leq P_i \leq 0.00$	d
Randiadumetorum	0.004	$0.00 \leq P_i \leq 0.00$	d	7.11E-05	0.002	$0.0007 \leq P_i \leq 0.00$	a	0.001	0	$0.00 \leq P_i \leq 0.00$	d
Schleicheraoleosa	0.001	$0.00 \leq P_i \leq 0.00$	d	0.001	0	$0.00 \leq P_i \leq 0.00$	d	0.001	0	$0.00 \leq P_i \leq 0.00$	d
Schreberaswietenioides	1.00E-04	$0.00 \leq P_i \leq 0.00$	a	3.56E-05	0	$0.00 \leq P_i \leq 0.00$	d	1.00E-04	0	$0.00 \leq P_i \leq 0.00$	d
Semecarpusanacardium	0.002	$0.00 \leq P_i \leq 0.00$	d	0.001	0	$0.00 \leq P_i \leq 0.00$	d	0.001	0	$0.00 \leq P_i \leq 0.00$	d
Soymidafebrifuga	0.001	$0.00 \leq P_i \leq 0.00$	c	0.001	0	$0.00 \leq P_i \leq 0.00$	d	0.001	0.012	$0.009 \leq P_i \leq 0.014$	a
Sterculiaurens	5.62E-05	$0.00 \leq P_i \leq 0.00$	b	3.56E-05	0	$0.00 \leq P_i \leq 0.00$	d	0.0001	0	$0.00 \leq P_i \leq 0.00$	d
Syzygiumcumini	0.02	$0.00 \leq P_i \leq 0.00$	d	0.00594	0.026	$0.023 \leq P_i \leq 0.029$	a	0.009	0.051	$0.046 \leq P_i \leq 0.055$	a
Tectonagrandis	0.058	$0.00 \leq P_i \leq 0.00$	d	0.032	0.005	$0.003 \leq P_i \leq 0.006$	b	0.046	0.018	$0.015 \leq P_i \leq 0.021$	b
Terminaliaarjuna	0.000	$0.00 \leq P_i \leq 0.00$	d	3.56E-05	0	$0.00 \leq P_i \leq 0.00$	d	1.00E-04	0	$0.00 \leq P_i \leq 0.00$	d
Terminaliatomentosa	0.006	$0.003 \leq P_i \leq 0.007$	d	0.004	0.007	$0.005 \leq P_i \leq 0.009$	c	0.005	0.026	$0.022 \leq P_i \leq 0.029$	a
Ziziphusxylopyra	0.002	$0.017 \leq P_i \leq 0.024$	d	0.002	0.015	$0.012 \leq P_i \leq 0.017$	a	0.002	0.01	$0.008 \leq P_i \leq 0.012$	a
Creeper											
Asparagus racemosum	0	0		0	0	0		0.001	0	$0.00 \leq P_i \leq 0.00$	d
Bauhinia vahlii	0.001	$0.028 \leq P_i \leq 0.037$	a	0.0002	0.018	$0.015 \leq P_i \leq 0.021$	a	0.0004	0.024	$0.020 \leq P_i \leq 0.027$	a
Buteaparviflora	5.62E-05	$0.00 \leq P_i \leq 0.00$	a	0.0003	0.002	$0.001 \leq P_i \leq 0.003$	a	0.001	0.005	$0.003 \leq P_i \leq 0.006$	a
Ziziphusoenoplia	0.007	$0.045 \leq P_i \leq 0.056$	a	0.005	0.031	$0.027 \leq P_i \leq 0.034$	a	0.005	0.026	$0.022 \leq P_i \leq 0.029$	a

(Continued)

Table 6.2 (Continued)

Plant species	Summer			Post-monsoon				Winter				R
	Pio	Pie	95% Bonferroni	R	Pio	Pie	95% bonferroni	R	Pio	Pie	95% Bonferroni	
Shrub												
Bamboo spp.	0.02	0	0.043≤Pi≤0.054	d	0.018	0.003	0.002≤Pi≤0.004	b	0.018	0.008	0.006≤Pi≤0.010	b
Grewia hirsute	0.005	0	0.00≤Pi≤0.00	d	0.009	0	0.00≤Pi≤0.00	d	0.01	0	0.00≤Pi≤0.00	d
Grewiaspp	0.003	0	0.00≤Pi≤0.00	d	0	0	0		0	0	0	
Hilectruxizora	0.008	0.049	0.00≤Pi≤0.00	a	0.006	0.009	0.007≤Pi≤0.02	c	0.005	0.011	0.008≤Pi≤0.013	a
Holarrhenaantidysenterica	0.001	0	0.00≤Pi≤0.00	d	0.001	0	0.00≤Pi≤0.00	d	0.0001	0	0.00≤Pi≤0.00	d
Lantana camara	0.065	0	0.00≤Pi≤0.00	d	0.041	0.011	0.009≤Pi≤0.013	b	0.042	0.014	0.011≤Pi≤0.017	b
Pheonixaquilis	0.008	0	0.00≤Pi≤0.00	d	0.009	0	0.00≤Pi≤0.00	d	0.007	0.006	0.004≤Pi≤0.007	c
Vitexnegundo	0	0	0		0.002	0	0.00≤Pi≤0.00	d	0	0	0	
Herb												
Casiatora	0.081	0	0.00≤Pi≤0.00	d	0.052	0	0.00≤Pi≤0.00	d	0.049	0	0.00≤Pi≤0.00	d
Desmodium spp.	0.000	0	0.00≤Pi≤0.00	d	0.002	0	0.00≤Pi≤0.00	d	0	0	0	
Desmodiumtriflorum					0.03	0.097	0.091≤Pi≤0.103	a	0	0	0	
Elephantopusscaber	0.023	0	0.00≤Pi≤0.00	d	0.023	0.003	0.002≤Pi≤0.004	b	0.016	0.035	0.031≤Pi≤0.039	a
Guizotiaabyssinica	0.053	0.067	0.06≤Pi≤0.07	a	0.037	0.11	0.103≤Pi≤0.116	a	0.049	0.079	0.00≤Pi≤0.00	d
Marsileaquadrifolia	0	0	0		0.057	0.079	0.074≤Pi≤0.084	a	0.117	0.024	0.021≤Pi≤0.028	b
Ocimumbasilicum	0	0	0		0.006	0	0.00≤Pi≤0.00	d	0	0	0	
Ocimumcanum	0.009	0	0.00≤Pi≤0.00	d	0.016	0.038	0.003≤Pi≤0.041	b	0.016	0.089	0.0827≤Pi≤0.0954	a
Phyllanthusamarus	0.007	0	0.00≤Pi≤0.00	d	0.028	0	0.00≤Pi≤0.00	d	0.009	0.012	0.010≤Pi≤0.015	a
Partheniumhysterophorus	0	0	0		0.013	0	0.00≤Pi≤0.00	d	0.001	0	0.00≤Pi≤0.00	d

Sida acuta	0.021	0.00≤Pi≤0.00	d 0.04	0	0.00≤Pi≤0.00	d 0.041	0	0.00≤Pi≤0.00	d
Sida spp.	0.001	0.00≤Pi≤0.00	d 0.031	0	0.00≤Pi≤0.00	d 0.046	0	0.00≤Pi≤0.00	d
Sida spp.	0.044	0.00≤Pi≤0.00	d 0.012	0.022	0.019≤Pi≤0.025	c 0.003	0.019	0.016≤Pi≤0.022	a
Spirodela polyrhiza	0.001	0.00≤Pi≤0.00	d 0.005	0	0.00≤Pi≤0.00	d 0	0	0	
Tribulus terrestris	0.015	0.00≤Pi≤0.00	d 0.002	0	0.00≤Pi≤0.00	d 0.001	0	0.00≤Pi≤0.00	d
Vallaris solanacea	0.009	0.00≤Pi≤0.00	d 0.011	0	0.00≤Pi≤0.00	d 0.009	0.009	0.007≤Pi≤0.011	c
Ventilagomaderaspatana	0.001	0.006≤Pi≤0.011	a 0	0	0	0	0	0	
Xanthium strumaxium	0.003	0.00≤Pi≤0.00	d 0	0	0	0	0	0	
Grasses									
Apluda mutica	0	0	0.014	0.018	0.015≤Pi≤0.020	c 0	0	0	
Chloris barbata	0.012	0.00≤Pi≤0.00	d 0.01	0	0.00≤Pi≤0.00	d 0.008	0.014	0.011≤Pi≤0.017	a
Cynodon dactylon	0.001	0.00≤Pi≤0.00	d 0.018	0.033	0.02≤Pi≤0.03	a 0.002	0.024	0.021≤Pi≤0.028	a
Cyperus cariosus	0.004	0.00≤Pi≤0.00	d 0.001	0	0.00≤Pi≤0.00	d 0.001	0	0.00≤Pi≤0.00	d
Dicanthium spp.	0.029	0.055≤Pi≤0.067	a 0.071	0.033	0.029≤Pi≤0.036	b 0.055	0.069	0.064≤Pi≤0.075	a
Eragrosti stenella	0.213	0.234≤Pi≤0.255	a 0.195	0.128	0.121≤Pi≤0.135	c 0.206	0.068	0.062≤Pi≤0.073	b
Eulaliopsis binata	0.014	0.008≤Pi≤0.014	c 0.014	0.016	0.013≤Pi≤0.018	c 0.002	0	0.00≤Pi≤0.00	d
Heteropogon contortus	0.073	0.126≤Pi≤0.143	a 0.064	0.043	0.038≤Pi≤0.047	c 0.078	0.052	0.047≤Pi≤0.057	b
Imperata cylindrical	0.003	0.001≤Pi≤0.002	b 0.007	0	0.00≤Pi≤0.00	d 0.004	0	0.00≤Pi≤0.00	d
Pennisetum pedicellatum	0.000	0.00≤Pi≤0.00	d 0.003	0	0.00≤Pi≤0.00	d 0.002	0.006	0.005≤Pi≤0.008	a
Themeda quadrivalvis	0.012	0.00≤Pi≤0.00	d 0.012	0	0.00≤Pi≤0.00	d 0.004	0.011	0.008≤Pi≤0.013	a

proportion of *Eragrostis tenella* was maximum (12.827%), and minimum (1.555%) for *Eulaliopsis binate* (Table 6.1). The result also reveals that in the diet of chousingha during post-monsoon season, the percentage of herb was found maximum (34.88%), followed by tree (30.78%), grass (27.02%), climber (5.08%), and minimum (2.24%) for shrub. The results show that during post-monsoon season, the diet of chousingha constitutes of 72.98% browse items and 27.02% grass items. Table 6.2 shows the values of 95% Bonferroni confidence intervals values for different plant species utilised by chousingha during post-monsoon season. A total of 89 plants were recorded in the chousingha's habitat. The results show that during post-monsoon chousingha preferred 19 tree, three climbers, three herbs, and one grass species. *Buchanania lanzan*, *Cassia fistula*, *Bauhinia variegata*, *Emblica offcinalis*, *Aegle marmelos*, *Chloroxylon swietenia*, *Albizia odoratissima*, *Grewia latifolia*, *Anogeissus latifolia*, *Zizyphus xylopyrus*, *Syzygium cumini*, *Flacourtia indica*, *Milosa tomentosa*, *Ixora arborea*, *Madhuca indica*, *Randia dumetorum*, *Lannea coromandalica*, *Mitragyna parvifolia*, *Butea monosperma*, *Bauhinia vahlii*, *Butea parviflora*, *Ziziphus oenoplia*, *Marsilea quadrifolia*, *Guizotia abyssinica*, *Desmodium triflorum*, and *Cynodon dactylon* are the plants which were utilised more than their availability. The species which were utilised less than their availability were *Tectona grandis*, *Diospyros melanoxylon*, *Bamboo spp.*, *Lantana camara*, *Elephantopus scaber*, *Ocimum canum*, and *Dicanthium spp.* Utilised in proportion to their availability were *Gardinialatifolia*, *Gymnosporia spinosa*, *Lagerstroemia parviflora*, *Terminalia tomentosa*, *Hilectrux izora*, *Sida spp.*, *Eragrostis tenella*, *Apluda mutica*, *Eulaliopsis binata*, and *Heteropogon contortus*. Except these, 46 species were also sampled but not recorded from the diet, which suggests total avoidance by the chousingha during post-monsoon season.

During winters a total of 3,658 plant fragments were examined to investigate food and feeding habit of chousingha in winter season. In winter, 48 species (41 browse and seven grass species) of plants were identified from their diet. Out of all identified browse species from the pellet group of chousingha, the proportion of *Ocimum canum* was maximum (8.911%), and minimum (0.191%) for *Annona squamosa*. Similarly, for grasses, were utilised in proportion to their availability the proportion of *Dicanthium spp.* was maximum (6.943%), and minimum (0.628%) for *Pennisetum pedicellatum* (Table 6.1). In the diet of chousingha during winter, the percentage of trees was maximum (39.61%), followed by herb (26.79%), grass (24.38%), climber (5.41%), and minimum (3.81%) for shrubs. During winter, the diet of chousingha constitutes of 75.62% of browse items and 24.38% of grass items. Table 6.2 represents the values of 95% Bonferroni confidence intervals values for different plant species utilised by chousingha in winters. During winter, 54 trees, six shrubs, four climbers, 12 herbs, and 10 grass species were recorded with 23,692 individuals of these species. The results suggest that in winters chousingha preferred 21 trees, three climbers, one shrub, four herbs, and five grass species and completely avoided other species. *Cassia fistula*, *Bauhinia variegata*, *Emblica officinalis*, *Gardinia latifolia*, *Aegle marmelos*, *Kydia calycina*, *Albizia odoratissima*, *Grewia latifolia*, *Cleistanthus collinus*, *Ziziphus xylopyrus*, *Syzygium cumini*, *Milosa tomentosa*, *Bridelia retusa*, *Acacia catechu*, *Ixora arborea*, *Madhuca indica*, *Mitragyna parvi folia, Butea monosperma, Soymida febrifuga, Terminalia tomentosa, Annona squa-mosa, Bauhinia vahlii, Ziziphus oenoplia, Butea parviflora, Hilectrux izora, Phyllanthus amarus, Elephantopus scaber, Sida spp., Ocimum canum, Pennisetum pedicellatum, Cynodon dactylon, Themeda quadrivalvis, Chloris barbata,* and *Dicanthium spp.* are the plants which were utilised more than their availability. The species which were utilised less than their availability were *Lagerstroemia parviflora, Tectona grandis, Diospyros melanoxylon, Bamboo spp., Lantana camara, Marsilea quadrifo-lia, Eragrostis tenella* and *Heteropogon contortus.* Utilised in proportion to their availability were *Chloroxylon swietenia, Lannea coromandalica, Pheonix aquilis,* and *Vallaris solanacea.* Except these, another 40 species were sampled but not recorded from the diet, which suggests that they were totally avoided by the chousingha during winter season.

Apart from the identified plant fragments, 1,524, 1,143, and 467 unidenti-fied fragments of monocot and dicot were found in summer, post-monsoon, and winter, respectively, and their proportion was 30.12% and 69.88% in sum-mer, 28.60% and 71.39% in post-monsoon, and 76.01% and 23.98% in winter, respectively. It was also observed that *Lantana camara,* which is a weed and exotic species, is eaten by some of the ungulate species present in Pench. However, chousingha was not found to feed on *Lantana spp.* during summer, but it was reported during the post-monsoon season and winter season (1.10% and 1.39% respectively).

Discussion

Conservation of the species requires a good understanding of its ecology. Ungulates are the best indicator of the health of a forest, and understanding their food preferences and changes resulting from biotic influences are important in interpreting relationships between the environment and the consumer (Hanley, 1984; Leopold & Krausman, 1987; Ilyas & Khan, 2004).

Consumer-resource interactions are the core motif of ecological food chains or food webs (Bascompte, 2009) and are an umbrella term for a variety of more specialised types of biological species interactions, including the plant-herbivore. The study and modelling of these kinds of interactions is comparatively new in population ecology (Murdoch et al., 2003; Turchin, 2003). According to Gotelli (2001), the relationship between herbivores and plants is cyclic. When palat-able species of plants are abundant, herbivores increase in numbers, reducing the plant density and causing a decline in the number of herbivores, which eventu-ally results in the recovery of plants and starts the same cycle again. This sug-gests that the population of herbivores fluctuates around the carrying capacity of wildlife habitats. Thus, it is important to study the feeding ecology of herbivores. Keystone herbivores also keep vegetation populations in check and allow for a greater diversity of both herbivores and plants (Smith & Smith, 2001).

Because of the problem in following ungulates in wildlife habitats, it is dif-ficult to have direct observations of the animals to study their feeding ecology. Moreover, animals cannot be sacrificed for applying the oesophageal and rumen fistula technique for collecting rumen content. Therefore, it is most relevant to rely on indirect evidence (pellets) to study the feeding ecology of the species. Micro-histology was first described by Baumgartner and Martin (1939) and reviewed by Holechek et al. (1982). It is undoubtedly the only technique to be able to study the feeding ecology of small secretive species such as chousingha. However, this method for determining the food habits of small herbivores can be biased by differential digestibility of ingested plant species (Holechek et al., 1982). Forbs are usually highly digestible and as a result, underestimated by faecal analyses (Vavra et al., 1978; Vavra & Holechek, 1980; McInnis et al., 1983). The content of pellets by faecal analysis can be overestimated in some grass and browse species, while others are underestimated (Dearden et al., 1975; Leslie et al., 1983). Research on food habits of ungulates has progressed as much in India as it has in Africa (e.g. Talbot & Talbot, 1962; Gwynne & Bell, 1968; Leuthold, 1970; Bell, 1971; Jarman & Sinclair, 1979). In India, Schaller (1967) listed the food plants of wild ungulates, in Kanha, and Berwick (1976) discussed the field-feeding traits in Gir. Feeding patterns of different mountain ungulates were first reported by Schaller (1977), but the study did not provide their detailed dietary profile, whereas Mishra and Johnsingh (1996), Ilyas and Khan (2004), and Ilyas and Syed (2014) studied the diet of temperate and sub-alpine ungulates. Furthermore, Haleem et al. (2014) studied food and feeding habits of chousingha in Pench Tiger Reserve. Different ungulate species were found to have varying food and feeding habits. Some may be primarily grazers and occasionally browsers, and on the other hand others may be primarily browsers but occasionally grazers.

Results suggested that chousingha preferred dicot over monocot in general, but the proportion of grasses was found maximum in the diet, which is similar to the study of Baskaran et al. (2011). Among all identified plants species from the diet of chousingha, the proportion of *Eragrostis tenella* was found maximum throughout different seasons, which suggested that the animal is largely a grazer, which may be due to the high availability of *Eragrostis tenella* grass. However, the result of diet preference analysis indicated that chousingha mostly prefer *Albizia odoratissima* in summer, *Guizotia abyssinica* in post-monsoon, and *Ocimum canum* in winter, which may be due to the availability of leaves at the accessible height of tree seedling and sapling in summer. It was also observed that *Guizotia abyssinica* and *Ocimum canum* sprout in large numbers after the rainy season and are available throughout the year for chousingha. The outcome of the study reveals that despite the availability of a variety of plant food items, chousingha prefer to feed on some of the selected plant materials and avoid the rest of the plant species, and therefore they are considered to be specialist, also supported by Levine's Index for niche breath (Haleem et al., 2014). Since chousingha is a selective feeder, for the long term conservation of the declining population of the species, managers have to improve the existing habitat, giving more emphasis to maintaining the preferred plant species inside as well as outside the protected areas.

Acknowledgment: The authors thank the SERB-DST for providing funding support to carry out the work. The authors are also thankful to the Chairman,

Department of Wildlife Sciences, for providing permission to carry out the work in Pench Tiger Reserve. The authors are very thankful to the Chief Wildlife Warden and all the forest staff of Madhya Pradesh, especially Pench Tiger Reserve, for not only giving the permission to work but also for providing all logistic support for smooth functioning of the project in Pench Tiger Reserve for three years.

References

Alipayo, D., Valdez, R., Holechek, J. L., & Cardenas, M. (1992). Evaluation of micro histological analysis for determining ruminant diet botanical composition. *Journal of Range Management, 45*(2), 148–152.

Bascompte, J. (2009). Disentangling the web of life. *Science, 325*(5939), 416–419.

Baskaran, N., & Desai, A. A. (1999). An ecological investigation on four-horned antelope (*Tetracerus quadricornis*) in Mudumalai Wildlife Sanctuary, Tamil Nadu. *Technical report, Bombay Natural History Society*. Bombay.

Baskaran, N., Desai, A., & Udhayan, A. (2009). Population distribution and conservation of the four-horned antelope (*Tetracerus quadricornis*) in the tropical forest of Southern India. *Journal of Science and Environment Technology, 2*(3), 139–144.

Baskaran, N., Kannan, V., Thiyagesan, K., & Desai, A. (2011). Behavioural ecology of four-horned antelope (*Tetracerus quadricornis* de Blainville, 1816) in the tropical forests of southern India. *Mammalian Biology, 76*(1), 741–747.

Baumgartner, L. L., & Martin, A. C. (1939). Plant histology as an aid in squirrel food-habits studies. *Journal of Wildlife Management, 3*(3), 266–268.

Bell, R. H. V. (1970). The use of the herb layer by grazing ungulates in the Serengeti. In A. Watson (Ed.), *Animal populations in relation to their food resources* (pp. 111–124). Blackwell Science Publication.

Bell, R. H. V. (1971). A grazing ecosystem in Serengeti. *Science America, 225*(1), 86–93.

Berwick, S. H. (1974). *The community of wild ruminants in Gir Forest ecosystem, India* [Ph.D. Thesis]. Yale University.

Berwick, S. H. (1976). The Gir Forest: An endangered ecosystem. *American Science, 64*, 28–40.

Dearden, B. L., Pegau, R. E., & Hansen, R. M. (1975). Precision of microhistological estimates of ruminant food habits. *Journal of Wildlife Management, 39*(2), 402–407.

Fowler, J., & Cohen, L. (1986). *Statistics for ornithologist*. BTO Guide No. 22. British Trust of Ornithology.

Gotelli, N. J. (2001). *A primer of ecology* (3rd ed.). Sinauer Associates, pp. 265.

Green, M. J. B. (1985). Ecological separation in Himalayan ungulates. *Journal of Zoology, 1*(4), 693–719.

Gyynne, M. O., & Bell, R. H. (1968). Selection of vegetation components by grazing ungulates in Serengeti National Park. *Nature, 220*, 390–393.

Haleem, A., Ilyas, O., Syed, Z., Arya, S. K., & Ekwal, I. (2014). Distribution, status and aspect of ecology of mammalian species in Kedarnath Wildlife Sanctuary, Uttarakhand Himalayas, India. *Journal of Materials and Environmental Science, 5*(3), 683–692.

Hanley, T. A. (1984). Habitat patches and their selection by wapiti and black tailed deer in a coastal montane coniferous forest. *Journal of Applied Ecology, 21*(2), 423–436.

Holechek, J. L., Vavra, M., & Pieper, R. D. (1982). Botanical composition determination of range herbivore diets: A review. *Journal of Range Management, 35*(3), 309–315.

Ilyas, O., Khan, J. A., & Khan, A. (1998). Status, distribution and conservation of deer populations of oak forests in Kumaon Himalayas, India. *Proceeding of 3rd International deer biology congress* (pp. 18–24). Hungary.

Ilyas, O., & Khan, J. A. (2004). Food habits of barking deer (*Muntiacus muntjak*) and goral (*Nemorhaedus goral* bedfordi) in Binsar Wildlife Sanctuary, India. *Mammalia, 67*, 521–531.

Ilyas, O., & Syed, Z. (2014). Habitat use and overlap of Himalayan musk deer (*Moschus chrysogaster*) with reference to habitat quality and human use in Kedarnath Wildlife Sanctuary, Uttarakhand Himalayas, India. *UGC-Major Research Project Report*, AMU Aligarh.

Jarman, P. (1974). The social organization of antelopes in relation to their ecology. *Behaviour, 58*, 215–267.

Jarman, P. J., & Sinclair, A. R. E. (1979). Feeding strategy and the pattern of resource partitioning in ungulates. In A. R. E. Sinclair & M. Norton-Griffiths (Eds.), *Serengeti: Dynamics of an ecosystem* (pp. 130–163). University of Chicago Press.

Karanth, K. U., & Sunquist, M. E. (1992). Population structure, density and biomass of large herbivores in the tropical forests of Nagarhole, India. *Journal of Tropical Ecology, 8*(1), 21–35.

Kleiber, M. (1961). *The fire of life: An introduction to animal energetic*. John Wiley.

Krishna, Y. C. (2006). *Estimating occupancy and assessing the influence of covariates on the distribution of the four horned antelope (Tetracerus quadricornis) in a South Indian tropical Forest* [M.Sc Thesis]. National Center for Biological Science.

Krishna, Y. C., Clyne, P. J., Krishnaswamy, J., & Kumar, N. S. (2009). Distributional and ecological review of the four horned antelope, *Tetracerus quadricornis*. *Mammalia, 73*, 1–6.

Krishna, Y. C., Krishnaswamy, J., & Kumar, N. S. (2008). Habitat factors affecting site occupancy and relative abundance of four-horned antelope. *Journal of Zoology, 276*(1) 63–70.

Krishnan, M. (1972). An ecological survey of the larger mammals of peninsular India. *Journal of the Bombay Natural History Society, 69*, 469–501.

Leopold, B. D., & Krausman, P. R. (1987). Diurnal activity patterns of desert mule deer in relation to temperature. *Texas Journal of Science, 39*, 49–53.

Leslie, D. M., Vavra, M., Edward, E. S., & Slater, R. C. (1983). Correcting for differential digestibility in micro-histological analyses involving common coastal forages of the pacific northwest. *Journal of Range Management, 36*(6), 730–732.

Leuthold, W. (1970). Preliminary observations on food habits of gerenuk in national park Kenya. *Ecology of African Wildlife Journal, 8*(1), 73–84.

Mc Farland, D. (1981). *The Oxford companion to animal behavior*. Oxford University Press.

Mclnnis, M. L., Vavra, M., & Krueger, W. C. (1983). A comparison of four methods used to determine the diets of large herbivores. *Journal of Range Management, 36*(3), 302–306.

Mishra, C., & Johnsingh, A. J. T. (1996). On habitat selection by the goral *Nemorhaedus goral* Bedfordi. *Journal of Zoology, 240*(3), 573–580.

Murdoch, W. W., Briggs, C. J., & Nisbet, R. M. (2003). *Consumer–resource dynamics*. Princeton University Press.

Neu, W. C., Byers, C. R., & Peek, J. M. (1974). A technique for analysis of utilization availability data. *Journal of Wildlife Management, 38*(3), 541–545.

Prater, S. H. (1971). *The book of Indian animals (Bombay Natural History Society)* (3rd ed.). Oxford University Press, pp. 324.

Rahmani, A. R. (2001). Antelopes : global survey and regional action plans, part 4 : North Africa, the Middle East, and Asia. S. C. Kingswood & D. P. Mallon (Eds). Antelope Specialist Group, IUCN.

Rice, C. G. (1990). The status of four-horned antelope *Tetracerus quadricornis*. *Journal of the Bombay Natural History Society*, 88, 63–66.

Sankar, K., Qureshi, Q., Pasha, M. K., & Areendran, G. (2000). Ecology of gaur (*Bos gaurus*) in Pench Tiger Reserve, Madhya Pradesh, final report. Wildlife Institute of India.

Satkopan, S. (1972). Key to identification of plant remain in animal dropping. *Journal of Bombay Natural History Society*, 69(1), 139–150.

Schaller, G. (1977). *Mountain monarches: Wild sheep and goat of Himalaya*. University of Chicago Press, pp. 423.

Schaller, G. B. (1967). *The deer and the tiger*. The University of Chicago Press.

Sharatchandra, H. C., & Gadgil, M. (1975). A year of Bandipur. *Journal of the Bomay Natural History Society*, 72(3), 623–647.

Sharma, K., Rahmani, A. R., & Chundawat, R. S. (2005). Ecology and distribution of four-horned antelope *Tetracerus quadricornis* in India. *Final Report-DST*, Bombay Natural History Society.

Shukla, R. (1990). *An ecological study of interactions between wild animals and vegetation in Pench Sanctuary and environments* [PhD Thesis]. Dept. of Botany. Sagar, India: Harisingh Gaur Vishwavidyalaya, p. 249.

Singh, L. A. K., & Swain, D. (2003). The four horned antelope (*Tetracerus quadricornis*) in Simlipal. *Zoos' Print Journal*, 18(9), 1197–1198.

Smith, R. L., & Smith, T. M. (2001). *Ecology and field biology* (6th ed.). Benjamin Cummings, pp. 720.

Sparks, D. R., & Malechek, J. C. (1968). Estimating percentage dry weight in diets using a microscopic technique. *Journal of Range Management*, 21(4), 264–285.

Syed, Z., & Ilyas, O. (2012). Status, distribution and aspects of ecology of Alpine musk deer (*Moschus chrysogaster*) in Uttarakhand Himalayas, India. *IUCN - DSG Newsletter*, No 24, pp. 4–12.

Talbot, L. M., & Talbot, M. H. (1962). Food preferences of some East African wild ungulates. *East African Agriculture and Forestry Journal*, 27(3), 131–138.

Turchin, P. (2003). *Complex population dynamics*. Princeton University Press, pp. 450.

Vavra, M., & Holechek, J. L. (1980). Factors influencing micro-histological analysis of herbivores diet. *Journal of Range Management*, 33(5), 371–374.

Vavra, M., Rice, R. W., & Hansen, R. M. (1978). A comparison of esophageal fistula and fecal material to determine steer diets. *Journal of Range Management*, 31(1), L1–L13.

Zar, J. H. (1999). *Bio statistical analysis* (4th ed.). Pearson Education Inc., pp. 663.

7 Do the Niches of Sympatric Sparrows Differ?

Anukul Nath and Hilloljyoti Singha

Introduction

On a global scale humans are shifting the composition of habitat, and with ever increasing human population, the importance of incorporating the impact of human activities into the study of ecology is becoming more widely recognised (Palmer et al., 2004). Very recently, urban ecosystems have attracted the attention of ecologists. It is only since the 1970s that urban ecosystems have been examined more broadly, revealing that in spite of severe urbanisation they retained a variety of vegetative structures and supported several wildlife species (Gilbert, 1989, Adams, 1994). However, as landscape becomes more urbanised, the distribution and abundance of species tend to homogenise urban avifauna by reducing both species richness and evenness (Marzluff et al., 2001).

The tree sparrow *Passer montanus* and house sparrow *Passer domesticus* are widely distributed sympatric species between which hybrids been reported. Both these species originated from a common ancestor in the Pleistocene—the tree sparrow in China, the house sparrow in the Middle East. Later expansion has led to extensive overlapped distribution (Summers-Smith, 1988). Traditionally, house sparrows form commensal relationships with mankind allowing them to invade numerous habitats in various parts of the world (Summers-Smith, 1963). While they initially thrived well in response to urbanisation, in recent decades this species has suddenly suffered massive declines, most pronounced in highly urbanised city centres (Chamberlain et al., 2007; De Laet & Summers-Smith, 2007). Although the house sparrow is still common in many cities, such as Berlin (Witt, 2005) and Paris (Mahler, 2006), studies also show that there has been a dramatic decline of the species in many parts of the world. Decline in sparrow population is also reported from India (Daniels, 2008), but there is no authentic historical data on the species to compare population trends (Dhanya & Azeez, 2010). However, Rajashekhar and Venkatesha (2008), Khera et al. (2009), and Dhanya and Azeez (2010) made an effort to record the population of house sparrows in Bangalore, Delhi, and Aaraku, Andhra Pradesh. In the absence of sound data on population dynamics, it has not been possible to assert that there have indeed been declines in sparrow numbers anywhere in India. Lack of scientific information has also stood in the way of attributing apparent declines to any

DOI: 10.4324/9781003321422-8

specific cause for declining popularly cited. However, in 2013 a country-wide citizen science project in India, known as "Citizen Sparrow," evaluated the status of the house sparrow in India and revealed that

> Based on over 10,000 reports from across the country and from different years, some patterns are clear. Sparrow occurrence is reported to be lower at present than in the past, and this is consistent across the country. Sparrow occurrence is lower in cities compared with towns and villages, and this is again consistent in different parts of the country. Still, there is large variation in the occurrence of sparrows from city to city. For example, sparrows are reported to be widespread in Mumbai and Coimbatore, but are missing in most localities in Bangalore and Chennai, while Delhi is intermediate.
>
> (www.citizensparrow.in)

In north-eastern parts of India, both the house sparrow and tree sparrow have occupied various human habitations. Nath et al. (2019) reported the presence of both the species in the different habitats of the rapidly urbanising capital city (Guwahati) of Assam. Here we summarise the studies that were carried out on sympatric sparrows, emphasising the niche of the above-mentioned sympatric species on a macroscale spatial distribution and at a microscale habitat selection.

Study Area

Guwahati, the capital city of Assam, is one of the most important cities in terms of population size, connectivity of transport, and strategic location. The landscape is characterised by the river Brahmaputra and fragmented hill slopes. The city is the gateway to north-east India and is important, being capital of the state Assam. With new emerging socio-economic forces, the city is growing very fast, responding to diverse developments, both internal and external. Guwahati is said to be the city of Eastern light, the legendary Pragjyotishpur. The city has a rich historical past and finds frequent mention in medieval historical sources and also in Mahabharata, Ramayana, and Raghuvansham of Kalidasa. The city of Guwahati is located on the southern bank of the river Brahmaputra extending from 26° 10′ north latitude and 92° 49′ east longitude. Guwahati is situated towards the south-eastern side of Kamrup district bounded by Nalbari district in the north, Darrang and Marigaon districts in the east, Meghalaya state in the south, and Goalpara and Barpeta districts in the west. The city is surrounded by hillocks on the southern and eastern sides on undulating plains, with varying altitudes of 49.5 m to 55.5 m above mean sea level (MSL). The central part of the city has small hillocks, namely Chunsali Hill (293 m), Nilachal Hill (193 m), Nabagraha Hill (217 m), and Sarania Hill (193 m). As per provisional reports of Census India, the population of Guwahati in 2011 was 957,352. The population of Guwahati has risen from 8,934 in 1891 to around 100,000 in 1961, and according to the census record of 2001 the population rose up to 809,895. The city holds an adequate portion of the area encompassing hills and water bodies. Due to rapid urbanisation, these areas are under tremendous pressure. Rapidly increasing urbanisation on the hills of Guwahati has resulted in the partial removal of vegetation cover in the forested area.

Methodology

The study was carried out at macroscale to know the distribution of the species at microscale and to investigate the micro-habitat use. The survey on the sparrow species was carried out in different human habitation areas of Guwahati city. The data were collected during 2014–2015 in 500 m × 500 m grids, and in each grid one or two sampling points were placed at an interval of at least 150–250 m apart, yielding a final set of 572 sampling locations (Figure 7.1). Five-minute point counts were conducted between 6.00 a.m. and 9.00 a.m. We used kriging methods to map the spatial distribution of the species (see Nath et al., 2019). For the micro-habitat use, we sampled different urban habitats. We also sampled 45-point count stations at each point and visited ten times between September 2013 and February 2014, and September 2014 to February 2015, and counted sparrows within a 30 m radius. In each of these locations, 12 ecological covariates were collected to investigate the micro-habitat use. The covariates are as follows: (a) Distance to nearest green patch, (b) green cover (including tree and grass), (c) grass cover, (d) plant diversity, (e) distance to nearest marketplace, (f) number of food shops, (g) number of rolling shutters people use for the front gate of shops and garages, (h) number of single storied houses, and (i) number of multi-storied houses, along with the roof types such as (j) tin and (k) concrete within the 30 m radius and (l) area opening of drains. We conducted Principal Component Analysis (PCA) and extracted orthogonal components representing similar ecological gradients (see Nath et al., 2019). After that, we built a generalised linear model for both the sparrow species, with a combination of final covariates (degree of urbanisation, greenness, housing structure). The analysis was carried out in program R v 3.2.5.

Results

The house sparrow count ranged from 0 to 22 individuals per point count station with a mean of 3.17 ± 4.05SD, whereas the tree sparrow varied from 0 to 12 individuals per station with a mean of 0.76 ± 1.97SD. Interpolation of point count data across Guwahati city varying in different urban gradients showed significant trends in the distribution of sparrows. The crowded zones of the city predominantly in Paltan Bazar, Fency Bazar, and the area adjacent to Maligaon had the highest probability of occurrence of house sparrows (Figure 7.2). The results of indicator kriging showed that the probability of occurrence of house sparrows declines in the areas with low settlement densities located far from the urban core (Figure 7.3). In contrast, tree sparrows had a high probability of occurrence in the eastern part of Guwahati city, in and around the Beltola area, and very low occurrence probability of distribution was found in the commercial

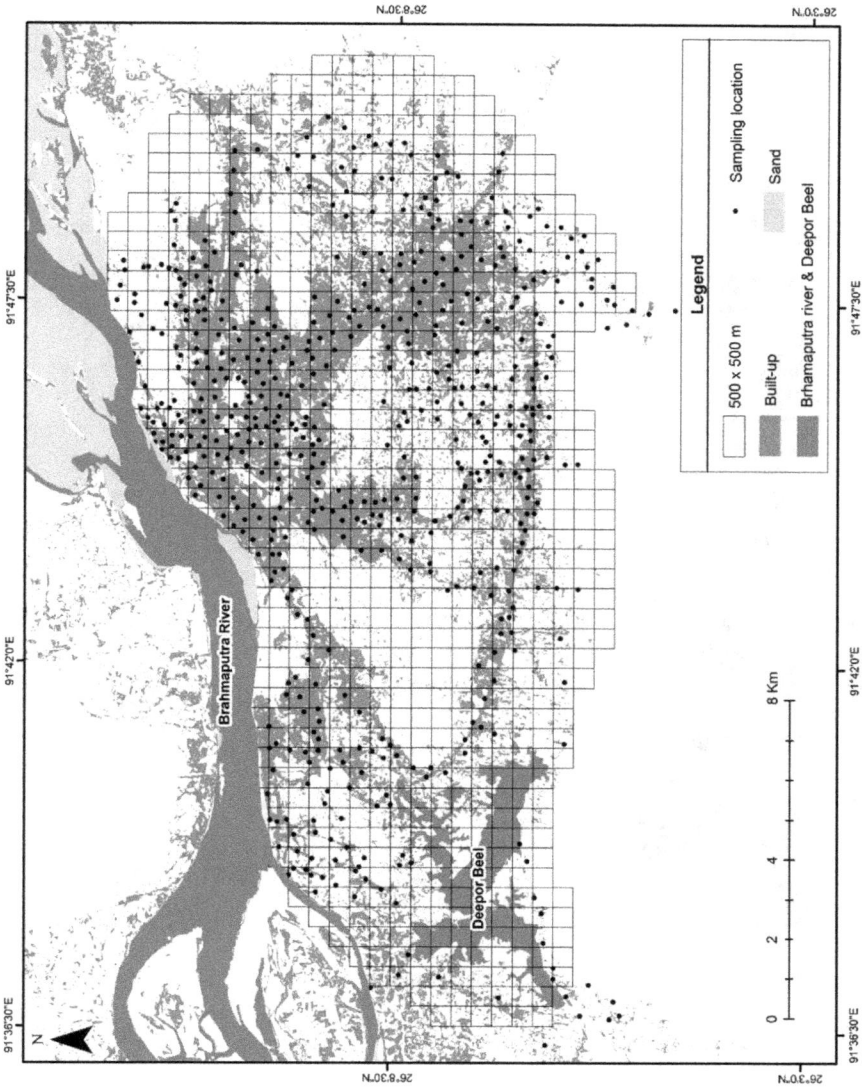

Figure 7.1 Study area showing the point count locations in Guwahati city.

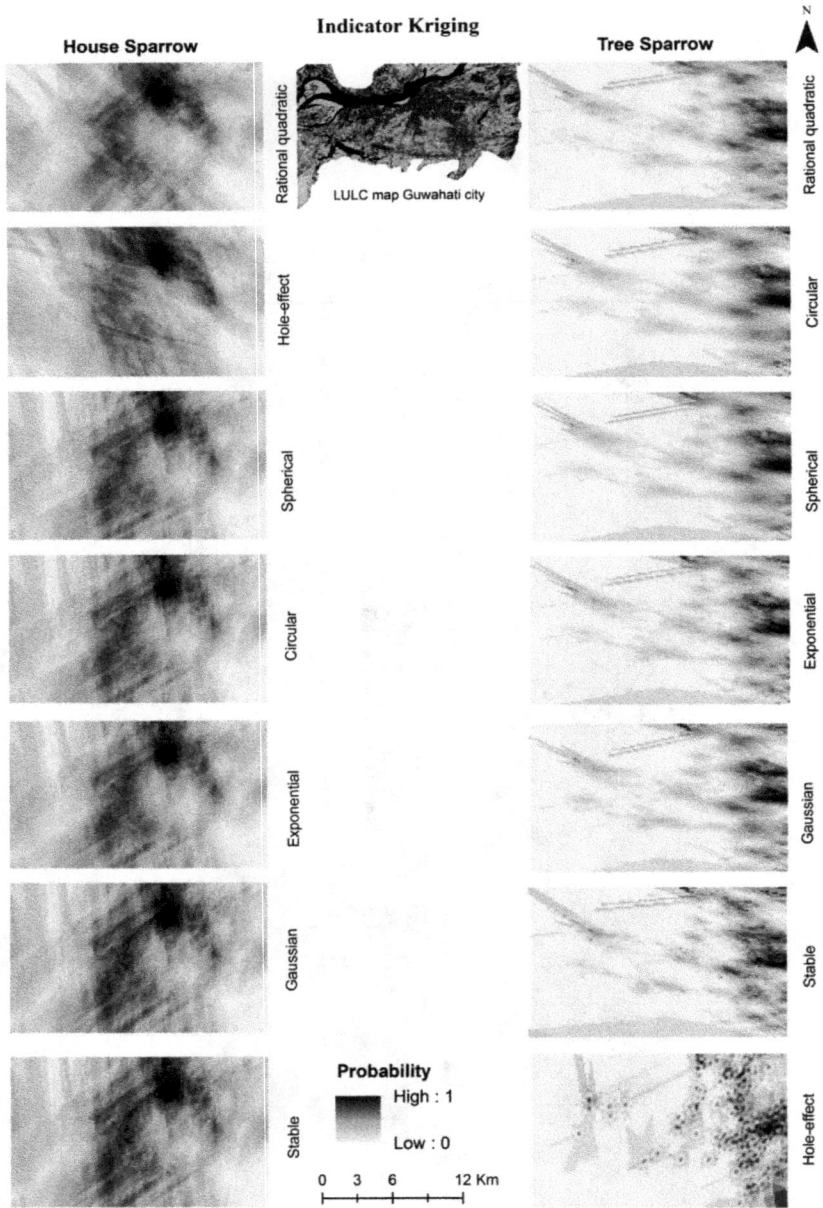

Figure 7.2 Output of semi variograms model with decreasing order of RMS value.

Figure 7.3 Representation of house sparrow (top) and tree sparrow (bottom) and LULC
(bottom) of Guwahati city, Assam, India.

Table 7.1 Summary of top three candidate regression model explaining house
sparrow habitat usage

Model	Wi	AICc	Δ AICc	Df	Log L
(Deg. of urbanisation + deg. of urbanisation²)	0.593	192.6	0	4	-91.807
(Deg. of urbanisation + deg. of urbanisation² + house type)	0.214	194.7	2.04	5	-91.558
(Deg. of urbanisation)	0.075	196.7	4.13	3	-95.081

zones, except for the Maligaon Railway station, which was comparatively less
crowded compared to other key commercial centres. Degree of urbanisation
was found to be most important predictor (quadratic) in determining the house
sparrow abundance (Tables 7.1 and 7.2). On the other hand, tree sparrow
were found to be associated with greenness and housing structure (interaction:
Greenness * housing structure) and negatively influenced by degree of urbanisation (Tables 7.3 and 7.4).

Discussion

The grade of urbanisation played an important role in shaping the abundance of
house sparrows. The number of restaurants and grocery (food) shops and proximity

Table 7.2 Summary of model (house sparrow ~ deg. of urbanisation + deg. of urbanisation^2)

Coefficients	β estimate	Std. Error	t value	Pr(>|t|)
(Intercept)	0.3556	1.208	0.294	0.76994
Deg. of urbanisation	3.6389	1.206	3.017	0.00432 **
Deg. of urbanisation^2	-0.6737	0.2627	-2.565	0.01400 *

Significant codes: '**' 0.01 '*' 0.05

Table 7.3 Summary statistics top three candidate regression model explaining tree sparrow habitat usage

Models	Wi	AICc	Δ AICc	Df	logLik	init.theta
(Greenness*house type)	0.425	125	0	5	-56.75	0.808
(Greenness*house type + deg. of urbanisation)	0.176	126.8	1.76	6	-56.29	0.883
(Greenness*house type + open drain)	0.132	127.4	2.34	6	-56.58	0.826

Table 7.4 Summary of model (tree sparrow ~ greenness * house type)

Coefficients:

	β estimate	Std. Error	z value	Pr(>|z|)
(Intercept)	-2.3438	1.0129	-2.314	0.020666*
Greenness	2.0314	0.5703	3.562	0.000368***
House type	1.2849	0.5656	2.272	0.023096*
Greenness * house type	-1.0971	0.3122	-3.515	0.00044***

Significant. codes: 0 '***' 0.001 '**' 0.01 '*' 0.05 '.' 0.1 " 1

to market places directly associated with food resources and front rolling shutters of shops offer nesting sites. However, it was also found that the species avoid areas where developmental pressure was at a peak, where there are very few options to place a nest. Henceforth, the expansion of modern buildings with box-type structures will be a key risk to the breeding of sparrows in the near future. House sparrows largely nest in holes and gaps in soffit boards and under tiles; thus, this tendency may have an effect on the accessibility of nest sites (Nath et al., 2016). A country-wide examination held in the UK showed that house sparrows avoided newer structures (built after 1985) or those that had experienced extensive roof repairs in the last decade (Shaw et al., 2008). This may result in fewer nesting opportunities, and there may also be a possibility of adverse respiratory effects from airborne fiberglass on breeding birds and chicks (Crick et al., 2002). Murgui (2008) found that in Valencia, Spain, build-up had a quadratic relationship with house sparrow density. Subsequently, Evans et al. (2009) reported that in the

UK, the high numbers of buildings have a negative influence on the densities of house sparrow and collared dove. Marzluff (2001) found that house sparrows were at maximum abundance at medium levels of built-up and landcover, suggesting that they could benefit from a balanced mixture of habitats (Murgui, 2008) which would provide a nest site (in manmade structures) and feeding resources (gardens or parks). An adequate amount of landscape diversity is a moderate predictor of the abundance of this species Murgui (2008).

Conversely, in the present study the tree sparrow selected habitation in which multi-storied building had a good extent of green cover, were high in plant diversity, and were spaces near to green patches or parks. Šálek et al. (2015) also observed similar results, where residential areas had a higher percentage of green habitats for tree sparrows. Subsequently, tree sparrows were also found to avoid places with highly commercial and crowded city centres. Zhang and Zheng (2010) also noted that in Beijing the number of tree sparrows declined along urbanisation gradients. Highly urbanised areas, such as commercial areas, main roads, and high-rise buildings, are not fit for inhabitation by tree sparrows. Though tree sparrows show some adaptation to human environments, they are unable to adapt to the overly rapid growth of urban development (Zhang & Zheng, 2010). Studies indicated that a decline in the house sparrow population was due to the decrease in the suitability of urban parks and gardens in Hamburg, Germany (Mitschke & Mulsow, 2003). Hence, the key tool for increasing urban sparrow populations would be the security of a mosaic-type of habitat, prioritising suitable design and management of private gardens and allotments. In cities, green areas such as parks, gardens, and lawns are vital habitats for foraging birds. Studies have already showed the role of urban parks as food resources for sparrows in urbanised landscapes (Murgui, 2008; Chamberlain et al., 2007). Compared to other large cities in India, Guwahati has not yet been completely urbanised, and quite a few parks are still there for recreational activities.

Closely related and ecologically similar species only differ substantially in one or a very small number of niche dimensions. This limited differentiation reflects evolutionary constrains on morphology, physiology, and behaviour as a result of descent from a common ancestor (Brown, 1984). This leads to assumptions of the extent to which these species vary in their requirements. The same can be noticed in the case of the house sparrow and the Eurasian tree sparrow, which also have same common ancestor and are ecologically similar too; often hybrids are reported within the distribution of the species where they overlap (Summers-Smith, 1988). Both species occupy almost one third of the earth's surface in varying densities in cities and towns within their distribution limit (Lack, 1971). Quite a few studies showed that there is a positive relationship between local abundance and geographic distribution among closely related ecologically similar species (Hanski, 1982a,b; Bock & Ricklefs, 1983). In general, the species that have the highest local population densities tend to inhibit a greater proportion of sample sites within a region and to have wider geographic ranges. In the present study as well, we found similar patterns in a much constrained city-wide scale. In Guwahati city, house sparrows occupied a much wider area than that of the tree sparrow. However, the house sparrow was found to be distributed throughout the city but concentrated in the urban core, whereas tree sparrow distribution was more scattered, and they were found to be residing in areas far from crowded zones of the city, mostly occupying eastern part of the city adjacent to hills. This may be the first order of niche differentiation by both species at a local level. In the major zones of allopathy, these two sparrows fill more or less similar ecological niches with some differences in fine scale habitat selection. In Lammi, south Finland, house sparrow and tree sparrow successfully colonised in the municipality (Vepsäläinen et al., 2005), and the study also suggested that house sparrows have a higher probability to disappear from those areas where tree sparrows had disappeared or were absent, compared to areas occupied by tree sparrows. Studies have shown that the tree sparrow plays the urban role successfully in those parts of the world where it is not in competition with the house sparrow (Marchant et al., 1990). Summers-Smith (1988) mentioned that the house sparrow and tree sparrow present a complicated picture. Newton (1998) also observes that the separation between these two species is "less clear cut": In some regions, the tree sparrow occupies the towns and the house sparrow the rural areas, while in other regions the situation is reversed. Henceforth, the habitat needs of tree sparrows and house sparrows at a local level look to be rather similar, although, with some overlaps, they inhabit different gradients of the urban environment.

References

Adams, L. W. (1994). *Urban wildlife habitats, a landscape perspective.* University of Minnesota Press.

Bock, C. E., & Ricklefs, R. E. (1983). Range size and local abundance of some North American songbirds: A positive correlation. *American Naturalists, 122*(2), 295–299.

Brown, J. H. (1984). On the relationship between abundance and distribution of species. *American Naturalist, 124*(2), 255–279.

Chamberlain, D. E., Toms, M. P., Cleary-Mcharg, R., & Banks, A. N. (2007). House sparrow (*Passer domesticus*) habitat use in urbanized landscapes. *Journal of Ornithology, 148*(4), 453–462.

Crick, H., Robinson, R., Appleton, G., Clark, N., & Richard, A. (2002). *Investigation into the causes of the decline of starlings and house sparrows in Great Britain.* BTO Report Number 290, (BTO/DEFRA).

Daniels, R. J. R. (2008). Can we save the sparrow? *Current Science, 95*(11), 1527–1578.

De Laet, J., & Summers-Smith, J. D. (2007). The status of the urban house sparrow *Passer domesticus* in North-Western Europe: A review. *Journal of Ornithology, 148*(S2), 275–278.

Dhanya, R., & Azeez, P. A. (2010). The house sparrow *Passer domesticus* population of arakku township, Andhra Pradesh, India. *Indian Bird, 5*, 180–181.

Evans, K., Newson, S., & Gaston, K. J. (2009). Habitat influences on urban avian assemblages. *Ibis, 151*(1), 19–39.

Gilbert, O. L. (1989). *The ecology of urban habitats.* Chapman & Hall.

Hanski, I. (1982a). Communities of bumblebees: Testing the core-satellite species hypothesis. *Annales Zoologicifennici, 19*, 65–73.

Hanski, I. (1982b). Distributional ecology of anthropochorus plants in villages surrounded by forest. *Annales Botanicifennici, 19*, 1–15.

Kheera, N., Das, A., Srivastava, S., & Jain, S. (2009). Habitat-wise distribution of the house sparrow (*Passer domesticus*) in Delhi, India. *Urban Ecosystem, 13*(1), 147–153.

Lack, D. (1971). *Ecological isolation in birds.* Blackwell.

Mahler, F. (2006). The house sparrow in Paris: A centre of persistence? Poster presented at the international congress of ornithology (Hamburg). *Journal of Ornithology, 147*, 207.

Marchant, J. H., Hudson, R., Carter, S. P., & Whittington, P. A. (1990). *Population trends in British breeding birds.* British Trust for Ornithology, Tring.

Marzluff, J. M., Bowman, R., & Donelly, R. (2001). *Avian ecology and conservation in an urbanizing world.* Kluwer Academic Publishers.

Mitschke, A., & Mulsow, R. (2003). Düstere Aussichten Für Einen Häufigen Stadtvogel-Vorkommen Und Bestandsentwicklung des Haussperlings in Hamburg. *Artenschutzreport, (Sonderheft), 14*, 4–12.

Murgui, E. (2008). Seasonal pattern of habitat selection of house Sparrow *Passer domesticus* in the urban landscape of Valencia (Spain). *Journal of Ornithology.* https://doi.org/10.1007/s10336-008-0320-z

Narth, A., Singha, H., Haque, M., & Lahkar, P. (2019). Sparrows in urban complexity: Macro and micro-scale habitat use of sympatric sparrows in Guwahati City, India. *Urban Ecosystem, 22*(6), 1047–1060.

Nath, A., Singha, H., Deb, P., Das, A. K., & Lahakar, B. P. (2016). Nesting in a crowd: Response of house sparrow towards proximity to spatial cues in commercial zones of Guwahati City. *Proceedings of the Zoological Society, 69*(2), 249–254.

Newton, I. (1998). *Population limitation in birds.* Academic Press.

Palmer, M. A. E., Bernhardt, E., Chornesky, S., Collins, A., Dobson, C., Duke, B., Gold, R., Jacobson, S., Kingsland, R., Kranz, M., Mappin, M. L., Martinez, F., Micheli, J., Morse, M., Pace, M., Pascual, S., Palumbi, O. J. Reichman, A. Simons, A., Townsend, A., & Turner, M. (2004). Ecology for a crowded planet. *Science, 304*(5675), 1251–1252.

Rajashekar, S., & Venkatesha, M. G. (2008). Occurrence of house sparrow, *Passer domesticus indicus* in and around Bangalore. *Current Science, 94*(4), 446–449.

Šálek, M., Riegert, J., & Grill, S. (2015). House sparrows *Passer domesticus* and tree sparrows *Passer montanus*: Fine-scale distribution, population densities and habitat selection in a central European city. *Acta Ornithologica, 50*(2), 221–232.

Shaw, L. M., Chamberlain, D., & Evans, M. (2008). The house sparrow *Passer domesticus* in urban areas: Reviewing a possible link between post-decline distribution and human socioeconomic status. *Journal of Ornithology, 149*(3), 293–299.

Summers-Smith, J. D. (1963). *The house Sparrow.* Collins.

Summers-Smith, J. D. (1988). *The sparrows.* T. and A.D. Poyser.

Vepsäläinen, V., Pakkala, T., & Andtiainen, J. (2005). Population increase and aspects of colonization of the tree sparrow *Passer montanus*, and its relationships with the house sparrow *Passer domesticus*, in the agricultural landscapes of Southern Finland. *Ornis Fennica, 82*, 117–128.

Witt, K. (2005). *Birds in European cities* (J. G. Kelcey & G. Rheinwald (Eds.) (pp. 17–40). Ginster Verlag.

Zhang, S., & Zheng, G. (2010). Effect of urbanization on the abundance and distribution of tree sparrows (*Passer montanus*) in Beijing. *Chinese Birds, 1*(3), 188–197.

8 Avifauna of Suru Valley

Species Composition and Habitats

Iqbal Ali Khan and Anil Kumar

Introduction

Ladakh is a high-altitude desert with an arid landscape, located in the western Himalaya. In the north, it is connected by the international border with China and in the east with Tibet. The southern boundary lies partly with Jammu and Himachal Pradesh and on the west with Pakistan Occupied Kashmir (POK) and the Kashmir region. Ladakh harbours a unique diversity of birds. The Kargil district of Ladakh is located at the crossroads of several biogeographic zones, and it is home to species native to Central Asia, Tibet, and alpine Himalayan ecosystems. The district is comprised of numerous valleys/nallah such as Suru Valley, Drass Valley, Zanskar Valley, Wakha Nallah, and Kanji Nallah and is peppered by alpine meadows, mountain steppes, snow-covered valleys, rocky canyons, and high-altitude cold deserts. Suru Valley including Rangdum is categorised in the A3 criterion of Important Bird and Biodiversity Area (Rahmani et al., 2016), indicating that the valley holds a significant component of the group of species whose distributions are largely confined to one biome. Based on the published literature as well as his field observations, Pfister (2004) compiled a checklist of 310 avian species from Ladakh. The present study is an attempt to document the avifauna of the Suru Valley based on the published literature and observations made during a period of three years, i.e., from 2018 to 2020.

Study Area

Suru Valley extends from Kargil town to Zanskar Valley. It is one of the biodiversity-rich, fertile, and productive areas of the Ladakh region and receives higher levels of moisture than other adjoining areas (Figure 8.1). Vegetation in the valley consists of herbs, shrubs, and trees, which offer favourable habitats for bird species. Tree species such as Willow (*Salix alba*), Juniper (*Juniperus communis*), and Poplar (*Populus nigra*) are found near the river and village areas. Herb species like Wild iris (*Iris lacteal*), Himalayan may apple (*Podophullum hexandrum*), Yellow colchicum (*Colchicum luteum*), Himalayan meadow crane's-bill (*Geranium pratense*), and Spotted heart orchid (*Dactylorhiza hatagirea*) are found in Suru Valley. Prickly Wild rose (*Rosa acicularis*), Sea-buck thorn (*Hippophaer hamnoides*), and Caragana (*Caragana arborescens*) are the main shrubs in this region. The valley is mostly drained by the Suru river and its tributaries which rise from the Pensi La Glacier.

DOI: 10.4324/9781003321422-9

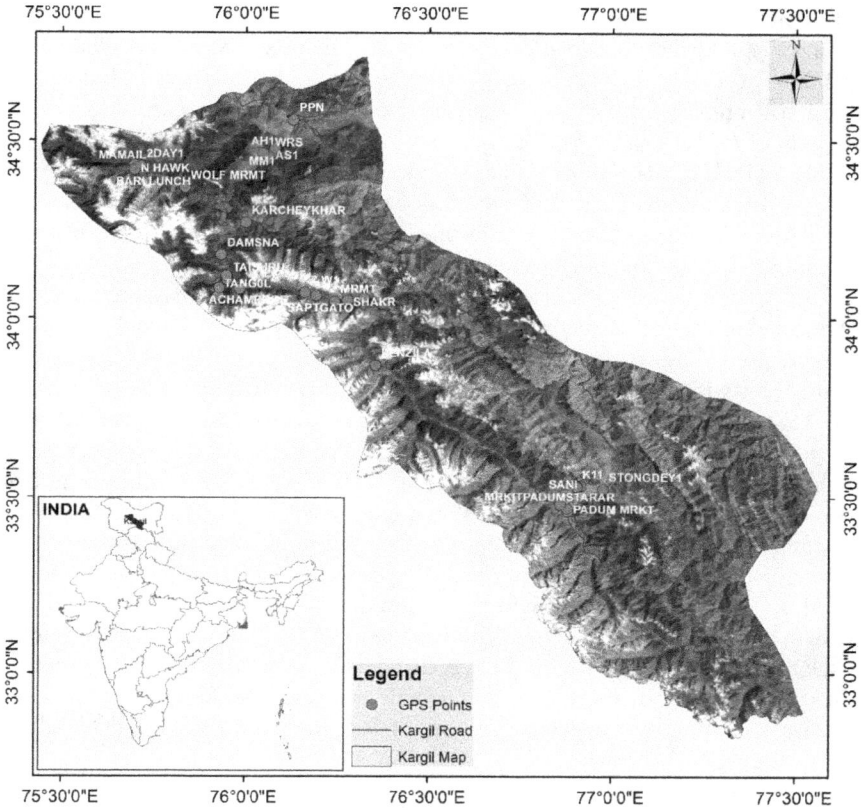

Figure 8.1 Map showing the surveyed localities in study area in Suru Valley, Kargil District, Ladakh.

Besides the diverse assemblage of birds, the valley harbours large mammals such as the Himalayan brown bear *Ursus arctos isabellinus*, Tibetan wolf *Canis lupus*, Asiatic ibex *Capra ibex*, Snow leopard *Panthera uncia*, and smaller mammals such as the Himalayan red fox *Vulpes vulpes*, Mountain weasel *Mustela altaica*, Stoat *Mustela ermine*, Silvery mountain vole *Alticola argentatus*, Long-tailed marmot *Marmota caudate*, Royle's pika *Ochotona roylei*, and Plateau pika *Ochotona curzoniae* (Ahmad et al., 2015).

Methodology

The present article is based on the published literature and information generated through field surveys of various habitats of Suru Valley undertaken from

March 2018 to June 2019. Some interesting species recorded during 2020 were also included in the checklist. Five surveys were conducted within three years. During the fieldwork, observations on birds were made every day from 6.00 a.m. to 4.00 p.m. (with few exceptions), with the help of prismatic field binoculars and a still-photography camera. Identification of species was carried out with the help of *Pocket Guide of the Birds of the Indian Subcontinent* (Grimmett et al., 2003) and *Field Guide of Birds of India* (Grewal et al., 2017). The taxonomic order and nomenclature by Clements were followed (Clements et al., 2018). *Birds of South Asia: The Ripley Guide* (Rasmussen & Anderton, 2005) and *Birds of India* (Kazmierczak & Perlo, 2000) were also considered.

The occurrence of some species, such as Arctic tern (*Sterna paradisaea*) as mentioned in the earlier records, was debatable, as they do not have distribution in the region. While some of species, such as Long-billed bush warbler (*Locustella major*), Saker falcon (*Falco cherrug*), Eurasian nightjar (*Caprimulgus europaeus*), Forest wagtail (*Dendronanthus indicus*), Eurasian jackdaw (*Coloeus monedula*), and Barred warbler (*Sylvia nisoria*) were reported in past, they have not been seen in the valley for decades. Conservation status (IUCN, 2018) of the species has been incorporated in the study. Based on migratory status, birds were categorised into four groups, namely winter migrants (long-distance migrants winter in the region, often stay as passage migrants during onward and return journey in some areas), summer migrants (arrive in the region during the summer season and few of them behave as passage migrants during the journey), passage migrants (purely passage migrants seen only for a short period during onward and return journeys), and residents (stay in the region throughout the year, including a few of them migrating locally owing to decreasing temperature in winter and food availability, within the region) (Kumar, 2015).

Results and Discussions

Species Composition

A total of 140 avian species was recorded from Suru Valley based on previous studies and including new sightings. In the present study we recorded a total of 97 species (Table 8.1), as well as a few species whose identification was not confirmed. We added 12 interesting and/or new records for the area, such as Indian spot-billed duck (*Anus poecilorhyncha*), Greylag goose (*Anser anser*), Great crested grebe (*Podiceps cristatus*), Eurasian coot (*Fulica atra*), Common moorhen (*Gallinula chloropus*), Tickell's thrush (*Turdus unicolor*), Himalayan bluetail (*Tarsiger rufilatus*), Blue-fronted redstart (*Phoenicurus frontalis*), Spotted dove (*Spilopelia chinensis*), Laughing dove (*Spilopelia senegalensis*), Bar-tailed treecreeper (*Certhia himalayana*), and Rufous-naped tit (*Periparus rufonuchalis*).

New Records of Birds

Following migrant waterfowls and terrestrial birds were recorded for the first time in Suru Valley.

Table 8.1 Checklist of the bird species, IUCN status, new records, month, and area of sightings from the study area

Common name	Scientific name	Migratory status	IUCN status	Earlier records	Sighting details (recent most sighting)	Number/flock Size	Area of sighting	Behaviour
Order: Anseriformes								
Family: Anatidae								
Bar-headed goose	Anser indicus	SM	LC	AC	June 2018	6	Rangdum	S
Greylag goose	Anser anser	SM	LC	C*	April 2020	1	Minjee	S
Ruddy shelduck	Tadorna ferruginea	PM/R	LC	AC	July 2018	8	Rangdum	FF
Mallard	Anas platyrhynchos	PM/R	LC	BC	March 2019	4	Minjee	S
Indian spot-billed duck	Anas poecilorhyncha	SM/PM	LC	C*	May 2019	6	Damsna	FF
Northern shoveler	Spatula clypeata	SM/PM	LC	AC	Sept 2018	2	Damsna	S
Northern pintail	Anas acuta	SM/PM	LC	AC	July 2018	4	Rangdum	P
Garganey	Spatula querquedula	SM	LC	AC	June 2019	7	Rangdum	FF
Green-winged teal	Anas crecca	SM	LC	AC	September 1983	1	Sankoo	FG
Common merganser	Mergus merganser	PM/R	LC	ABC	September 2018	2	Damsna	P
Order: Galliformes								
Family: Phasianidae								
Chukar	Alectoris chukar	R	LC	ABC	June 2018	7	Minjee	FF
Himalayan snowcock	Tetraogallus himalayensis	R	LC	ABC	July 2018	2	Parkachik	FF
Rain quail	Coturnix coromandelica	SM	LC	A	July 1980	1	Sankoo	C
Common quail	Coturnix coturnix	SM	LC	A	August 1925	1	Minjee	C
Order: Podicipediformes								
Family: Podicipedidae								
Great crested grebe	Podiceps cristatus	SM	LC	C*	June 2018	2	Taisuru	S
Order: Accipitriformes								
Family: Accipitridae								
Lammergeier	Gypaetus barbatus	R	NT	ABC	April 2018	2	Sankoo	Ho

(Continued)

Table 8.1 (Continued)

Common name	Scientific name	Migratory status	IUCN status	Earlier records	Sighting details (recent most sighting)	Number/flock Size	Area of sighting	Behaviour
Himalayan griffon	Gyps himalayensis	R	NT	ABC	July 2018	1	Karchey khar	Ho
Golden eagle	Aquila chrysaetos	R	LC	ABC	September 2018	2	Karchey khar	F
Booted eagle	Hieraaetus pennatus	SM	LC	AC	April 2019	1	Minjee	F
Eurasian marsh-harrier	Circus aeruginosus	PM	LC	A	September 1984	2	Rangdum	F
Eurasian sparrowhawk	Accipiter nisus	R	LC	ABC	May 2019	1	Minjee	H
Northern goshawk	Accipiter gentilis	R/PM	LC	AC	October 2018	1	Minjee	H
Black-eared kite	Milvus (migrans) lineatus	SM	LC	AC	August 2018	1	Tangol	P
Order: Charadriiformes **Family: Ibidorhynchidae**								
Ibisbill	Ibidorhyncha struthersii	R/SM	LC	AB	August 2018	2	Rangdum	F
Family: Recurvirostridae								
Black-winged stilt	Himantopus himantopus	R/PM	LC	ABC	September 2018	4	Minjee	FG
Family: Charadriidae								
Lesser sand-plover	Charadrius mongolus	SM	LC	AB	June 2011	2	Rangdum	FG
Family: Jacanidae								
Pheasant-tailed jacana	Hrdrophasianus chirurgus	SM	LC	A	July 1980	1	Panikhar	FG
Family: Scolopacidae								
Common sandpiper	Actitis hypoleucos	SM	LC	AC	August 2018	2	Rangdum	FG
Green sandpiper	Tringa ochropus	SM/WM	LC	AC	August 2018	3	Rangdum	FG
Common greenshank	Tringa nebularia	WM/PM	LC	ABC	August 2018	2	Rangdum	FG
Wood sandpiper	Tringa glareola	SM	LC	A	September 1984	2	Rangdum	FG
Common redshank	Tringa totanus	SM/PM	LC	ABC	August 2018	4	Rangdum	FG
Ruff	Calidris pugnax	PM	LC	A	July 1980	2	Rangdum	FG
Eurasian curlew	Numenius arquata	SM/PM	NT	A	August 1983	2	Rangdum	FG

Common name	Scientific name							
Curlew sandpiper	*Calidris ferruginea*	PM	NT	A	July 1980	1	Rangdum	FG
Temminck's stint	*Calidris temminckii*	WM/PM	LC	A	July 1980	X	Rangdum	FF
Little stint	*Calidris minuta*	SM/PM	LC	A	September 1931	1	Rangdum	FG
Solitary snipe	*Gallinago solitaria*	R/WM	LC	A	August 1984	1	Rangdum	FG
Family: Laridae								
Black-headed gull	*Chroicocephalus ridibundus*	WM/PM	LC	AC	July 1980	1	Rangdum	F
Common tern	*Sterna hirundo*	SM	LC	ABC	April 2020	1	Minjee	F
Arctic tern	*Sterna paradisaea*	V	LC	A	July 1928	1	Rangdum	F
Order: Columbiformes								
Family: Columbidae								
Rock pigeon	*Columba livia*	R	LC	ABC	June	X	Minjee	FG
Hill pigeon	*Columba rupestris*	R	LC	ABC	October	X	Minjee	FG
Snow pigeon	*Columba leuconota*	R/SM	LC	AB	September 2018	X	Parkachik	PG
Oriental turtle-dove	*Streptopelia orientalis*	R/SM	LC	ABC	April 2018	X	Minjee	FG
Spotted dove	*Spilopelia chinensis*	PM	LC	C*	September 2020	2	Titi-chumik	FG
Laughing dove	*Spilopelia senegalensis*	PM	LC	C*	October 2020	4	Titi-chumik	FG
Order: Gruiformes								
Family: Rallidae								
Common moorhen	*Gallinule chloropus*	SM/PM	LC	C*	April 2020	2	Minjee	FG
Eurasian coot	*Fulica atra*	WM/PM	LC	C*	November 2018	4	Chutuk	S
Order: Cuculiformes								
Family: Cuculidae								
Common cuckoo	*Cuculus canorus*	SM	LC	ABC	April 2019	1	Minjee	P
Order: Strigiformes								
Family: Strigidae								
Little owl	*Athene noctua*	R	LC	AB	July 2011	1	Parkachik	P
Order: Caprimulgiformes								

(Continued)

Table 8.1 (Continued)

Common name	Scientific name	Migratory status	IUCN status	Earlier records	Sighting details (recent most sighting)	Number/flock Size	Area of sighting	Behaviour
Family: Caprimulgidae								
Eurasian nightjar	Caprimulgis europaeus	SM	LC	A	September 1983	1	Parkachik	P
Family: Apodidae								
Alpine swift	Apus melba	PM	LC	AC	July 2018	4	Damsna	F
Common swift	Apus apus	SM	LC	AB	June 2015	2	Trespone	F
Little swift	Apus affinis	SM	LC	B	June 2011	1	Sankoo	F
Order: Coraciformes								
Family: Alcedinidae								
Common kingfisher	Alcedo atthis	R/SM	LC	AC	July 2018	1	Minjee	P
Family: Coraciidae								
European roller	Coracias garrulus	V	LC	A	July 1977	1	Sankoo	p
Order: Bucerotiformes								
Family: Upupidae								
Eurasian hoopoe	Upupa epops	SM	LC	ABC	July 2018	2	Minjee	FG
Order: Piciformes								
Family: Picidae								
Eurasian wryneck	Jynx torquilla	SM	LC	AC	April 2020	1	Parkachik	Pk
Scaly-bellied woodpecker	Picus squamatus	R/PM	LC	AC	May 2018	2	Minjee	Pk
Order: Falconiformes								
Family: Falconidae								
Common kestrel	Falco tinnunculus	R/SM	LC	ABC	May 2018	3	Minjee	H
Eurasian hobby	Falco subbuteo	SM	LC	AC	July 2018	1	Sankoo	Pr
Saker falcon	Falco cherrug	SM	EN	A	July 1977	1	Panikhar	H
Peregrine falcon	Falco peregrinus	R/PM	LC	A	August 1982	2	Parkachik	H
Order: Passeriformes								
Family: Laniidae								
Long-tailed shrike	Lanius schach	R/SM	LC	ABC	May 2019	2	Minjee	P
Grey-backed shrike	Lanius tephronotus	SM	LC	A	July 1972	1	Sankoo	P

Family: Oriolidae								
Indian golden oriole	*Oriolus kundoo*	SM	LC	AC	May 2019	1	Minjee	P
Family: Corvidae								
Eurasian magpie	*Pica pica bactriana*	R	LC	ABC	Seen all over months	X	Minjee	FG
Red-billed chough	*Pyrrhocorax pyrrhocorax*	R	LC	ABC	Seen all over months	X	Sankoo	FF
Yellow-billed chough	*Pyrrhocorax graculus*	R	LC	ABC	March 2019	X	Damsna	FF
Eurasian jackdaw	*Corvus monedula*	SM	LC	A	May 1923	X	Titichumik	FF
Carrion crow	*Corvus corone*	R/PM	LC	ABC	September 2018	3	Minjee	M
Large-billed crow	*Corvus macrorhynchos*	R/PM	LC	A	October 1977	X	Titi-chumik	FF
Common raven	*Corvus corax*	R/PM	LC	ABC	June 2019	4	Trespone	M
Family: Alaudidae								
Asian short-toed lark	*Calandrella cheleensis*	SM	LC	A	July 1982	1	Rangdum	FG
Hume's lark	*Calandrella acutirostris*	SM	LC	A	July 1982	2	Parkachik	FG
Oriental skylark	*Alauda gulgula*	SM	LC	ABC	May 2018	6	Minjee	Sky
Horned lark	*Eremophila alpestris*	R	LC	AB	June 2011	4	Rangdum	FG
Family: Hirundinidae								
Eurasian crag-martin	*Ptyonoprogne rupestris*	SM	LC	ABC	August 1982	2	Minjee	F
Barn swallow	*Hirundo rustica*	SM	LC	A	August 1982	1	Parkachik	F
Common house-martin	*Delichon urbicum*	SM	LC	AB	June 2011	2	Sankoo	F
Family: Paridae								
Cinereous tit	*Parus cinereus*	R/PM	LC	ABC	April 2018	6	Minjee	P
Rufous-naped tit	*Parus rufonuchalis*	WM	LC	C*	December 2020	1	Goma-Minjee	F
Fire-capped tit	*Cephalopyrus flammiceps*	SM	LC	A	September 1931	4	Rangdum	F
Family: Aegithalidae								
White-browed tit-warbler	*Leptopoecile sophiae*	SM	LC	AC	October 1979	2	Titi-chumik	Pr

(Continued)

Table 8.1 (Continued)

Common name	Scientific name	Migratory status	IUCN status	Earlier records	Sighting details (recent most sighting)	Number/flock Size	Area of sighting	Behaviour
Family: Tichodromidae								
Wallcreeper	Tichodroma muraria	SM	LC	AC	August 1983	1	Minjee	P
Family: Cathiidae								
Bar-tailed treecreeper	Certhia himalayana	WV	LC	C*	November 2020	1	Goma-Minjee	P
Family: Troglodytidae								
Eurasian wren	Troglodytes troglodytes	R/PM	LC	AC	July 2018	1	Minjee	P
Family: Cinclidae								
White-throated dipper	Cinclus cinclus	R	LC	ABC	July 2018	1	Minjee	F
Brown dipper	Cinclus pallasii	R	LC	ABC	June 2018	2	Minjee	F
Family: Regulidae								
Goldcrest	Regulus regulus	R/PM	LC	AC	August 2018	1	Sankoo	P
Family: Phylloscopidae								
Mountain chiffchaff	Phylloscopus sindianus	SM	LC	ABC	August 2018	2	Minjee	Pr
Plain leaf-warbler	Phylloscopus neglectus	SM	LC	AB	June 2011	4	Damsna	FG
Tickell's leaf-warbler	Phylloscopus affinis	SM	LC	ABC	May 2018	2	Sankoo	Pr
Sulphur-bellied warbler	Phylloscopus griseolus	SM	LC	AB	June 2011	2	Parkachik	Pr
Greenish warbler	Phylloscopus trochiloides	SM	LC	ABC	July 2018	2	Minjee	FG
Yellow-browed warbler	Phylloscopus inornatus	SM	LC	A	September 1983	4	Panikhar	FG
Family: Acrocephalidae								
Blyth's reed-warbler	Acrocephalus dumetorum	PM	LC	AC	August 2018	2	Sankoo	P
Family: Locustellidae								
Long-billed bush-warbler	Locustella major	SM	NT	A	July 1977	1	Sankoo	P
Family: Sylvidae								
Lesser whitethroat	Sylvia curruca	WM	LC	ABC	May 2018	2	Sankoo	P
Common whitethroat	Sylvia communis	PM	LC	A	July 1979	4	Titichumik	P
Barred warbler	Sylvia nisoria	SM	LC	A	September 1983	1	Parkachik	P

Common name	Scientific name				Date		Location	
Family: Muscicapidae								
Bluethroat	*Luscinia svecica*	SM/PM	LC	ABC	April 2018	1	Sankoo	P
White-tailed rubythroat	*Calliope pectoralis*	SM/WM	LC	AB	June 2011	1	Panikhar	P
Blue whistling-thrush	*Myophonus caeruleus*	R	LC	ABC	April 2019	2	Minjee	P
Little forktail	*Enicurus scouleri*	V	LC	AC	July 1977	2	Sankoo	W
Himalayan bluetail	*Tarsiger rufilatus*	SM	LC	C*	June 2018	1	Gramthang	FG
White-capped redstart	*Phoenicurus leucocephalus*	R/SM	LC	ABC	May 2018	4	Minjee	W
White-winged redstart	*Phoenicurus erythrogastrus*	R/SM	LC	ABC	July 2018	2	Parkachik	W
Black redstart	*Phoenicurus ochruros*	SM/PM	LC	ABC	May 2019	X	Minjee	FF
Blue-Fronted redstart	*Phoenicurus frontalis*	SM/PM	LC	C*	July 2018	1	Sankoo	P
Blue Rock-thrush	*Monticola solitarius*	SM	LC	ABC	August 2018	1	Karchey Khar	Pr
Siberian stonechat	*Saxicola maurus*	SM	LC	AC	May 2019	2	Damsna	P
Variable wheatear	*Oenanthe picata*	SM	LC	AC	June 2018	2	Minjee	P
Pied wheatear	*Oenanthe pleschanka*	SM	LC	AC	July 2018	1	Minjee	P
Desert wheatear	*Oenanthe deserti*	SM	LC	AC	September 2018	2	Minjee	P
Family: Turdidae								
Black-throated thrush	*Turdus atrogularis*	WM	LC	AC	June 2019	2	Minjee	F
Tickell's thrush	*Turdus unicolor*	SM	LC	C*	May 2019	1	Minjee	p
Family: Sturnidae								
Rosy starling	*Pastor roseus*	PM	LC	A	September 1984	4	Panikhar	FG
Brahminy starling	*Sturnia pagodarum*	V	LC	B	June 2011	2	Rangdum	FG
Family: Prunellidae								
Robin accentor	*Prunella rubeculoides*	R	LC	ABC	September 2018	6	Minjee	FF
Rufous-breasted accentor	*Prunella strophiata*	R/SM	LC	AC	October 2018	2	Minjee	Pr
Brown accentor	*Prunella fulvescens*	R/PM	LC	AC	October 2018	X	Minjee	FG
Black-throated accentor	*Prunella atrogularis*	WM	LC	AC	October 2018	2	Minjee	Pr

(Continued)

Table 8.1 (Continued)

Common name	Scientific name	Migratory status	IUCN status	Earlier records	Sighting details (recent most sighting)	Number/flock Size	Area of sighting	Behaviour
Family: Motacillidae								
Western yellow wagtail	Motacilla flava	WM/PM	LC	AB	September 2018	2		W
Citrine wagtail	Motacilla citreola	SM	LC	ABC	April 2019	3	Minjee	W
Gray wagtail	Motacilla cinerea	SM	LC	ABC	June 2011	2	Sankoo	W
White wagtail	Motacilla alba	SM/PM	LC	ABC	April 2019	2	Minjee	W
Forest wagtail	Dendronanthus indicus	SM/PM	LC	A	August 1868	1	Suru	P
Rosy pipit	Anthus roseatus	SM	LC	AC	May 2019	2	Minjee	P
Tree pipit	Anthus trivialis	PM/SM	LC	A	June 1927	2	Sankoo	P
Family: Emberizidae								
Rock bunting	Emberiza cia	SM/R	LC	ABC	April 2018	X	Minjee	FF
Family: Fringillidae								
Plain mountain-finch	Leucosticte nemoricola	R/WM	LC	ABC	May 2018	7	Parkachik	FF
Black-headed mountain-finch	Leucosticte brandti	R	LC	ABC	May 2018	2	Rangdum	Pr
Common rosefinch	Carpodacus erythrinus	SM/PM	LC	ABC	June 2019	2	Minjee	P
Red-fronted rosefinch	Carpodacus puniceus	R	LC	ABC	July 2018	1	Parkachik	P
Great rosefinch	Carpodacus rubicilla	R	LC	ABC	May 2019	1	Parkachik	P
European goldfinch	Carduelis carduelis	R/WM	LC	ABC	September 2018	6	Minjee	FG
Twite	Carduelis flavirostris	R/SM	LC	A	September 1984	7	Rangdum	FG
Fire-fronted serin	Serinus pusillus	R/WM	LC	ABC	May 2019	X	Minjee	FG
Family: Passeridae								
House sparrow	Passer domesticus	R/SM	LC	ABC	April 2019	X	Minjee	FG

(**Abbreviations:** Migratory Status: WM—winter migrants; SM—summer migrants; PM—passage migrants; R—residents; V—Vagrants. IUCN status: LC—Least Concern; NT—Near Threatened; VU—Vulnerable; ED—Endangered. Earlier records: A—Holmes, 1983; B—Ahmad et al. 2015; C—Birds recorded by author; *—New bird record. Number or flock Size: X—more than ten individuals. Behaviour: S—swimming; C—calling; F—flight; FF—foraging in flock; FG—foraging on ground; P—perching; Pr—preening; H—hawking; Ho—hovering; M—mobbing; Pk—pecking; Sky—skylarking; W—wiggling).

Indian spot-billed duck *Anus poecilorhyncha*: A flock of six individuals was seen near the shoreline of Suru River at Damsna Village (34° 09' 20.79"N; 75° 57' 18.94"E) feeding in the grass patch on 25 March 2019.

Eurasian coot *Fulica atra*: Two individuals were recorded swimming in Suru river under the bridge of Chutuk Village (34° 29' 45.30"N; 76° 07' 09.37"E) on 6 November 2018.

Great crested grebe *Podiceps cristatus*: A pair was seen at Taisuru Village (34° 08' 12. 74"N; 75° 57' 29.79"E) swimming near the river bridge on 13 June 2018.

Greylag goose *Anser anser*: One individual was recorded swimming in the River Suru at Minjee area (34° 09' 29.82"N; 76° 55' 06.39"E) on 19 April 2020, which was photographed by Mr. Karamjeet Singh (a local photographer) on our request, and provided to us.

Common moorhen *Gallinula chloropus*: One individual was also recorded foraging near the river shore at Minjee (34° 09' 29.82"N; 76° 55' 06.39"E) in the same month on 24 April 2020, which was again photographed by Mr. Karamjeet Singh on our request, who provided the photograph to us.

Himalayan blue-tail *Tarsiger rufilatus*: One confirmed sighting of male foraging in the cultivated Pea field at Gramthang Village (34° 28' 27.58"N; 76° 04' 51.39"E) on 22 June 2018. The bird flew away when approached for photography.

Tickell's thrush *Turdus unicolor*: One male individual was recorded, perched on the top branch of poplar tree and delivering a song. It was photographed at the Minjee area (34° 28' 54.51"N; 76° 05' 31.64"E) on 14 May 2019.

Blue-fronted redstart *Phoenicurus frontalis*: Sighting of an individual in non-breeding plumage was confirmed and photographed at Sankoo Village (34° 17' 42.14"N; 75° 57' 37.96"E) on 13 July 2018.

Spotted dove *Spilopelia chinensis*: A pair was sighted near Titi-chumik area, foraging solitarily on the barren field on 10 September 2020 (34° 52' 20.17"N; 076° 13' 17.24"E).

Laughing dove *Spilopelia senegalensis*: Two pairs were seen and photographed, foraging with the flock of hill pigeons *Columba rupestris* just near the same area where the spotted dove *Spilopeliachinensis* had been sighted the previous month (34° 52' 20.17"N; 076° 13' 17.24"E) on 23 October 2020.

Bar-tailed treecreeper *Certhia himalayana*: One individual was seen creeping over a tall poplar tree searching for invertebrate larvae hiding under the bark. It was photographed in Goma-Minjee (34° 28' 50.27"N; 076° 04' 53.12"E) on 4 November 2020.

Rufous-naped tit *Periparus rufonuchalis*: One individual was seen foraging in thick branches of a willow tree at Goma-Minjee (34° 28' 55.87"N; 076° 04' 55.19"E) on 3 December 2020.

Globally Threatened Species

Based on previous studies and records, Suru Valley supports the occurrence of some of globally threatened and near-threatened birds like Saker falcon (*Falco cherrug*), categorised as endangered by IUCN and rarely seen in the valley. Five species, namely, Lammergier (*Gypaetus barbatus*), Himalayan griffon (*Gyps himalayensis*), Eurasian curlew (*Numenius arquata*), Curlew sandpiper (*Calidris ferruginea*), and Long-billed bush warbler (*Locustella major*) comes under the category of "near threatened" category of IUCN.

Migratory Species

About of 74% of avian species are migratory in Suru Valley. They showed different patterns of migration for breeding and foraging. Some species were winter migrants (7%) which had travelled a long distance, mainly comprising of waterfowl, shorebirds, raptors, and some passerines. Some species were passage migrants (22%) which spend a few days in the area during fall and spring. Some species (43%) arrived in the region during summer, usually treated as summer migrants. The rest of the species (26%) were residents in the area. Some of the species known to migrate partially from high-altitude areas to low altitudes generally move to a higher elevation in summer and come back to lower elevations or valleys in winter owing to snowfall. The status of local migrants is still uncertain and undocumented. Some accidental/vagrant (2%) species were also seen in the valley.

Major Conservation Issues

In earlier days, tourism was very limited in the area owing to the rugged, hostile climate and poor infrastructure, and tourists used to prefer to visit Padum and Zanskar Valley for exploration owing to relatively better facilities and convenience. However, nowadays, due to large-scale infrastructure (especially transportation and accommodation), the Suru Valley has become the main destination for most visitors. Some events/activities like mountaineering, rafting, and rock climbing are organised by the Kargil administration. Moreover, local travel agencies provide quality services at affordable costs, which allures tourists and consequently has led to a manifold increase in tourism in the Suru Valley. However, it harms avian diversity owing to habitat degradation and disturbance. Conversion of wasteland and agriculture expansion is seen in the valley, because local people are earning from tourists staying locally and in guest houses, and this activity is increasing year by year.

Habitat loss and overgrazing by livestock are the main threats being faced by the avifauna in the valley. Poplar trees are used in construction at a large scale, but such trees are the preferred nesting sites for birds such as Eurasian magpie (*Pica pica*) and Northern goshawk (*Accipiter gentilis*). Local people and nomadic herders of the valley graze their livestock in mountains, and there are no specific sites reserved for grazing by the administration, which leads to uncontrolled/unorganised grazing and creating an imbalance in food proportion of wild ungulates like the Asiatic ibex *Capra ibex* and birds like the Himalayan snowcock (*Tetraogallus himalayensis*) and Chukar partridge (*Alectoris chukkar*). At the moment, the habitats are relatively healthy and mostly undisturbed. However, due to unregulated grazing of livestock, rapid developmental activities, and tourist influx, it is predicted that in the near future the scenario will change drastically as in other parts of the country. So, to secure and conserve the biodiversity of the valley, protective measures should be adopted in advance. Livestock grazing needs to be regulated to minimise habitat sharing with wild animals. Such measures may be useful to maintain the heterogeneous properties of the biome and increase habitat richness. It has been suggested that tourist inflow and unnecessary events in wildlife habitats near the wetlands be minimised for the breeding period of migratory birds.

Conclusion

This article increases information on the avifauna of the Suru Valley, Ladakh. This region comprises various habitats which are fragile and vulnerable to climate change. Analysis of data revealed that the valley is no doubt a bird paradise that sustains a sizeable number of avian species; many are summer migrants, winter migrants, and passage migrants. Yet information about local resident birds is scanty; only an opportunistic study has been done in this area. The population of many avian species is declining, resulting in biodiversity loss. Therefore, long-term monitoring and regular studies on avifauna are needed to understand the occurrence of species, populations, and emerging conservation threats. There is a need to adopt preventive measures to sustain the avian diversity in the valley.

Acknowledgments: We are grateful to Dr. Dhriti Banerjee, Director, Zoological Survey of India, Kolkata, for encouragement and support, and thankful to Dr. Gaurav Sharma, Scientist-E and Officer-in-Charge, Northern Regional Centre, ZSI, Dehradun for departmental facilities. The first author, i.e., I.K., is grateful to the second author, i.e., A.K., for research guidance. Valuable inputs were given by Mr. Raza Ali Abidi, Wildlife Warden, and the team department of Wildlife Protection Division Kargil is also gratefully acknowledged.

References

Ahmed, T., Khan, A., & Chandan, P. (2015). A pilot survey of the avifauna of Rangdum Valley, Kargil, Ladakh (Indian trans-Himalaya). *Journal of Threatened Taxa, 7*(6), 7274–7281. https://doi.org/10.11609/jott.o3965.7274-81

Clements, J. F., Schulenberg, T. S., Iliff, M. J., Roberson, D., Fredericks, T. A., Sullivan, B. L., & Wood, C. L. (2018). *The e-Bird/Clements checklist of birds of the world: v2018.*

Grewal, B., Sen, S., Singh, S., Devasar, N., & Bhatia, G. (2017). *A photographic field guide to the birds of India, Pakistan, Nepal, Bhutan, Sri Lanka, and Bangladesh.* Princeton University Press.

Grimmett, R., Inskipp, C., & Inskipp, T. (2003). *Pocket guide to the birds of the Indian subcontinent.* Oxford University Press.

Holmes, P. R. (1983). Avifauna of the Suru river Valley. *Forktail, 2,* 21–41. IUCN (2018).

Kazmierczak, K., & Perlo, B. V. (2000). *A field guide to the birds of India.* Om Book Service, pp. 352.

Kumar, A. (2015). Birds of Himachal Pradesh: Species composition and conservation issues. *Journal of Experimental Zoology*, India, *18*(2), 505–529.

Pfister, O. (2004). *Birds and mammals of Ladakh*. Oxford University Press.

Rahmani, A. R., Islam, M. Z., & Kasambe, R. M. (2016). Important bird and biodiversity areas in India: Priority sites for conservation (revised and updated). Bombay natural history society, Indian bird conservation network, royal society for the protection of birds and birdlife international (U.K.). pp. 1992 + xii.

Rasmussen, P. C., & Anderton, J. C. (2005). *Birds of South Asia: The ripley guide* (Vol. 1, pp. 1–378). Smithsonian Institution & Lynx Editions.

9 Indian Robin

Song Characteristics and Functions

Poornima Nailwal, Subhechha Tapaswini, and Anil Kumar

Introduction

Bioacoustic studies play a crucial role in behavioural ecology, socio-biology, and the evolutionary and neurobiology of birds. However, in India, scientific studies on avian vocalisations are confined to limited species (Kumar, 2012a, 2016, Dadwal et al., 2017). Indian birds are known to use a varied range of species-specific vocalisations composed of simple, context-specific, monosyllabic, monotonous calls to highly varied, frequency-modulated, and complex songs (Kumar, 2012a). A review of literature revealed that physical characteristics of song, mode of production, temporal organisation, and structural complexity (variability of elements) are highly diverse among oscine birds (Marler & Slabbekoorn, 2004). Some species use only one category of songs while others are known to sing two categories of song (Spector, 1992). The first category songs are often simple and highly stereotyped, sung at relatively low rates for male-male competition. In contrast, second category songs are usually more complex and variable, and are sung at higher rates for inter-sexual communication (Highsmith, 1989; Kroodsma et al., 1989; Staicer 1989; Bolsinger, 2000). The mode of singing also varies among birds. Mostly oscine birds use two types of singing modes i.e., "repeat" or "eventual variety" (singer produces one song type/strophe repeatedly before switching to another type of strophe), and "serial" or "immediate variety" mode (singer produces a variety of song strophes/ phrases without any apparent rule) (Molles & Vehrencamp, 1999).

Robins use both simple and complex vocalisations for communication (Kumar, 2011). For example, the European robin (*Erithacus rubecula*) uses complex vocalisations for territory defence (Hoelzel, 1986; Scriba & Goymann, 2010) and four species of African forest robins (Genus Stiphrornis) have been reported to use two types of vocalisation (Beresford & Cracraft, 1999). Oriental magpie robin *Copsychus saularis* also uses two categories of songs (Kumar, 2013; Singh et al., 2019). Complex songs were delivered predominantly during courtship in the presence of neighbouring male rivalry and for female attention. Complex songs are an indicator of male quality. Discrete songs are simple, highly stereotyped, and sung at a low rate. Discrete songs are delivered mainly for territory advertisements throughout the breeding season and are spontaneous and common. The Indian robin *Copsychus fulicatus* (Family: Muscicapidae) is an endemic,

DOI: 10.4324/9781003321422-10

small-sized (19 cm), passerine species, distributed throughout the Indian sub-continent except north-eastern region of India, higher Himalaya, and the Thar desert, and is divided into five distinct races (Collar & Bonan, 2018). It is a common, resident, territorial, and sexually dimorphic bird. It prefers dry, stony areas with sparse scrub, arid stony ridges, low rocky hills outcrops, edges of cultivation, and deserted buildings, gardens and groves (Ali & Ripley, 1973; Grimmett et al., 1998). Males use discrete songs for inter-sexual (mate attraction) and intra-sexual (male-male competition for territory) contexts (Kumar, 2011). Information on singing behaviour and socio-biology of the species is limited (Ali & Ripley, 1973, Nirmala & Vijayan, 2003, Kumar, 2012b).

Study Area

Chandrapur is located between 19° 30' and 20° 45'N latitude and 78° 46' and 80° 00'E longitude in the Vidharbha region (Figure 9.1), spreading over 11,443 km^2 (Shende, 2013). On the northern side, this district is bounded by Nagpur, Bhandara, and Wardha on the western side, and Yavatmal to Gadchiroli on the eastern side. Chandrapur district is confined within the basin of the two rivers, the Wainganga and the Wardha, which are tributaries of the Godavari river. This district is bestowed with diverse flora and fauna as well as valuable mineral resources such as limestone, bauxite, iron, and chromite (Shende, 2013).

Figure 9.1 Map of the Chandrapur district showing the song recording locations of eight individuals of Indian robin (No. 1 to 8 in circles).

Mode of singing

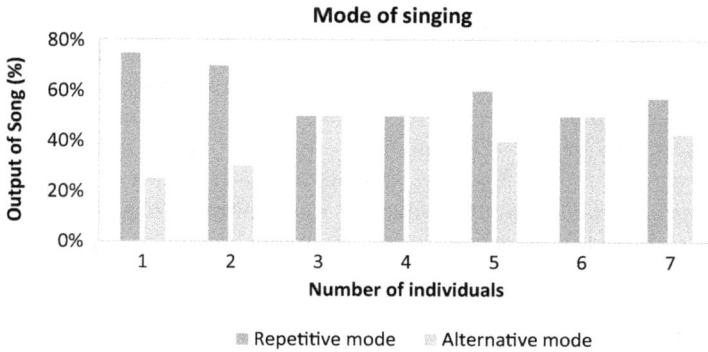

Figure 9.2 Percentage of mode of singing in Indian robin.

The climate of the district falls under tropical hot with a high range of temperatures throughout the year. Blistering hot summer and moderate winter are two major and distinct seasons prevail entirely the year. The southwest monsoon brings rainfall in late September to this district. The mean maximum temperature during summer is 46 °C, and the mean maximum temperature during winters is 28 °C. This district faces average annual rainfall of 1,420 mm. Different kinds of soil are found following the geographical features of this district. Black, regular loam, and clay loam are found along the Wardha and Wainganga valleys. The district is covered with 3,651 km² of forested area. The vegetation type of this area is classified as a southern tropical dry deciduous forest. The dominant species is teak with other associated species, i.e., bamboo (*Dendrocalamus strictus*), semal (*Bombax ceiba*), tendu (*Diospyros melanoxylon*), etc. Nearby open city-side areas like Durgapur, Astabhuja ward, Lohara village, and Ballarpur are dominated by the scrubby thorny vegetation of *Prosopis juliflora*.

Methodology

Field Recordings

The recording of vocalisations was carried out from May to June 2019 in the open scrubland of Chandrapur, Maharashtra (Figure 9.1). Song behaviour of a total of ten individuals was observed, of which the songs of eight individuals were considered for bioacoustic analysis (recordings of two individuals were excluded due to poor sound quality). All birds recorded were unbanded. To avoid recording the same individual twice, one of us attentively inspected the bird and the other one recorded the song. To avoid signal mixing, we took due care to record one individual every day. A total of 93 song samples were recorded mostly in the morning hours from 5.00 a.m. to 9.00 a.m. Signals were recorded with the help of SONY D100 digital audio-recorder, and individuals were observed using 12×50 Nikon field binoculars.

Data Analysis

Signals were recorded at a sampling frequency of 48 kHz. Recordings were saved in the computer as wav files. After preliminary analyses at 48 kHz, signals were customised at 24 kHz sampling rate and 16 bit resolution for better spectrograms. After sample frequency conversion, signals were analysed with the help of Avisoft SAS Lab Pro software. In the software signals appeared in the form of a spectrogram, the X-axis represents time in seconds and Y-axis denotes frequency in kHz. Maximum frequency and minimum frequency were measured using time and frequency cursors. Dominant frequency of strophes were measured after creating the power spectrum of the strophes. For the analysis of types of strophes per individual, different strophes were printed on A4 size paper. All spectrograms were quantified using following settings on Avisoft SAS Lab Pro: 512 FFT length, 50% frame, flat top window, and 87.5% time-window overlap. In the present study, minimum frequency, maximum frequency, bandwidth, dominant frequency, duration of strophes, followed gap, types of elements per strophe, and number of elements per strophe were quantified. Number of strophes, types of strophes, song bout duration, singing rate, and mode of singing were registered to stipulate the song characters and behavioural correlates were used to infer the biological functions of song. The presence of whistles and trills were also noted.

Results

Songs were sung by individual male robins mostly during dawn (5.00–9.00 a.m.) and dusk (5.00–6.30 p.m.) hours for territory advertisement. Individuals preferably choose tall, exposed branches of trees, barren rocks, walls, boundaries of houses, or electric wires as their song posts. The song of the Indian robin is composed of strophes, sporadic whistles, and trills. Strophes are comprised of various types of elements arranged in irregular fashion. Song repertoire was comprised of 2 to 11 types of strophes per individual. The frequency of song ranges from 1.2 kHz to 9.37 kHz (n=150). The minimum frequency, maximum frequency, range of frequency (bandwidth), and dominant frequency of songs were 2.69±0.04, 7.36±0.06, 4.7±0.08, and 5.00±0.04 kHz (n=150), respectively (Figures 9.3, 9.4 and 9.5). Frequency bandwidth shows maximum deviation from its mean ranging from minimum 2.34 to maximum 7.26 kHz, and the dominant frequency ranges from 3.09 to 6.23 kHz (Tables 9.1 & 9.2). The average duration of the strophes was 0.68±0.02sec (n=150). Followed gap between strophes was 8.77±0.52sec (n=150). Songs were sung at the rate of 8.73±0.57 strophes per minute. Analysis of the data showed that duration of strophes was almost consistent with mean duration of strophes, while gap between strophes was relatively varied (Figure 9.4). The Indian robin delivers short duration song bouts ranging from 1.00 to 3.00 min duration. Different kinds of singing modes were observed based on strophe arrangement in a song bout. They were mainly stereotyped and alternative (Figure 9.2).

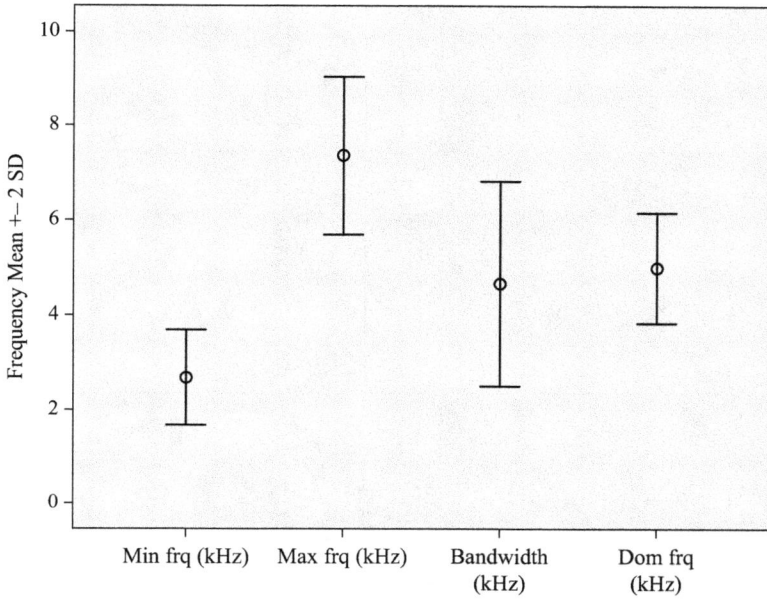

Figure 9.3 Error bar plot of variability of different physical parameters of song in Indian robin.

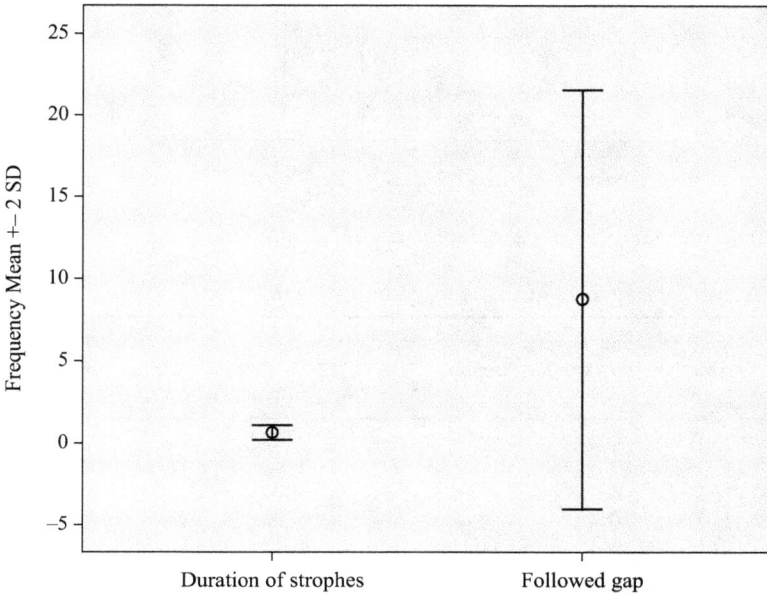

Figure 9.4 Error bar plot of duration of strophes and followed gap in Indian robin.

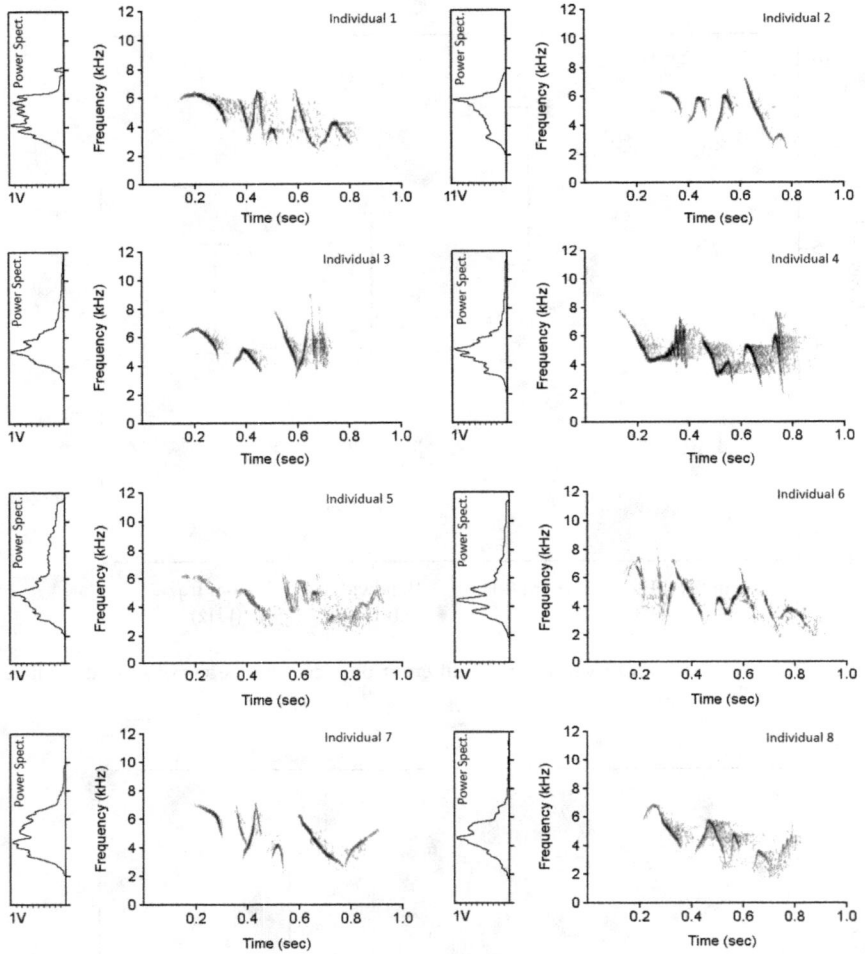

Figure 9.5 Spectrograms of the songs of Indian robin C. *f. intermedius.*

Table 9.1 Acoustical features of discrete song in Indian robin

No. of strophes analysed (n)	Song bout duration (sec)	Singing rate (strophes/ min)	Types of strophes	Mode of singing		Presence of trills	Presence of whistles
				Repetitive	Alternative		
20	150	9	6	75%	25%	Yes	Yes
20	120	7	6	70%	30%	Yes	Yes
21	162	6	9	50%	50%	Yes	Yes
23	120	10	8	50%	50%	Yes	Yes
21	72	6.2	5	60%	40%	Yes	Yes
16	102	7	6	50%	50%	Yes	Yes
3	60	12	1	100%	Nil	No	Yes
26	90	13	14	57%	43%	Yes	Yes

Table 9.2 Physical characteristics of strophes in Indian robin

Individual	Minimum frequency strophe (kHz)	Maximum frequency strophes (kHz)	Bandwidth (kHz)	Dominant frequency (kHz)	Duration of strophes (sec)	Followed gap (sec)	Types of element per strophe	Number of elements per strophe
1	2.65±0.03 (n=20)	7.21±0.14 (n=20)	4.55±0.14 (n=20)	4.96±.16 (n=20)	0.61±0.02 (n=20)	6.65±0.74 (n=20)	2.1±0.1 (n=20)	3.2±0.11 (n=20)
2	2.89±0.09 (n=21)	7.4±0.2 (n=21)	4.51±0.25 (n=21)	4.93±.16 (n=21)	0.63±0.04 (n=21)	8.96±1.05 (n=21)	2.35±0.15 (n=21)	4.2±0.11 (n=21)
3	2.53±0.11 (n=21)	8.28±0.15 (n=21)	5.75±0.2 (n=21)	5.03±.13 (n=21)	0.86±0.05 (n=21)	9.04±1.48 (n=21)	2.52±0.14 (n=21)	3.90±0.24 (n=21)
4	2.74±0.15 (n=23)	7.54±0.11 (n=23)	4.8±0.23 (n=23)	5.16±.06 (n=23)	0.66±0.03 (n=23)	7.74±1.01 (n=23)	2.39±0.17 (n=23)	3.13±0.27 (n=23)
5	2.76±0.07 (n=21)	6.47±0.06 (n=21)	3.69±0.07 (n=21)	5.12±0.11 (n=21)	0.71±0.03 (n=21)	11.98±1.52 (n=21)	2.47±0.11 (n=21)	3.71±0.10 (n=21)
6	2.72±0.17 (n=16)	7.63±0.13 (n=16)	4.90±0.24 (n=16)	5.13±.09 (n=16)	0.75±0.08 (n=16)	10.43±2.43 (n=16)	2.37±0.17 (n=16)	4.43±0.32 (n=16)
7	2.39±0 (n=3)	7.17±0.05 (n=3)	4.78±0.05 (n=3)	4.03±0 (n=3)	0.74±0.01 (n=3)	0.12±0.64 (n=3)	3±0 (n=3)	3±0 (n=3)
8	2.61±0.08 (n=26)	7.12±0.16 (n=26)	4.50±0.21 (n=26)	4.84±.10 (n=26)	0.58±0.02 (n=26)	7.65±1.33 (n=26)	2.46±0.13 (n=26)	3.92±0.17 (n=26)
Total	2.69±0.04 (150)	7.36±0.06 (150)	4.66±0.08 (150)	5±0.04 (150)	0.68±0.01 (150)	8.77±0.52 (150)	2.4±0.05 (150)	3.75±0.08 (150)

Mode of Singing

Two types of singing patterns were identified in the recordings. In some song samples, the same strophe was repeated again and again while in some others, more than one type of strophe was uttered several time before switch over to another type of strophe. Alternatively, several types of strophes were used in a song bout.

Discussion

Results revealed that songs in the population of Indian robin (*C. f. intermedius*) were comprised of strophes (2 to 14), delivered in two modes namely "repetitive mode" and "alternative mode." When compared to a previous study (Kumar, 2011) carried out on the song characteristics of Indian robin (*C. f. cambaiensis*) in northern India, it was found that both sub-species used almost the same pattern of singing, except some minor structural variations. The minimum frequency and maximum frequency of songs were slightly lower than the north Indian population. A review of the literature revealed that bird song may be almost consistent in structure in some species throughout the distribution ranges, while in some others, it may vary substantially (Catchpole & Slater, 2008). Song variations have been documented in various species (Hamao et al., 2018). Both on a continental scale (Kroodsma et al., 1999) and small island level (Schottler, 1995; Baker, 1996) have been documented. For the Indian robin, the variations in song characteristics at a sub-species level appear minor in structure. However, the present study is conducted at the preliminary level, and further evidence is yet to be gathered and investigated to document the sub-species level variations and possible driving forces.

Acknowledgements: We are grateful to Dr. Dhriti Banerjee, Director, ZSI, Kolkata, for kind support and encouragement, and Officer-in-Charge, ZSI, NRC, Dehradun, for departmental facilities. The first two authors would like to convey sincere gratitude to Dr. Anil Kumar, Scientist-E and Principal Investigator, SERB funded project (CRG/2018/003508), ZSI, NRC, Dehradun, for his kind support and guidance throughout the fieldwork. We would like to express our special thanks to the Forest Department, Chandrapur Forest Division for providing accommodation and valuable support.

References

Ali, S., & Ripley, S. D. (1973). *Hand book of the birds of India and Pakistan. Vol. 9, Robins to Wagtails*. Oxford University Press.

Baker, M. C. (1996). Depauperate meme pool of vocal signals in an island population of singing honeyeaters. *Animal Behaviour, 51*(4), 853–858.

Beresford, P., & Cracraft, J. (1999). Speciation in African forest robins (Stiphrornis): Species limits, phylogenetic relationships, and molecular biogeography. *American Museum Novitates, 3270*, 1–22.

Bolsinger, J. S. (2000). Use of two song categories by golden-cheeked warblers. *Condor*, *102*(3), 539–552.

Catchpole, C. K., & Slater, P. J. (2008). *Bird song: Biological themes and variations* (2nd ed.). Cambridge University Press.

Collar, N., & Bonan, A. (2018). Indian robin (*Saxicoloides fulicatus*). In J. Delhoyo, A. Elliott, J. Sargatal, D. A. Christie & E. De Juana (Eds.), *Handbook of the birds of the world alive*. Lynx Edicions. Retrieved June 18, 2018, from https://www.hbw.com/node /58491.

Dadwal, N., Bhatt, D., & Singh, A. (2017). Singing patterns of pied bush chat (*Saxicola caprata*) across years and nesting cycles. *Wilson Journal of Ornithology*, *129*(4), 713–726.

Grimmett, R., Inskipp, C., & Inskipp, T. (1998). *Birds of the Indian subcontinent*. Oxford University Press.

Hamao, S., Komatsu, H., & Shinohara, M. (2018). Geographic variation in yellow bunting songs. *Ornithological Science*, *17*(2), 159–164.

Highsmith, R. T. (1989). The singing behavior of golden-winged warblers. *Wilson Bulletin*, *101*, 36–50.

Hoelzel, A. R. (1986). Song characteristics and response to playback of male and female robins (*Erithacus rubecula*). *Ibis*, *128*, 115–127.

Kroodsma, D. E., Bereson, R. C., Byers, B. E., & Minear, E. (1989). Use of song types by the chestnut-sided warbler: Evidence for both intra- and inter-sexual functions. *Canadian Journal of Zoology*, *67*(2), 447–456.

Kroodsma, D. E., Byers, B. E., Halkin, S. L., Hill, C., Minis, D., Bolsinger, J. R., & Wilda, K. (1999). Geographic variation in black-capped chickadee songs and singing behaviour. *Auk*, *116*(2), 387–402.

Kumar, A. (2011). Physical characteristics, categories and functions of song in Indian robin *Saxicoloides fulicata* (Aves: Muscicapidae). *Journal of Threatened Taxa*, *3*(7), 1909–1918.

Kumar, A. (2012a). Songs and calls of Indian birds: Implications for behavioral studies, systematic and conservation. *Journal of Bombay Natural History Society*, *109*(1&2), 60–71.

Kumar, A. (2012b). Breeding biology of Indian robin *Saxicoloides fulicata* (Aves: Muscicapidae) in Northern India. *Journal of Experimental Zoology*, India, *15*(1), 57–61.

Kumar, A. (2013). Song system in an ecologically important songbird the oriental magpie robin *Copsychus saularis*. In R. Jayapal, S. Babu, G. Quadros, P. R. Arun, P. Pramod, H. N. Kumara, & P. A. Azeez (Eds.), *Ecosystem services and functions of birds: Proceedings of the second international conference on Indian ornithology* (pp. 25–27). SACON, Coimbatore, India, 19–23 November 2013, pp. 232.

Kumar, A. (2016). Why do birds sing? Secrets behind the melody of birdsong. *Hornbill*, *4*, 16–22.

Marler, P., & Slabbekoorn, H. (2004). *Nature's music: The science of birdsong*. Elsevier Academic Press.

Molles, L. E., & Vehrencamp, S. L. (1999). Repertoire size, repertoire overlap, and singing modes in the banded wren (*Thryothorus pleurostictus*). *Auk*, *116*(3), 677–689.

Nirmala, Sr., T., & Vijayan, L. (2003). Breeding behavior of the Indian robin *Saxicoloides fulicata* in the Anaikatty Hills, Coimbatore. In *Proceedings of the 28th ESI conference, February 7–8 2003* (pp. 43–46). KMTR.

Schottler, B. (1995). Songs of blue tit *Parus caeruleus palmensis* from La Palma (Canary Islands): A test of hypothesis. *Bioacoustics*, *6*(2), 135–152.

Scriba, M. F., & Goymann, W. (2010). European robins (*Erithacus rubecula*) lack an increase in testosterone during simulated territorial intrusions. *Journal of Ornithology*, *151*(3), 607–614.

Shende, R. R. (2013). *Ground water in formation, Chandrapur district, Maharashtra*. Central Ground Water Board, Ministry of Water Resources, Govt. of India, 1759/DBR/2013, pp. 24.

Singh, A., Bhatt, D. & Sethi, V. K. (2019). Singing behaviour of the oriental magpie robin (*Copsychus saularis*). *Journal of Ornithology*, *160*(1), 185–193.

Spector, D. A. (1992). Wood-warbler song systems. A review of paruline singing behaviors. In M. D. Power (Ed.), *Current ornithology* (pp. 199–238). Plenum Press.

Staicer, C. A. (1989). Characteristics, use and significance of two singing behaviors in Grace's warbler (*Dendroica graciae*). *Auk*, *106*(1), 49–63.

Part II
Human-Wildlife Interaction

10 Human-Tiger Conflict in Corbett Tiger Reserve

Sharad Kumar, Jamal Ahmad Khan, and Afifullah Khan

Introduction

Human-carnivore conflict is one of the major issues in wildlife conservation worldwide, because conflict often puts human communities against carnivores and those who seek to preserve or restore their populations (Torres et al., 1996; Bangs et al., 1998; Berg, 1998; Karanth & Madusudan, 2002; Treves & Karanth, 2003; Loe & Roskaft, 2004). Large carnivores are always considered a threat to human beings (Brantingham, 1998; Treves & Naughton-Treves, 1999), and problems of large carnivore-human conflict have historically occurred wherever wild predators have co-existed with humans (Davenport, 1953; Schaefer et al., 1981; Bibikov, 1982; Boitani, 1982; Kaczensky, 1996; Sekhar, 1998; Vijayan & Pati, 2002; Madhusudan, 2003; Mishra et al., 2003). Attacks on livestock and humans by large carnivores and retaliatory killings by humans have persisted for centuries (Boitani, 1995) and are a major reason behind extirpating them in parts of their range or reducing them to small rampant populations (Servheen, 1990; Promberger & Schroder, 1993). Under a variety of demographic, economic, and social pressures, human alteration of carnivore habitats or exploitation of carnivores has led to escalated conflicts (Mladenoff et al., 1997; Liu et al., 2001; Treves & Karanth, 2003).

In India, after the commencement of the Wildlife (Protection) Act, 1972, populations of large carnivores have increased, and the frequency and economic cost of conflict between humans and carnivores appear to have increased (Treves & Karanth, 2003). The intensity of the problem has also increased with the increase in human and livestock populations. Increasing human and livestock populations puts immense pressure on natural habitats of wild animals and poses a threat to wildlife conservation (Mishra, 1997). Intensive grazing by domestic cattle and other activities of local people have been degrading tiger habitats in terms of retarded growth of vegetation, increase in abundance of weeds, and ultimately depletion of natural prey base (Madhusudan, 2000). As a consequence of increase in livestock and depletion of prey base, large carnivores such as the tiger are compelled to prey on domestic livestock. Persecution by human beings in response to tiger-human conflict, along with decline of prey base, degradation of habitat, and poaching, are the major causes of the decline of tigers in the recent

DOI: 10.4324/9781003321422-12

past (Nyhus & Tilson, 2004). Over most of its range, the tiger shares its habitat with humans, and the problem of human-tiger conflict is posing a threat to the long-term conservation of tiger over its entire range.

Corbett Tiger Reserve (CTR) is a significant tiger conservation unit under the Project Tiger Scheme of the Government of India and plays a crucial role in the conservation of tigers. In the Corbett Tiger Reserve, with a high density of tigers (Jhala et al., 2015), large carnivores (tiger and leopard) attack a significant number of cattle, creating apathy among the local people toward conservation of large carnivores and posing a threat to the long-term conservation of tigers in the Corbett landscape. Therefore, mitigation of human-tiger conflict is essential to ensure the long-term survival of tigers in the Corbett landscape, and to tailor the strategies to mitigate the problem, it is imperative to have a good understanding of the problem. Therefore, we investigate the problem of human-tiger conflict in and around the Corbett Tiger Reserve from 2002 to 2006.

Study Area

The study was conducted in and around the Corbett Tiger Reserve, located in the civil districts of Nainital, Pauri Garhwal, and Almora in the Uttarakhand state of India (Figure 10.1). The Corbett Tiger Reserve, within its boundary, incorporates areas of Corbett National Park and Sonanadi Wildlife Sanctuary and is spread over an area of 1,288.34 km². The terrain of the study area is hilly and consists of a number of ridges and valleys. The geological formation of the study area is divided into the recent and the Siwalik series. The altitude ranges from 350 m to 1,210 m. The climate is tropical with three distinct seasons: summer, monsoon, and winter. The average temperature varies from 13 °C in January to 31 °C in May. The vegetation of the area is quite heterogeneous. According to Champion and Seth (1968), eight forest types (moist Siwalik sal forest; moist Bhabar Dun sal forest; western Gangetic moist mixed deciduous forest; alluvial Savannah woodland forest; dry Siwalik sal forest; northern dry mixed deciduous forest; Khair-Sissoo forest; and lower or Siwalik chir pine forest) are found in the study area. A major portion of the area is covered by sal (*Shorea robusta*) along with several related species.

Methods

To get a preliminary understanding of problem of the human-tiger conflict, information on livestock attacks by tigers during 2001 and 2002 was collected from the forest administration of the CTR. Analysis of the forest department data indicated that the problem was more severe at the south-eastern limit of the study area. On the basis of this, the area of the CTR was divided into two zones, viz intensive (south-eastern portion) and extensive (northern portion) study.

Due to logistical limitations, it was not possible to reach every reported carnivore attack site on time. So, we gathered secondary information about attacks on a monthly basis from uncovered areas of the northern zone (extensive area).

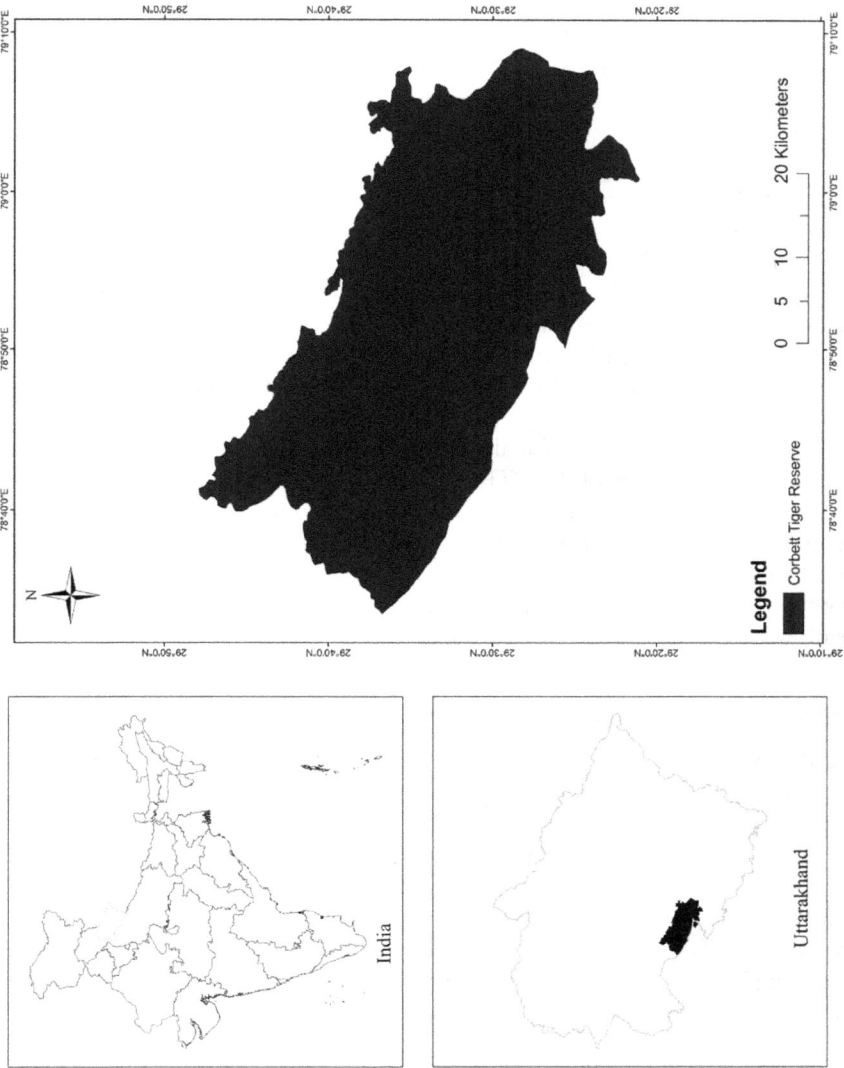

Figure 10.1 Location map of the Corbett Tiger Reserve (CTR).

To study carnivore-human conflict in detail, carnivore attacks on livestock and human beings reported by the villagers and forest department were inspected physically to establish the identity of the predator involved and to understand the factors responsible for cattle attacks. The predator was identified on the basis of patterns of feeding, pugmarks, hairs, and other indirect evidences. Data on location, predator species, and species killed, as well as health, weight, age, and sex of species was recorded. Economic loss incurred to locals due to livestock killed by tigers was estimated based on the current market price of the species killed. Livestock species were categorised in different age classes as adult (> 5 years), sub-adult (3 to 5 years), calf (1.5 to 3 years), and weaner (< 1.5 years). Chi square goodness of fit test was used to test the significance difference in livestock attacked by tigers among various seasons and livestock categories.

Results

Human Attacks

Over a period of five years, from 2002 to 2006, 47 humans were attacked by wild animals in and around the CTR. Of those, only four were killed and the rest were injured by wild animals. Tiger contributed 40.2% (n=19) to attacks on humans and emerged as the most conflicting species, while sloth bear and wild pig each contributed 21.3% (n=10) followed by leopard 10.6% (n=5), elephant 4.3% (n=2), and blue bull 2.1% (n=1). Tigers killed three humans and injured 16 others during the five-year period. Out of 19 cases of tiger attacks, 14 victims were male and the rest were female. All the tiger attacks were reported from the south-eastern boundary of the CTR.

In 2002, no attacks on humans by tigers were reported, whereas the maximum number of attacks (n=7) was reported during 2005 and minimum number of attacks (n=3) was reported during 2006. Tigers injured four, three, seven, and two humans during 2003, 2004, 2005, and 2006 respectively, whereas one human death was reported in each year during 2003, 2004, and 2006.

Repercussions of Tiger Attack on Humans

Local people react very aggressively after an attack by a tiger on humans, and managers of the protected areas face the most challenging jobs in such a situation. They are in a very difficult situation and face tremendous pressure from local people to eliminate or capture the tiger. Situations become more strenuous when attacks on human lead to death and people assume that the tiger has become a man-eater. During the study period, three tigers were eliminated from in and around the CTR in response to tiger-human conflict. No case of persecution or poisoning by local communities in response to human-tiger conflict was recorded during the study period.

Livestock Depredation

Number of Attacks and Livestock Killed

During a five-year period (2002–2006), 3,027 incidents of carnivore (tiger—1,924 and leopard—1,103) attacks on livestock were recorded in and around the periphery of the CTR. Tigers killed 1,418 cattle with an average rate of 284 livestock per year and injured 506 cattle with an average rate of 101 livestock per year. Tigers killed 241, 245, 306, 293, and 333 cattle while injuring 80, 55, 109, 136, and 126 cattle in 2002, 2003, 2004, 2005, and 2006 respectively. The livestock attacked by tigers during a five-year period (2002–2006) were from 166 villages. Tigers attacked 1,924 animals, with an average rate of 12 animals per village. The settlement of *Belghatti gujjar khatta*, located in the southern part of the Corbett Tiger Reserve, saw the most livestock attacks.

Composition of Age and Sex of the Species

Table 10.1 provides the characteristics of livestock attacked by tigers in and around the CTR. The distribution of livestock attacked by tigers differed significantly among the different livestock classes ($\chi2 = 2484.79$, d.f. = 7, P<0.001). Tigers attacked more cows (37.7%), followed by buffaloes (35%), whereas they attacked fewer bullocks (11.4%), buffalo calves (6.7%), cow calves (5%), horses (0.2%), and mules (0.1%). On comparison of livestock killed and injured separately, it was found that tigers killed more cows in comparison to buffaloes but injured more buffaloes in comparison to cows.

Figure 10.2 provides the distribution of livestock killed and injured in different months over a five-year period (2002–2006). The distribution of tiger attacks on livestock differed significantly in different months ($\chi2 = 153.73$, d.f. = 11, P<0.001). The majority (57%) of cattle attacks were recorded during the months of August (n= 335), September (n= 237), October (n= 236), and July (n= 234).

Table 10.1 Numbers and characteristics of livestock attacked by tiger in and around the CTR

S. No.	Livestock class	Killed	Injured	Total
1	Cow	555	171	726
2	Bullock	174	46	220
3	Cow calves	78	27	105
4	Buffalo	469	207	673
5	He-Buffalo	51	14	65
6	Buffalo calves	85	44	129
7	Horse	4	0	4
8	Mule	2	0	2
Total		1418	506	1924

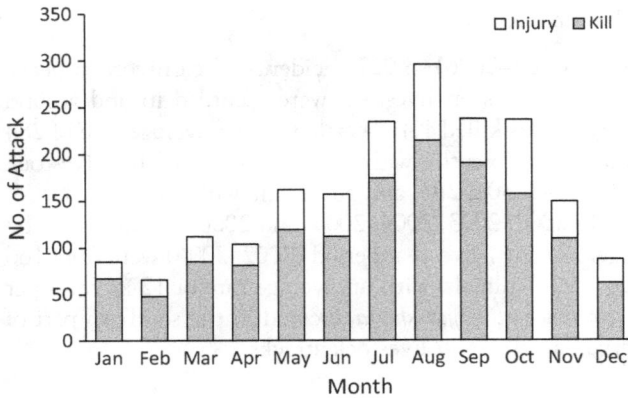

Figure 10.2 Month-wise distribution of livestock attacked by tiger in and around the CTR.

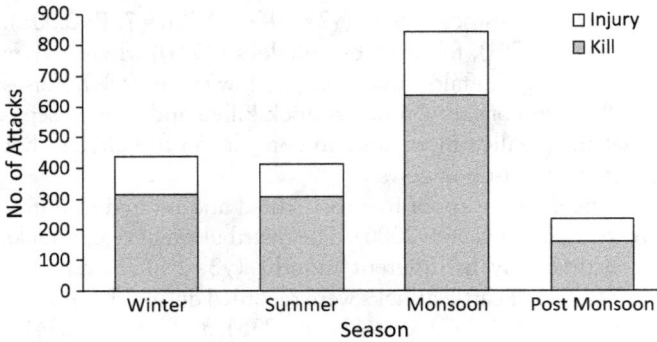

Figure 10.3 Season-wise distribution of livestock attacked by tigers in and around the CTR.

The distribution of tiger attacks on livestock differed significantly between seasons ($\chi2 = 104.51$, d.f. = 3, P<0.001). The frequency of a tiger's attack on livestock was highest (n=839) in the monsoon. There were 412 and 438 incidents of tiger attacks during summer and winter respectively, while only 235 incidents of tiger attacks on cattle were recorded during the post monsoon (Figure 10.3).

Economic Loss

From 2002 to 2006, 3,027 attacks of large carnivores on livestock were recorded, with an average of 12 livestock per week. Tigers attacked seven livestock per week. Out of 506 livestock injured by tigers, 137 died afterwards. To determine the impact of depredation in terms of economic loss, incidents of livestock dying due to injury caused by tigers were also included in the calculation of total loss. Table 10.2 provides a presentation of various age and sex classes of livestock species and their associated economic costs. Cows, buffaloes, and bullocks were most frequently depredated by tigers and formed the largest component of economic loss. From 2002 to 2006, tigers were responsible for estimated economic loss of INR 35,550,000 (USD 500,704) to people sharing range with tigers in and around the CTR. Over a period of five years, the maximum loss was recorded in 2006 (INR 8,232,000) and the minimum in 2002 (INR 6,002,000).

Discussion and Conclusion

Humans dominate the landscapes of most parts of the world, and human impact and movement go far beyond the edges of cities and into wild habitats. This incursion of humans into wild habitats, plus habituation of some wild species, has resulted in an increased potential for human wildlife encounters, including wildlife attacks on humans (Quigley & Herrero, 2005). Though there has been a recent increase in our appreciation and understanding of the importance of wildlife conservation, attacks on humans and their property tend to illicit a strong negative reaction towards the conservation of carnivores. Remaining tiger habitats in Indian sub-continent are like islands surrounded by sea of human-dominated landscapes, and the future of tigers depends on the goodwill and support of people living around tiger habitats. Therefore, unless there are strong incentives to conserve tigers for local people, it is impossible to ensure the future survival of tigers. Conflict between the local people and tigers creates hindrances in the motivation of local people for the conservation of tigers. For conservation measures to be successful, it is imperative that different forms of tiger-human conflicts be managed. An effective conservation programme needs a constant process of mitigation of conflict and negotiation over the long term. For conservation to be effective, we must constantly engage in multifaceted exploration of our relationship with the natural environment and with each other.

Out of 19 incidents of tiger attacks on humans, tigers killed only three victims, which indicated that tigers are not attacking humans with the motive to consume. It is clear that tigers do not consider humans as prey, and attacks on humans are purely accidental. Attacks on human beings have been considered an aberrant form of behaviour for tigers (Chauhan, 2005). In CTR, a large number of people enter forests daily for the collection of NTFPs (non-timber forest products) and to graze livestock. The presence of a large number of humans in tiger range gives rise to a high probability of encounters with tigers. Therefore, incidents of tiger attacks on humans would consequently be higher if tigers considered humans as occasional prey. In the study period, only three cases of deaths were reported during a five-year time span. Three human deaths over a period of five years is quite low in comparison to the Sundarbans where Hendrich (1975) reported 24.3 deaths per year over a period of 15 years of study. Therefore, these insignificant incidences of human attacks clearly

Table 10.2 Economic loss incurred to livestock owners over a period of 2002–2006

Livestock class	Estimated cost	Individuals killed					
		2002	2003	2004	2005	2006	Overall
Buffalo	50000	69	70	85	66	81	371
Buffalo (SA)	30000	3	11	11	21	18	64
He-Buffalo	50000	1	0	3	5	3	12
He-Buffalo (SA)	30000	1	0	0	2	3	6
Buffalo calf	7000	23	19	22	32	15	111
Buffalo weaner	3000	0	1	4	0	4	9
Cow	15000	95	93	138	133	169	628
Cow (SA)	10000	0	5	12	14	2	33
Bullock	15000	40	44	38	38	42	202
Bullock (SA)	10000	3	1	5	8	4	21
Cow calf	4000	25	11	17	13	15	81
Cow weaner	2000	3	2	5	1	0	11
Horse	30000	2	0	0	0	0	2
Mule	25000	0	1	1	2	0	4
Total livestock killed		265	258	341	335	356	1555
Total loss (INR)		6,002,000	6,154,000	7,809,000	7,353,000	8,232,000	35,550,000
Total loss (USD)		84,535	86,676	109,986	103,563	115,944	500,704

indicated that tigers responsible for attacks were not man-eaters, and the occasional attack on humans was the result of a sudden close encounter of humans with tigers and in response to self-security or protection of cubs. But in spite of the low number of incidents, this is the most intolerable form of conflict and needs some strategies to reduce tiger attacks on humans. Though local people cannot be legally denied entry into the buffer areas, they can be given practical skills and trained to defend themselves to avoid any potential attack by tigers. Reducing attacks not only reduces injury and loss of human life but can also help in conserving wildlife populations, promoting good will towards wildlife, minimising economic loss, and improving the quality of life for humans (Quigley & Herrero, 2005).

Tigers attacked more males (n= 14) in comparison to females (n= 5), and a similar trend was also reported by Chauhan (2005) regarding Katerniaghat Wildlife Sanctuary, where he discussed that males go to forests more in comparison to females and are more vulnerable to tiger attacks. But the situation in the CTR is different; mostly women in groups go inside the forests to collect fodder, fuelwood, and other NTFPs. Human settlements are scattered in and around the forests, and local people (generally men) have to commute through tiger habitats when they move outside for work. Mostly, they go early in the morning for work and come late in the evening, and these timings coincide with the crepuscularity of the tiger, thus making them more vulnerable to attacks. Secondly, women always move in groups while men are mostly solitary or move in smaller groups, and this makes men more vulnerable to tiger attacks in comparison to women. But tigers killed more females in comparison to males. That is because males are generally stronger in comparison to females and have higher chances of survival after attack as compared to females. Females got frightened easily and ran away when a tiger attacked their fellow group member while males developed courage to save their fellow group member.

A case of human attack by a tigress at Dhikala tourism facility was believed to be due to the collective pressure of tourists and hunger. This tigress had a litter of cubs, and thus close approach of tourists disturbed the animal prior to attacking its natural prey. Finding food is difficult if females close to their cubs are disturbed by human presence. This might lead to the attack on the person when she encounters him by chance.

Livestock attacks leading to economic losses are probably the most important issue in the study area. Our findings are based on the monitoring of incidents reported by villagers. Results clearly indicated that tiger attacks on livestock are high and on humans are occasional. Findings of the study showed that tigers are inflicting substantial loss to local people in and around the buffer zone of the Corbett Tiger Reserve. Estimated annual economic loss due to livestock depredation is INR 7,110,000 year. The calculated annual economic loss due to livestock damage is INR 42,831 per village. It may be an acceptable figure at a community level, but the loss is very high at an individual level for those who lose their livestock. On some occasions, a tiger killed four to five livestock of a single household in a single day. These multiple killings by tigers ruined the economic status of the family. Since the majority of the population living in and around the tiger habitats is relatively poor and their livelihood depends on livestock husbandry, the high proportion of losses due to livestock depredation by tigers emerged as the main reason behind human-carnivore conflict, and its mitigation is the prime determinant in resolving tiger-human conflict in the study area.

According to Finn (1929), livestock depredation is an aberrant behaviour of a tiger when it becomes regular and obsessively habituated. But the Corbett area has had a long-standing problem of livestock depredation by tigers, and livestock contributed to 23% of the diet of tigers in the buffer zone of the CTR (Kumar, 2010). Livestock species, because of their reduced escape abilities compared to wild herbivores, are more prone to predation (Nowell & Jackson 1996). Livestock depredation in the area is not a recent phenomenon, but the frequency of the killing of livestock by carnivores has increased during the last two decades, which coincides with the rising population of tigers in Corbett Tiger Reserve.

Despite the fact that villagers have been living with tigers for centuries, they do not take any preventive measures to reduce the attacks of tigers on their livestock. They leave their cattle unattended to graze in the forest, and if they are attended to, the people do not carefully oversee their movement. Livestock left unattended to graze in forested areas becomes the dominant prey for carnivores in terms of available biomass (Schaller & Crawshaw, 1980; Schaller, 1983). Tigers killed more cows in comparison to buffalo because of two possible reasons: first, people living in hilly areas prefer to rear cows to buffaloes, as cows are more adaptable to grazing in hilly terrains due to their small size; and secondly, it was easy for tigers to handle a cow at the time of attack. The statement was also supported by data on injured livestock. More buffaloes were able to defend themselves from the tiger in comparison to cows which indicated that tigers find difficulty in handling large prey species such as buffalo. Most of the cases of livestock depredation occurred when the owner was not accompanying the livestock. Livestock are rarely accompanied by the owners during the daytime and are left ranging free to graze in the area. Care in herd attendance and active defence is totally lacking in the study area. Though conflict is very high in the CTR, people seem to give poor attention to their cattle.

The highest numbers of attacks on livestock by tigers were recorded in monsoons (July, August, September, and October). Patterson et al. (2004) have also reported a similar trend in the case of lion predation on livestock in Tsavo National Park in Kenya. Heightened attack in rainy seasons may also be a result of an increase in vegetation cover, since rainfall in the study area stimulated fast vegetative growth, which provides cover for predators to ambush prey. It might also be possible that during the rainy season, vegetative cover increases and predators move close to human settlements, increasing the chances of tigers' encounters with livestock. Another factor which contributed to the increase in attacks during the monsoon was the lower utilisation of killed prey during the monsoon. It was found that during the monsoon, meat petrified rapidly and maggots developed quickly on the carcass. However, Chauhan (2005) reports a high incidence of livestock attacks in winter in Dudhwa National Park, Uttar Pradesh.

Tigers attack on an average of nine livestock per week, and this is a considerable loss to local people living in and around the CTR. Belghatti gujjar settlement located in the Sanwalde Bhavar blocks has the maximum number of livestock attacked by tigers because of their habit of grazing livestock at night.

Compensation programmes increase tolerance of wildlife and promote more positive attitudes and support for conservation among local people who live closest to endangered and dangerous animals (Wagner et al., 1997). Currently, the livestock losses to local people are compensated by the Forest Department and The Corbett Foundation (NGO). But local people complain that the amount is low (50% of actual value); for smaller stocks, the amount one pays in transport and medical certificate costs to claim the compensation may negate the amount received. Another hindrance is the long bureaucratic procedure of securing compensation from the Forest Department.

In CTR, the most important factor raising antipathy among local people towards the conservation of tigers is the conflict between tiger and human beings as a result of livestock depredation by tigers, which threatens the economic livelihoods of the local people. Since a significant part of the tiger population in India lives outside protected areas, negative attitudes and perceptions by humans towards tigers are clearly the greatest imminent threat to the species survival. It is imperative to initiate programmes with local people at landscape level to maintain a viable population. It is therefore necessary that tiger conservation should be addressed within the context of local people's incentives, and appropriate social institutions to ensure the collective and not just individual interest in tiger conservation. Conservationists must pay attention to the feelings and incentives of local people and develop conservation initiatives with them. Producing guidelines for villagers on how to minimise conflict with tigers, through modification of tiger behaviour and livestock management practices, would be helpful in the conservation of tigers. Reduction in livestock depredation by carnivores could be achieved by various means, such as controlling predators (Stahl et al., 2001), exclusion of predators (Sillero-Zubivi & Laurenson, 2001), or by improving livestock management (Breitenmoser et al., 2005). But control and exclusion of predators such as tigers rarely produces a long-term decline in losses (Landa et al., 1999, Stahl et al., 2001). Therefore, measures for improvement in the practice of livestock husbandry and lifestyle of people sharing ranges with tigers could be implemented to reduce livestock depredation. Additionally, increasing the tolerance of local people towards tigers could be an optimal solution for the mitigation of tiger-human conflict in areas such as Corbett Tiger Reserve.

Acknowledgment: We would like to thank the Principal Chief Conservator of Forests (Wildlife), Uttarakhand for providing permission to conduct the study in the Corbett Tiger Reserve. We are grateful to the staff of Corbett Tiger Reserve for providing necessary support in the implementation of the study. We wish to express our sincere thanks to The Corbett Foundation for providing logistical support to the research team. We would like to thank our field assistants and staff at The Corbett Foundation for assistance in the implementation of this project. We would like to acknowledge the assistance of villagers for providing information about the incidences of carnivore attacks.

References

Bangs, E. E., Frits, S. H., Fontaine, J. A., Smith, D. W., Murphy, K. M., Mack, C. M., & Neimeyer, C. C. (1998). Status of grey wolf restoration in Montana, Idaho, and Wyoming. *Wildlife Society Bulletin, 26*, 785–798.

Berg, K. A. (1998). The future of the wolf in Minnesota: Control, sport or protection? In M. Fascione (Ed.), *Proceedings of the defenders of wildlife: Restoring the wolf conference* (pp. 40–44). Defenders of Wildlife.

Bibikov, D. I. (1982). Wolf ecology and management in the USSR. In F. H. Harrington & P. C. Paquet (Eds.), *Wolves of the world: Perspectives of behavior, ecology and conservation* (pp. 120–133). Noyes Publications.

Boitani, L. (1982). Wolf management in intensively used areas of Italy. In F. H. Harrington & P. C. Paquet (Eds.), *Wolves of the world: Perspectives of behavior, ecology and conservation* (pp. 158–172). Noyes Publications.

Boitani, L. (1995). Ecological and cultural diversities in the evolution of wolf-human relationship. In L. N. Carbyn, S. H. Fritts, & D. R. Seip (Eds.), *Ecology and conservation of wolves in changing world* (pp. 158–172). Canadian Circumpolar Institute, Occasional Publication No. 35.

Brantingham, P. J. (1998). Hominid–carnivore co-evolution and invasion of the predatory guild. *Journal of Anthropology and Archaeology, 17*(4), 327–353.

Breitenmoser, U., Angst, C., Landry, J. M., Breitenmoser-Wursten, C., Linnell, J. D. C., & Weber, J. M. (2005). Non-lethal techniques for reducing depredation. In R. Woodroffe, S. Thirgood, & A. Rabinowitz (Eds.), *People and wildlife: Conflict or coexistence?* (pp. 49–61). Cambridge University Press.

Champion, H. G., & Seth, S. K. (1968). *A revised survey of forest type of India.* Government of India, pp. 404.

Chauhan, N. P. S. (2005). Livestock depredation by tiger in and around Dudhwa National Park, Uttar Pradesh and mitigation strategies. *Indian Forester, 131*(10), 1319–1328.

Davenport, L. B. (1953). Agricultural depredation by black bear in Virginia. *Journal of Wildlife Management, 17*, 331–340.

Finn, F. (1929). *Stern dale's Mammalia.* Thacker and Co., pp. 347.

Hendrich, H. (1975). The status of the tiger Panthera tigris (Linne, 1758) in the Sundarbans mangrove forest (Bay of Bengal). *Saugetierkundliche Mitteilungen, 23*, 161–199.

Jhala, Y. V., Qureshi, Q., & Gopal, R. (Eds.). (2015). *The status of tigers, co-predators and prey in India 2014.* National Tiger Conservation Authority, New Delhi and Wildlife Institute of India, Dehradun. TR2015/021.

Kaczensky, P. (1996). *Large carnivore-livestock conflicts in Europe.* Munich Wildlife Society.

Karanth, K. U. & Madhusudan, M. D. (2002). Mitigating human-wildlife conflicts in Southern Asia. In J. Terborgh, C. van Schaik, L. Davenport and M. Rao (Eds), *Making parks work: strategies for protecting tropical nature* (pp. 250–264). Island Press.

Kumar, S. (2010). *Aspects of ecology of tiger (Panthera tigris) in Corbett Tiger Reserve (Uttaranchal)* [Ph.D. Thesis]. Aligarh Muslim University.

Land, A., Gudvangen, K., Swenson, J. E., & Roskaft, E. (1999). Factors associated with wolverine (Gulo gulo) predation on domestic sheep. *Journal of Applied Ecology, 36*(6), 963–973.

Liu, J. M., Linderman, Z., Ouyang, L. A., Yang, J., Zhang, J., & Zhang, H. (2001). Ecological degradation in protected areas: The case of Woolong nature reserve for giant pandas. *Science, 292*(5514), 98–101.

Löe, J., & Roskaft, E. (2004). Large carnivores and human safety: A review. *Ambio – A Journal of the Human Environment, 33*(6), 283–288.

Madhusudan, M. D. (2000). *Sacred cows and the protected forest: A study of livestock presence in Indian wildlife reserves.* CERC Technical Report #4, Centre for Ecological Research and Conservation, Mysore, pp. 24.

Madhusudan, M. D. (2003). Living amidst large wildlife: Livestock and crop depredation by large mammals in the interior villages of Bhadra Tiger Reserve, South India. *Environmental Management, 31*(4), 466–475.

Mishra, C. (1997). Livestock depredation by large carnivores in the Indian trans-Himalaya: Conflict perceptions and conservation prospects. *Environmental Conservation, 24*(4), 338–343.

Mishra, C., Allen, P., McCarthy, T., Madhusudan, M. D., Bayarjargal, A., & Prins, H. H. T. (2003). The role of incentive programs in conserving the snow leopard. *Conservation Biology, 17*(6), 1512–1520.

Mladenoff, D. J., Haight, R. G., Sickley, T. A., & Wydeveh, A. P. (1997). Cause and implications of species restoration in altered ecosystems. *Bio-Science, 47*(1), 21–31.

Nowell, K., & Jackson, P. (1996). *Wild cats: Status, survey and conservation action plan.* IUCN, pp. 382.

Nyhus, P. J., & Tilson, R. L. (2004). Characterizing human-tiger conflict in Sumatra, Indonesia: Implication for conservation. *Oryx, 38*(1), 68–74.

Patterson, B. D., Kasiki, S. M., Selempo, E., & Kays, R. W. (2004). Livestock predation by lions (Panthera leo) and other carnivores on ranches neighbouring Tsavo National Park, Kenya. *Biological Conservation, 119*(4), 507–516.

Promberger, C., & Schroder, W. (1993). *Wolves in Europe: Status and perspectives.* Munich Wildlife Society, pp. 136.

Quigley, H., & Herrero, S. (2005). Characterization and prevention of attacks on humans. In R. Woodroffe, S. Thirgood, & A. Rabinowitz (Eds.), *People and wildlife: Conflict or coexistence?* (pp. 27–48). Cambridge University Press.

Schaller, G. (1983). Mammals and their biomass on a Brazilian ranch. *Arquivos De Zoologia, 31*(1), 1–36. https://doi.org/10.11606/issn.2176-7793.v31i1p1-36

Schaller, G. B., & Crawshaw, P. G. Jr. (1980). Movement patterns of jaguar. *Biotropica, 12*(3), 161–168.

Schaefer, J. M., Andrew, R. D., & Dinsmore, J. J. (1981). An assessment of coyotes and dog predation on sheep in Southern Iowa. *Journal of Wildlife Management, 45*, 883–893.

Sekhar, N. U. (1998). Crop and livestock depredation caused by wild animals in protected areas: The case of Sariska tiger Reserve, Rajasthan, India. *Environmental Conservation, 25*(2), 160–171.

Servheen, C. (1990). The status and conservation of the bears of the world. International Conference on Bear Research and Management, Monograph Series 2, pp. 32.

Sillero-Zubiric, C., & Laurenson, M. K. (2001). Interaction between carnivores and local communities: Conflict or co-existence. In J. Gittleman, F. M. Funk, D. W. Macdonald, & R. K. Wayne (Eds.), *Carnivore conservation* (pp. 283–312). Cambridge University Press.

Stahl, P., Vande, J. M., Herrenschmidt, V., & Migot, P. (2001). The effect of removing lynx in reducing attack on sheep in the French Jura Mountains. *Biological Conservation*, *101*(1), 15–22.

Torres, S. G., Mansfield, T. M., Foley, J. E., Lupo, T., & Brinkhaus, A. (1996). Mountain lion and human activity in California: Testing speculations. *Wildlife Society Bulletin*, *24*, 457–460.

Treves, A., & Karanth, K. U. (2003). Human-carnivore conflict and perspectives on carnivore management worldwide. *Conservation Biology*, *17*(6), 1491–1499.

Treves, A., & Naughton-Treves, I. (1999). Risk and opportunity for humans coexisting with large carnivores. *Journal of Human Evolution*, *36*(3), 275–282.

Vijayan, S., & Pati, B. P. (2002). Impact of changing cropping patterns on man-animal conflicts around Gir protected area with specific reference to Talala sub-district, Gujarat, India. *Population and Environment*, *23*(6), 541–559.

Wagner, K. K., Schmidt, R. H., & Conover, M. R. (1997). Compensation programs for wildlife damage in North America. *Wildlife Society Bulletin*, *25*, 312–319.

11 Ecological Drivers of Livestock Depredation by Large Predators

Prashant Mahajan and Afifullah Khan

Introduction

Human activities over the last 200 years have transformed the planet and started an era now termed the Anthropocene. Conservation in the 21st century faces a fundamental challenge: To have a friendly nature between human activities and the needs of wildlife present in highly anthropogenic landscapes. Protected areas covers around 13% of the earth's surface and are the last hope for the conservation of threatened animals (Barua et al., 2013). Spreading of human activities and interventions into the protected areas has resulted in increased human-wildlife conflicts, and this interaction becomes more severe when it involves carnivores. This growing expansion by living alongside wildlife species can impose a variety of significant costs upon local people, which include depredation upon livestock, crop-raiding, attacks upon humans, disease transmission to stock or humans, and opportunity costs (Dickman, 2010).

Livestock depredation by large carnivores in and around the villages of India antagonises people and results in the development of a negative attitude towards the predators, which subsequently leads to retaliatory killing of the carnivores. Increasing cases of conflict among humans and large carnivores like leopards and tigers are a matter of great concern, especially for species conservation. Large carnivores in particular are predisposed to such conflicts due to their large home ranges and dietary requirements that increasingly overlap with people (Dar et al., 2009). As large carnivores are forced to live in increasing proximity to humans, competition for space and ungulate prey species leads to increases in human-carnivore interactions. Such conflict can be the most important cause of adult carnivore mortality in and around protected areas, and most conflict incidents occur when animals range around and beyond protected area borders into human-dominated landscapes (Gurung et al., 2008). Much of the livestock depredation by carnivore occurs at the edges of protected areas where carnivore and livestock overlap (Miller et al., 2016a).

Human-carnivore conflict is a complex multifaceted phenomenon and is governed by multiple factors. Conflicts could be influenced by local environmental conditions, prey abundance, habitat, or by socio-ecological factors i.e., livestock's husbandry practices (Singh et al., 2015). Differences between the

DOI: 10.4324/9781003321422-13

body sizes, anti-predator defences, and grazing requirements of livestock species result in distinct levels of vulnerability to wild carnivores (Seidensticker, 1976). For instance, in many areas smaller large carnivores like leopards, hyenas, and wild dogs primarily kill smaller-bodied livestock such as calves, sheep, and goats, whereas the largest carnivores like tigers target larger-bodied livestock such as adult cattle, buffalo, and horses (Sangay & Vernes, 2008). Likewise, large carnivore species employ unique hunting strategies and segregation tactics to avoid interspecific competition that results in risks for livestock at different times and locations. Also, large carnivore species require adequate vegetation cover to ambush its prey and are generally associated with dense forest and low visibility (Stahl et al., 2002; Thorn et al., 2012; Miller et al., 2015; Zarco-González et al., 2013). There are many studies worldwide which determine the various factors responsible for human-carnivore conflict but few of them have analysed the ecological factors responsible for conflict. Identifying priority human-carnivore conflict sites and ecological factors responsible for conflict at those sites can help managers to apply mitigation strategies at those sites and thus reduce conflict.

Our study was focused in Corbett Tiger Reserve (CTR), a protected area in Uttarakhand, India. CTR supports the high density of tigers with 9.4 tigers/100 km². The number of tigers has increased significantly in the landscape over the years, and currently there are 215 (169±261) tigers within CTR (Jhala et al., 2015). CTR also serves as the source site for tiger population, and often dispersing tigers come into conflict with humans (Jhala et al., 2015). Cattle are usually attacked both by the tigers and leopards. In an effort to prevent livestock owners from retaliating against carnivores, the Indian Forest Department as well as The Corbett Foundation financially compensates owners for domestic animals killed by the carnivores. To receive compensation, a livestock owner must locate and report the livestock carcass to the Forest Department as well to The Corbett Foundation within 48 hours, after which an officer visits the site to record the evidence of the death. Although not all the livestock owners choose to report lost livestock, many people living within the tiger reserve do. People do not report the loss of goats, sheep; or other small animals; as they are not compensated by the Forest Department or The Corbett Foundation. People of the villages usually go inside the forest in winter to collect fodder and fuelwood, as they collect for the summer season also.

Our objective is to assess the patterns of livestock depredation in and around the northern periphery of Corbett Tiger Reserve and to identify the certain vegetation structure variables associated with the vulnerability of livestock depredation by large carnivores. Tigers and leopards both are ambush predators, and vegetation structure can affect the hunting success of large predators that ambush prey. For identifying the priority conflict sites, our aim was to visualise the spatial hot spots of livestock kills across the northern periphery of Corbett Tiger Reserve. Therefore, through our study, we provide an ecological perspective on human livestock interaction and develop recommendations to reduce livestock vulnerability to carnivores.

Material and Methods

Study Area

The study was conducted in the buffer zone, at Northern Periphery of Corbett Tiger Reserve, Uttarakhand, India. The northern zone of CTR consist of three zones namely Rikhnikhal, Nainidanda, and Jaiharikhal. Jaiharikhal was the start of the study area and Nainidanda the end location of the study area (Figure 11.1). The study area lies between 29.46° N–29.40° N to 78.39° E to 78.58° E.

CTR is located in the foothills of the Himalayas in the districts of Nainital, Pauri (Uttrakhand), and Bijnore (Uttar Pradesh). CTR within its boundary includes areas of Corbett National Park, Sonanadi Wildlife Sanctuary, and buffer zones for both units. At present, the CTR covers an area of 1288.32 km^2, which includes an area of 520.82 km^2 of Corbett National Park, 301.81 km^2 of Sonandi Wildlife Sanctuary, and 466.32 km^2 of buffer area for both units. The entire CTR is divided into the core zone, the tourism zone, and the buffer zone.

The terrain of the study area is hilly and undulating consisting of a number of ridges and valleys. Plain and Mandal rivers formed major river valleys in the study area. The altitude ranges from 400 m in the southern part to 1,205 m in the north-eastern part of CTR. The geological formation of the area is divided into a) recent and b) Siwalik series. The general climate of the area is tropical with three distinct seasons i.e., summer, monsoon, and winter. June through September is a monsoon, followed by a post-monsoon season. Late November to March is a cool season. This is followed by a hot season from April to mid-June. Rains ranging from 5 to 15 cm can usually be expected in January and February, and thundershowers often accompanied by hail are not uncommon throughout the hot months of April and May.

Vegetation of study area is heterogeneous. Major portions of the buffer zone are covered by sal (*Shorea robusta*) along with related species. However, there is a wide variation in the species composition, mostly varying according to the altitude. Four major forest vegetation types were found in the study area: mixed, riverine, scrub, and plantation.

CTR has a high density of tigers, leopards, and others from the cat family. It carries a lustrous breed of deers—chital (*Axis Axis*), barking deer (*Muntiacus muntjak*), sambhar (*Rusa unicolor*), and hog deer (*Axis porcinus*). Among others are porcupine (*Hystrixindica*), yellow-throated marten (*Martes flavigula*), and Terai langur (*Semnopithecus hector*). The park has more than 580 species of birds, which include kingfishers, wagtails, forktails, pheasants, hornbill, eagles, vultures, and about 50 species of mammals. There are also 25 species of reptiles like crocodiles, king cobras, rock pythons, and monitor lizards, as well as hundreds of species of insects.

There are around 41 villages at the periphery of the KTR which has a human population of around 11,821 and supports a growing population of 5,729 cattle, 135 buffalo, 3,309 goats, 23 horses, and 21 sheep (The Corbett Foundation, Unpublished).

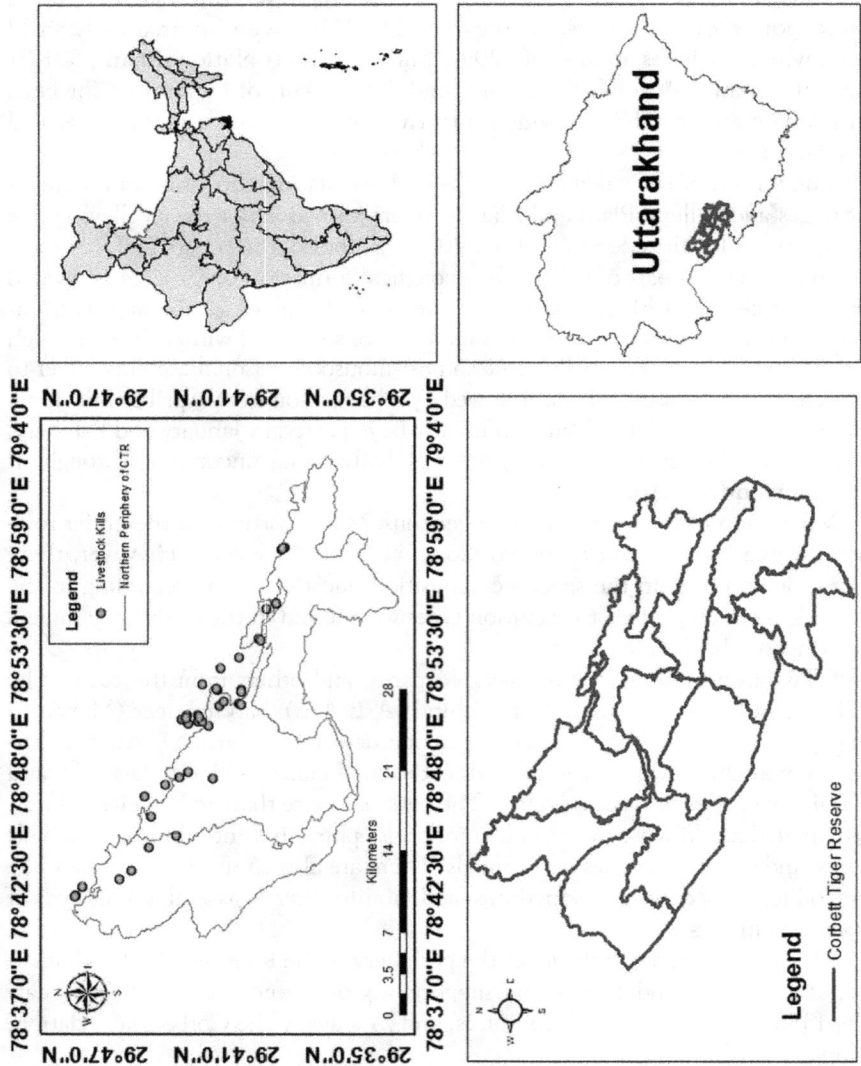

Figure 11.1 Study area of northern periphery of Corbett Tiger Reserve (CTR) showing kill locations.

Methodology

Data on livestock depredation was under an Interim Relief Scheme (IRS) run by The Corbett Foundation in collaboration with WWF India. Under this scheme, the villagers have to report the case of livestock injury or depredation within 48 hours.

Data Collection at Livestock Kill Sites

To obtain data on the ecological factors, we conducted ground surveys of freshly killed livestock reported for compensation from February 2018 to March 2018. We visited sites where livestock had been previous killed by collecting data from The Corbett Foundation database and visited those sites, asking the affected people about livestock mortality reported from November 2018 to January 2018. At each site we recorded the incident date and time, topography and elevation, livestock species and age, predator identity (based on direct sighting by the owner or indirect signs like pugmarks and scrapes which can be clearly distinguished between tigers and leopards), distance to human settlements, distance to water, and distance to forest and GPS coordinates. For the sites where livestock has been previously killed, the incident date and time, livestock species and age, predator, and GPS coordinates were collected from the database. Vegetative factors can affect the hunting success of large predators that ambush prey (Miller et al., 2015). We collected ground surveys to characterise vegetation variables at each kill site. At each kill site, a vegetation plot of 10 m was established to record data on: 1. mean shrub height; 2. percentage shrub cover; 3. percentage canopy cover; 4. density of trees; and 5. girth at breast height (GBH) of trees within the plot.

The shrub height was recorded at four places within the plot with a wooden stick of 1 m having markings on it. Canopy cover was recorded with ocular estimation and GBH of trees within the plots were recorded with the help of measuring tape. Percentage shrub cover and canopy cover was estimated by visual estimation, and shrub height was measured at different positions within the plot.

The same variables were quantified at the control site, which was an additional subplot at 400 m away from the kill site along a transect radiating out from the plot centre in a random direction. Therefore at each kill site, two vegetation plots were laid, comprising one kill site plot and one random plot.

The area around kill sites is frequently visited by the livestock, as pellets and dung were recorded from most of the sampled sites. This suggests that livestock graze widely throughout the area. The herders usually accompany the cattle to the pasture land inside the forest. Cattle mortality was measured at a maximum of 3 km from the village. All the kill sites were sampled within the study area.

Statistical Analysis

We examined five vegetation variables that are associated with kills of livestock by large carnivores. Since the data on the vegetation variables and landscape variables were not normally distributed, we used Wilcoxon matched-pairs tests to compare various vegetation structure variables among kill sites and random sites. Chi-square test was used to find the temporal differences in the frequencies of livestock and differences in various ages of cattle killed by leopards. These differences were not checked for tigers, as there were only five cases recorded of livestock killed by tigers during the study period. Chi-square goodness of fit test was used to find out the relationship between the topography and the frequencies of livestock killed by a leopard and also to delineate the forest blocks of the northern periphery of CTR.

To determine the prey preferences of leopard in the study area, Jacobs' index was used (Hayward, 2006).

$$D = (r - p)/(r + p - 2rp)$$

Jacobs' index standardises the relationship between the relative proportion that each species makes up of leopard kills r and prey relative abundance p (i.e., the proportion that each species makes up of the total abundance of all censured prey species in the study area). The standardised values range from +1 to −1, where +1 indicates maximum preference and -1 indicates maximum avoidance.

To map the hotspot of conflict, the natural neighbour interpolation technique was used. Natural neighbour (NN) interpolation finds the closest subset of input points to an unknown point and applies weights to them based on proportionate areas in order to interpolate a value (Sibson, 1981). Natural neighbours is local, using only one subset of points that surround the unknown point. It infers no trends and will produce no peaks, pits, ridges, or valleys not already represented by the input data.

Vegetation Structure Variables Associated with Livestock Depredation

We ran five non-spatial univariate models using the Generalized Linear Model. A binomial distribution was used for the response variable (presence and absence of attack) in the GLM using logit link function with kill and random sites as binary responses of 1 and 0. Akaike's information criterion (AICc) scores were used to access the relative fit of the models, which accounts for the small sample size. Logit link function uses logistic regression which forms a best fitting equation or function using the maximum likelihood method, which maximises the probability of classifying the observed data into the appropriate category given the regression coefficients. Before analysis, Spearman's rank correlation matrix was performed to confirm that none of the variables were inter-correlated using Spearman's rho ((r_s) >0.7 as a criterion for exclusion.

The probability graphs of carnivore depredation on livestock with the effect of different vegetation structure variables as predicted by the GLM were produced to find out the importance of the variables on livestock depredation. Importance ranges from 0 to 1, with importance values of 1 indicating that the variable made

a strong contribution to the model. We examined the relationship of each predictor variable to predict the probability of predation.

All the statistical analysis was carried out using R (v.3.5.0, R Project
Development Team, www.r-project.org) with the AIC modavag, car, cardata,
corplot, dismo, effects, ggplot2, and maptools packages.

Results

Patterns of Livestock Depredation

Data were collected for 42 sites where livestock depredation occurred from
November 2017 to March 2018. During the study period from February 2018
to March 2018, there were 25 cases of cattle reported for compensation, and 17
cases were from November 2017 to January 2018. Out of the total cases, 37 were
attributed to leopard (88%) while only five cases were attributed to tigers (12%).
Cases of cattle depredation occurred inside the village (12%) as well as inside the
forest (88%).

The northern periphery of CTR consists of three blocks, namely Rikhnikhal,
Jaiharikhal, and Nainidanda. There were significantly higher numbers of kills
in Rikhikhal block (74%) and lower number of kills in Jaiharikhal (19%) and
Nainidanda (7%) than expected by chance ($\chi2 = 31.5$, d.f. = 2, P<0.001). There
were a higher number of cattle killed in undulating topography (64%) as compared to hilly (22%) and plain topography (14%) than expected by chance
($\chi2 = 18.43$, d.f. = 2, P< 0.01). Leopard killed significantly higher numbers of cattle for the age group between four and six years as expected by chance ($\chi2 = 30.22$,
d.f. = 2, P<0.01, 76%) than for the other age groups. Tigers killed only one cow,
two calves, and two oxen during the study period. Two calves were of the age
four, one cow was of the age six, and the two oxen were of the ages five and eight,
respectively.

Cow was the most frequently killed prey species (59.5%), followed by calf
(21.6%), and ox (18.9%). Jacobs's index showed that cow is the preferred prey
(0.33 ± 0.06) species in the northern periphery of CTR, whereas, considering the
abundance of cow calf (-0.41 ± 0.07) and ox (-0.01 ± 0.02), these two species are
rather avoided.

More kills of cattle by leopards occurred in the afternoon and evening between
3 and 8 p.m. (70%) than in the morning (19%) and night (11%) as expected by
chance ($\chi2 = 60.53$, d.f. = 4, P<0.01). Out of the five cattle that were killed inside
the village, four kills occurred at night at 1.30 a.m., 4.00 a.m., 10 p.m., and 11
p.m., and one occurred at 6.00 p.m. Tigers killed three cattle in the afternoon—
one around 2 p.m. and two around 3 p.m., and also killed two cattle in the evening both around 5 p.m. All the cattle killed by tigers (n=5) occurred inside the
forest and none inside the village. Three cattle were killed deep inside the forest
at around 3 km distance from human habitation.

The data revealed that the number of kills by leopards generally decreases as
the distance from human habitation increases. The distribution of kills differed

significantly ($\chi2 = 47.45$, d.f. = 5, P<0.01). The majority of the cattle kills were located within a distance of 500 m (57%) from human habitation. 19% of the kills were located within the 500–1000 m of human habitation, and the rest (34%) were located 1,000–3,000 m from human habitation. The longest distance at which a cattle kill was recorded both by tigers and leopards was 3 km from human habitation. Field observations also revealed that most of the area around the sample sites contain cattle dung, suggesting that livestock graze widely throughout many parts of the area. Usually herders accompany the cattle when they take them inside the forest. Also pug marks and scats of the leopards were observed around the kill site, suggesting that they frequently visit those sites.

Economic Loss

Most losses were attributed to leopards followed by tigers. The total monetary loss due to tigers accounted for 34,000 INR, and in the case of leopards, the total monetary loss accounted for 272,000 INR. Therefore, the total economic loss caused by the large carnivores in the northern periphery of CTR during the study period was 306,000 INR. The cost incurred both by tigers and leopards for cows was 184,000 INR, for calves 50,000 INR, and for ox 72,000 INR.

Vegetation Structure Variables Governing Livestock Depredation by Large Carnivores

Wilcoxon matched-pairs test for measurements of vegetation structure revealed that vegetation structure variables differed between kill sites and random sites. Shrub cover was significantly higher in the vicinity of livestock kills than random sites, generating less visibility at kills (P< 0.001) (Table 11.1).

GLM models predicted the probability of large carnivores killing cattle given an encounter between both the species. The model with shrub cover was the top model according to the AICc scores (Table 11.2). Therefore, the risk

Table 11.1 Wilcoxon Signed-Rank test comparisons between mean values ± standard error of vegetative structure variables at the depredation sites (kill sites) and random sites across the northern periphery of Corbett Tiger Reserve; p-values in bold are significant

Variables	Random sites (n=42)	Cattle (n=42)	
		Kill sites	p-value
Tree density	204.76 ± 17.36	209.52 ± 23.07	0.565
Mean GBH (m)	0.74 ± 0.075	0.66 ± 0.07	0.288
Canopy cover (%)	36 ± 4.43	29.88 ± 4.13	0.314
Shrub cover (%)	39 ± 4.43	63.36 ± 4.64	0.003
Mean shrub height (m)	1.55 ± 0.11	2.09 ± 0.25	0.05

Table 11.2 Best GLM models explaining the relationship between the number of livestock killed by large carnivores and vegetation structure variables

	K	AICc	ΔAICc	Mod.Lik.	AICc wt.	Log.Lik	Cum.Wt.
Shrub cover	2	107.67	0.00	1.00	0.97	-51.76	0.97
Shrub height	2	115.63	7.95	0.01	0.01	-55.74	0.99
Canopy cover	2	119.42	11.74	0.00	0.00	-57.63	0.99
GBH	2	119.97	12.30	0.00	0.00	-57.91	0.99
Tree density	2	120.56	12.89	0.00	0.00	-58.21	1.00

to cattle increases with an increase in the scrub cover. Kill probability showed the linear relationship with the percentage shrub cover. Beta estimates of tree density and shrub height were positive; that is, they have a positive influence on the probability of kill but were not significant. Alternatively, canopy cover and girth at breast height showed a negative influence on the probability of kill, but again they were not significant. Therefore, only shrub cover was found to influence livestock depredation by large carnivores (p<0.01), suggesting that depredation risk was more closely associated to the areas where there is high shrub cover. The relationship between each percentage shrub cover to predictions of predation risk was measured with a 95% confidence interval (Figure 11.2).

Spatial Hotspot of Predation Risk

A predation hotspot map offered a visual insight into the spatial distribution of livestock depredations across the northern periphery of CTR. The hotspot map thus generated (Figure 11.3) can help managers to guide the villagers to avoid areas where there is high chance of livestock depredation occurring, hence reducing the chances of encounters between livestock and large carnivores.

Discussion

Pattern of Livestock Depredation

The rate of cattle depredation by leopards is higher in the northern zone than the southern zone of CTR (Bargali & Ahmed, 2018). The forest department compensates a very low amount for the loss of cattle by large carnivores, and it often involves long procedural delays in distribution. This often results in retaliatory measures among villagers like poisoning the carcass to kill the carnivore responsible for the livestock depredation. But in recent times the retaliation against depredating carnivores in CTR has decreased, largely due to the compensation program by The Corbett Foundation.

In the present study, only those sites were visited where cattle were killed and not those sites where other livestock were depredated, as people only get

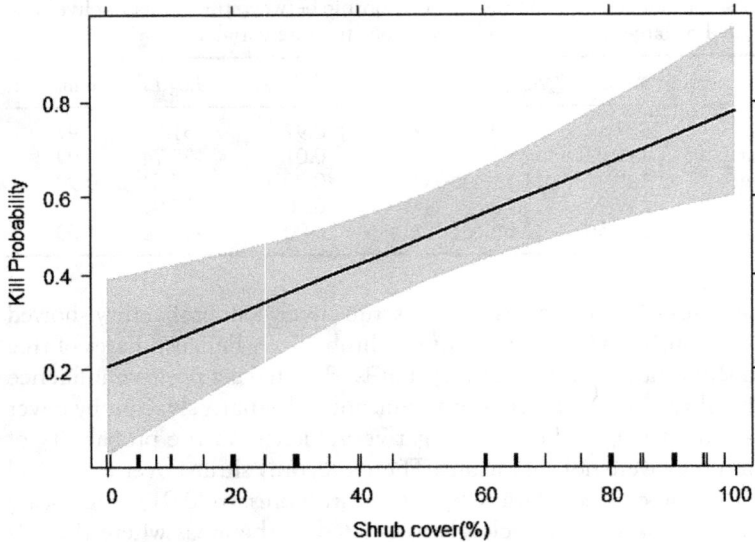

Figure 11.2 Relationship between percentage shrub cover and kill probability. The 95% confidence intervals are shown in blue.

compensation for cattle loss and therefore they only report the cattle that were depredated by carnivores and not any other livestock. During the study period, it was found that the leopard was the major carnivore responsible for killing cattle, as it killed more than tigers. This can be explained by many factors. First, the topography of the northern zone of CTR is more rugged and undulating where the leopard population can sustain well (Bargali & Ahmed, 2018) Also, this may be due to the fact that there is lower availability of free ranging cattle in CTR, where 30% of households stall-fed (Malviya & Ramesh, 2015). Second, this may be due to higher number of livestock holding inside the villages (Khan et al., 2016), which attract leopards, as wild prey may not be easily available. The cattle in CTR are of small size and are easy to handle, which provides easy catching probability than wild prey; wild prey show anti-predator behaviour which is often lacking in cattle having been attenuated by humans. Although the northern zone has sufficient cover, the area has low tiger prey densities due to human pressure (Jhala et al., 2011). Leopards usually kill wherever there is sufficient cover and adequately sized animals (Hayward et al., 2006). On the other hand, tigers require dense cover (Miller et al., 2015; Soh et al., 2014), and therefore cattle are at risk by tigers only when they graze far away from human habitation inside the forest. Our results also show that tiger killed cattle at a maximum of 3 km away from human habitation. Leopards, however, kill more cattle within a 0.5 km

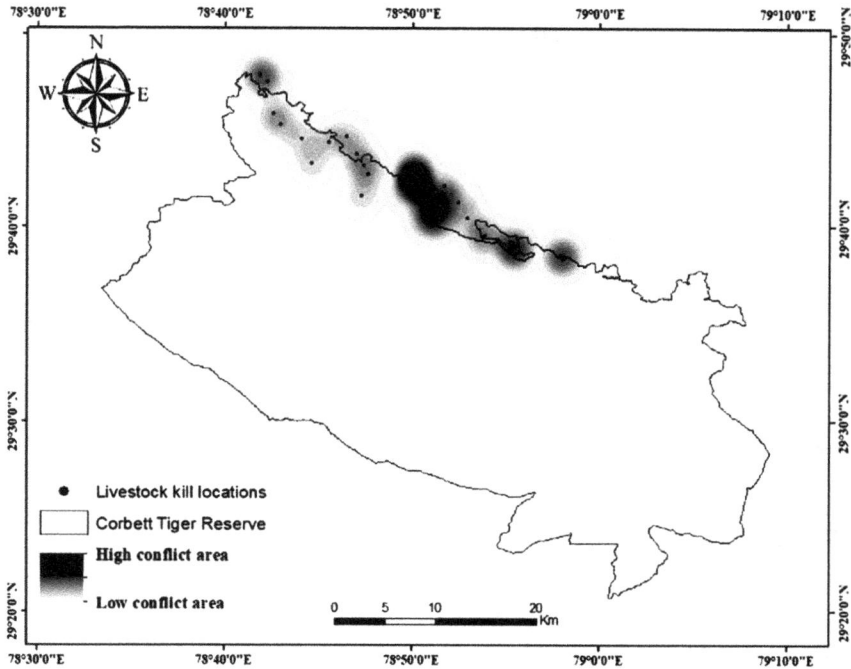

Figure 11.3 Livestock predation hotspot map based on the kill locations across the northern periphery of Corbett Tiger Reserve.

distance of human habitation, with only five cases inside the village and the rest within the fringes of the forest where there might be adequate cover and less human disturbance. Forest fringes also provide refuges so as to safely drag the animal, compared to areas near human habitation. Further, it may be possible that due to the high density of cattle (which falls in the preferred prey range of leopards) inside the village, more depredation occurred around the villages. Leopards are known to be a resilient species that live in forest fringes and are capable of surviving at very high densities in human-dominated landscapes (Athreya et al., 2013). Tigers generally use thick vegetation with heavy cover to ambush prey, and they are generally associated with dense forest and low visibility, whereas leopards generally rely on areas of enough vegetation cover to provide camouflage but not too dense to influence prey visibility and catchability (Miller, 2015; Abade et al., 2014). Leopards killed more cows between the ages of four and six years than oxen, as the size and weight of the cow is smaller than that of the ox and is therefore easier to handle. The prey size range of leopards is 10–40 kg (Hayward et al., 2006). This range of prey sizes provide easy handling opportunities not only to kill but also to drag prey to safer locations. Furthermore, leopards invest more energy in hunting medium-sized prey rather than smaller or excessively large suboptimal species (Hayward et al., 2006).

Leopards killed cattle at all hours of the day but particularly in late afternoon and evening, because this was the time when herders return from the forest after grazing the cattle and also the leopards are more active during that time, therefore the chance of encounters increases. Four out of five cases that occurred inside the village by leopards were at night at cowsheds, and one occurred in the evening. This may be due to the poor husbandry practices used by the villagers, as most of the types of cowsheds were open and without any protection. Our results show that out of five kills, three kills by tigers were in the afternoon between 2 and 3 p.m., likely because this was the time when cattle were farthest from the village and in dense vegetation where tigers often attack (Soh et al., 2014; Malviya & Ramesh, 2015; Miller et al., 2015).

In this study, the spatial and temporal separation between tigers and leopards were not taken, but from previous studies it was revealed that leopards have a higher tolerance for more open habitats than tigers, and both felid species move less during the midday period than in the morning and evening (Seidensticker, 1976; Miller et al., 2016a).

Vegetation Structure Variables

Our GLM models revealed that out of five univariate models selected, only one model with shrub cover was significant. The probability of cattle kills increases with an increase in shrub cover. Also, it should be noted that shrub cover may change seasonally. Since the data was only collected for a short term period, it may not provide a good estimation of conditions as they were at the time of the carnivore attack, as in a few cases, data was recorded after some months of the attack. But the results are also comparable with many previous studies on large carnivores that have shown that dense vegetation increases the cover available to carnivores, which reduces the visibility of approaching predators and hence increases the hunting success of carnivores, as most of the large carnivores are ambush predators and require appropriate cover to ambush prey (Stahl et al., 2002; Thorn et al., 2012; Zarco-González et al., 2013; Abade et al., 2014). Also, from previous studies, it was found that tigers commonly use thick vegetation with heavy cover to ambush prey. Tigers are generally associated with dense forest and low visibility, and the results also revealed that tigers killed cattle far from human habitation and in dense forest away from open habitats, whereas leopards generally rely on areas of enough vegetation cover to provide camouflage but not too dense to influence prey visibility and catchability. Further, the results showed that leopards mostly killed in areas within 500 m of human habitation, in areas having dense shrub cover around the fringes of forest (Miller, 2015; Miller et al., 2016a; Abade et al., 2014).

The data collected at kill sites offer more insight into ecological information about the sites where livestock are vulnerable, which is otherwise not possible by household surveys that have been the basis of many studies on human-carnivore conflict (Wang & Macdonald, 2006; Miller et al., 2016a). Different stages of hunt occur at different spatial scales; the final stages of the hunt—attacking, killing, and consumption—often occur over small areas (Gorini et al., 2012). Moreover, the probability of hunting success by felids is affected by the complexity of the vegetation involved. Felids usually employ dense cover to stalk or ambush prey, then often initiate their approach to prey from dense vegetation cover and remain concealed while hunting (Murray et al., 1995). Tigers and leopards use similar hunting strategies but different habitat domains (Miller et al., 2016b). Analysis of kill locations is the most useful tool for investigating the effect of habitat on predation, but it doesn't distinguish between the different predation stages nor take into account the failed hunting attempts. Therefore, kill locations can only be used to derive a relative measure of hunting success.

Conclusion

This study gave an insight into identifying when and where cattle are most likely to be depredated by leopards and tigers. Most cattle that are vulnerable to carnivore attack are of middle age (4–6 years) and are attacked primarily in the afternoon and evening (3–8 p.m.). Only five were killed inside the village, out of which four were attacked during the night time while most of the cattle were killed within 500 m of human habitation, inside the forest fringes having dense shrub cover. The number of cattle killed decreases as the distance from human habitation increases.

There are several potential limitations in the results. First, the study was for a very short period and only considered one season of data, therefore the results are not enough to apply to a year. Also, there were very few cases of tigers killing cattle, and hence the model was largely based on the leopard cases, which may interfere with the mitigation strategies used for the two different species. Only vegetation structure variables were considered, and no other variables associated with habitat and landscape, since landscape variables associated with prey accessibility can vary over different stages of hunt. Since we were only interested in the final stage of hunting, we only chose vegetation structure variables. This best explained the final hunting success of the felid, which often depends upon the complexity of the vegetation structure.

Understanding the temporal and vegetative factors that underlie ecological interactions between specific carnivore and cattle species will be essential for developing strong mitigation methods (Miller et al., 2016). Therefore, effective conservation requires means of coexistence of people and carnivores, which can be done by knowing the possible ecological correlates of conflict that can ultimately help to develop better mitigation strategies.

Acknowledgments: We would like to thank the Chairperson and faculty of the Department of Wildlife Sciences, Aligarh Muslim University for institutional assistance. We thank the Deputy Director, The Corbett Foundation for providing us with all the logistic support during the field. Finally, we would like to thank villagers for their kind cooperation in visiting the sites of livestock depredation, not harming wild animals, and extending their support in wildlife conservation.

References

Abade, L., Macdonald, D. W., & Dickman, A. J. (2014). Assessing the relative importance of landscape and husbandry factors in determining the large carnivore depredation risk in Tanzania's Ruaha landscape. *Biological Conservation, 180*, 241–248.

Athreya, V., Odden, M., Llinnell, J. D. C., Krishnaswamy, J., & Karanth, U. (2013). Big cats in our backyards: Persistence of large carnivores in a human dominated landscape in India. *PLOS ONE, 8*(3), e57872.

Bargali, H. S., & Ahmed, T. (2018). Patterns of livestock depredation by tiger (*Panthera tigris*) and leopard (*Panthera pardus*) in and around Corbett Tiger Reserve, Uttarakhand, India. *PLOS ONE, 13*(5), e0195612.

Barua, M., Bhagwat, S. A., & Jadhav, S. (2013). The hidden dimensions of human- wildlife conflict: Health impacts, opportunity and transaction costs. *Biological Conservation, 157*, 309–316.

Dar, N. I., Minhas, R. A., Zaman, Q., & Linkie, M. (2009). Predicting the patterns, perceptions and causes of human-carnivore conflict in and around Machiara National Park, Pakistan. *Biological Conservation, 142*(10), 2076–2082.

Dickman, A. J. (2010). Complexities of conflict: The importance of considering social factors for effectively resolving human-wildlife conflict. *Animal Conservation, 13*(5), 458–466.

Gorini, L., Linnell, J. D. C., May, R., Panzacchi, M., Boitani, L., Odden, M., & Nilsen, E. B. (2012). Habitat heterogeneity and mammalian predator-prey interactions. *Mammalian Review, 42*(1), 55–77.

Gurung, B., Smith, J., McDougal, C., Karki, J. B., & Barlow, A. (2008). Factors associated with human-killing Tigers in Chitwan National Park, Nepal. *Biological Conservation, 141*(12), 3069–3078.

Hayward, M. W., O'Brien, P. H. J., Hofmeyr, M., Balme, G., & Kerley, G. I. H. (2006). Prey preferences of the leopard (*Panthera pardus*). *Journal of Zoology, 270*(2), 298–313.

Jhala, Y. V., Qureshi, Q., & Gopal, R. (Eds.), (2015). The status of tigers, co-predators & prey in India 2014. National tiger conservation authority, New Delhi and the wildlife institute of India, Dehradun. TR2015/021.

Jhala, Y. V., Qureshi, Q., Gopal, R., & Sinha, P. R. (2011). Status of the tigers, co-predators, and prey in India. National tiger conservation authority, Govt. of India, New Delhi, and wildlife institute of India, Dehradun, pp. 302.

Khan, J. A., Khan, A., Kumar, S., Quadri, A., Khati, D. V. S., Malik, P. K., Kushwaha, S. P. S., Pandey, V., Sarin, G. D., & Musavi, A. (2016). Corbett Tiger Reserve: Local communities and livestock depredation by tigers, a case study from buffer zone. Chapter 4. In A. Musavi (Ed.), *Tiger abundance and cattle depredation* (pp. 119–144). Authorspress.

Malviya, M., & Ramesh, K. (2015). Human-felid conflict in corridor habitats: Implications for tiger and leopard conservation in Terai arc landscape, India. *Human-Wildlife Interactions, 9*(1), 48–57.

Miller, J. R. B. (2015). Mapping attack hotspots to mitigate human-carnivore conflict: Approaches and applications of spatial predation risk modeling. *Biodiversity and Conservation*, 24(12), 2887–2911.

Miller, J. R. B., Jhala, Y. B., & Schmitz, O. J. (2016b). Human perceptions mirror realities of carnivore attack risk for livestock: Implications for mitigating human-carnivore conflict. *PLOS ONE*, 11(9), e0162685.

Miller, J. R. B., Jhala, Y. V., & Jena, J. (2016a). Spatial and temporal patterns of livestock losses and hotspots of attack from tigers and leopards in Kanha Tiger Reserve, Central India. *Regional Environmental Change*, 16(1), 17–29.

Miller, J. R. B., Jhala, Y. V., Jena, J., & Schmitz, O. J. (2015). Landscape-scale accessibility of livestock to tigers: Implications of spatial grain for modeling predation risk to mitigate human-carnivore conflict. Ecology and Evolution, 5(6), 1354–1367.

Murray, D. L., Boutin, S., O'Donoghue, M., & Nams, V. O. (1995). Hunting behaviour of a sympatric felid and canid in relation to vegetative cover. *Animal Behaviour*, 50(5), 1203–1210.

Sangay, T., & Vernes, K. (2008). Human-wildlife conflict in the Kingdom of Bhutan: Patterns of livestock predation by large mammalian carnivores. *Biological Conservation*, 141(5), 1272–1282.

Seidensticker, J. (1976). On the ecological separation between tigers and leopards. Biotropica, 8(4), 225–234.

Sibson, R. (1981). A brief description of natural neighbour interpolation. In V. Barnett (Ed.), *Interpreting multivariate data* (pp. 21–36). John Wiley & Sons.

Singh, R., Nigam, P., Qureshi, Q., Sankar, K., Krausman, P. K., Goyal, S. P., & Nicholoson, K. L. (2015). Characterizing human-tiger conflict in and around Ranthambhore Tiger Reserve, Western India. *European Journal of Wildlife Research*, 61(2), 255–261.

Soh, Y. H., Carrasco, L. R., Miquelle, D. G., Jiang, J., Yang, J., Stokes, E. J., Tang, J., Kang, A., Liu, P., & Rao, M. (2014). Spatial correlates of livestock depredation by Amur tigers in Hunchun, China: Relevance of prey density and implications for protected area management. *Biological Conservation*, 169, 117–127.

Stahl, P., Vandel, J. M., Ruette, S., Coat, L., Coat, Y., & Balestra, L. (2002). Factors affecting lynx predation on sheep in French Jura. *Journal of Applied Ecology*, 39(2), 204–216.

Thorn, M., Gree, M., Dalerum, F., Bateman, P. W., & Scott, M. (2012). What drives human-carnivore conflict in the North West Province of South Africa? *Biological Conservation*, 150(1), 23–32.

Wang, S. W., & Macdonald, D. W. (2006). Livestock predation by carnivores in Jigme Singye Wangchuck National Park, Bhutan. *Biological Conservation*, 129(4), 558–565.

Zacro-González, M. M., Vilchis, O. M., & Alaníz, J. (2013). Spatial model of livestock predation by jaguar and puma in Mexico: Conservation planning. *Biological Conservation*, 159, 80–87.

12 Is Resource Sharing Leading to Human–Bear Interaction?

Kruti Patel, Arzoo Malik, and Nishith Dharaiya

Introduction

The sloth bear (*Melursus ursinus*) is the most widely distributed species in Indian sub-continent (Dharaiya et al., 2016). Around 85% of the sloth bear population is found in India and ranges from the foothills of Himalayas to the southern end of Western Ghats. Its distribution in the western part of India is limited by the desert of Rajasthan. However, their abundance varies in different habitats, probably depending on availability of resources. In 2016, the IUCN classified the sloth bear as vulnerable (Dharaiya et al., 2016). According to the Indian Wildlife (Protection) Act, 1972, the sloth bear is included in schedule I species. In Gujarat, they are found in five protected areas: Two in the north i.e., Jessore and Balaram Ambaji Wildlife Sanctuary, two in the central part i.e., Jambughoda and Ratanmahal Wildlife Sanctuary, and one in the southern part i.e., Shoolpaneshwar Wildlife Sanctuary. Apart from these five protected areas sloth bear is patchily distributed in un-protected forest patches of Gujarat.

Food and feeding ecology of the sloth bear in its range has been studied by several researchers: Bandipur Tiger Reserve (Johnsingh, 1981), Mudumalai (Baskaran et al., 1997; Desai et al., 1997), Neyyar Wildlife Sanctuary (Sreekumar & Balakrishnan, 2002), North Bilaspur forest division (Bargali et al., 2004), Panna Tiger Reserve (Yoganand et al., 2006), and Jessore Wildlife Sanctuary (Sukhadiya et al., 2013). Earlier studies of the sloth bear in countries like Nepal, India, and Sri Lanka suggested that the diets of sloth bears vary seasonally and depend on the availability of fruits and colonies of termite mounds (Schaller, 1967; Norris, 1969; Laurie & Seidensticker, 1997; Johnshingh, 1981; Baskaran, 1990; Gokula et al., 1991). It was recorded that they feed on large amount of fruits and insects (Seidensticker, 1976; Johnsing 1981; Joshi et al., 1997; Baskaran et al., 1997; Bargali et al., 2004).

With changing geographical conditions, the sloth bear's diet changes invariably. Studies have reported insects forming the major part of their annual diet (Joshi et al., 1997), whereas fruits were found to be consumed more in other areas (Baskaran et al., 1997; Bargali et al., 2004; Mewada, 2010; Sukhadiya et al., 2013).

DOI: 10.4324/9781003321422-14

Habitat degradation has also affected the food resource availability (Yoganand et al., 2006). It was observed that due to lack of food resources in forests, sloth bears are driven towards agricultural crops (Iswarish, 1984; Garshelis et al., 1999; Baragali et al., 2004; Dharaiya & Ratnayeke, 2009). According to Rajpurohit and Krausman (2000) food availability is an important aspect of habitat quality, playing a significant role in reducing conflicts. Generally, the sloth bear at pre-adult stage feeds mostly on insects, because they are rich in nutrition (Swenson et al., 1999; Mattson, 2001). Bears have a large home range, and they move a lot during fruiting season (Garshelis & Pelton, 1981; Peek et al., 1987; Yoganand et al., 2006). It was also reported that the sloth bear was found moving in different habitats in search of food in Chitwan National Park (Joshi et al., 1995, 1997). Data on the food biomass is required to compare the food habits among the study areas (Baskaran et al., 1997). These studies clearly indicate that sloth bear movement is largely dependent on the availability of food resources; hence utilisation and sharing of food resources may drive human–sloth bear conflicts. Keeping this in mind, this study was designed with the objectives of studying the availability of food for sloth bears and understanding the pattern of food resources shared between humans and sloth bears.

Study Area

The study was carried out in 2019 in Jessore Wildlife Sanctuary (JWLS), located in the northern part of Gujarat state. Jessore is unique in terms of having diversified flora and fauna of the western Aravalli mountain ranges. It is located between 24.20° N latitude and 72.3°E longitude and covers an area of 180.66 sq km (Figures 12.3 and 12.5).

Administratively, the sanctuary is divided into two forest ranges (Iqbalgadh and Dantiwada ranges) which are further divided into six rounds and 13 forest beats. Two major rivers, Banas and Sipu, pass through the sanctuary; there are also various tributaries of these rivers which are mostly dried beds. The hills are covered with dry deciduous mix forest. Dry deciduous habitat with rocky boulders and undulating plains is considered as one of the most suitable habitats for sloth bears (Sukhadiya et al., 2013).

The climate of JWLS is typically tropical to sub-tropical with three distinct seasons. This region experiences low rainfall during the monsoon season, with annual rainfall recorded at 460.66 mm in 2018. The highest intensity of rainfall was observed in July 2018 at 382.51 mm which was significantly below than the previous record of average rainfall observed in 2008 (765 mm) (Dharaiya, 2009). Winters are unexpectedly cold with minimum temperatures of 6°C, whereas summer is extremely hot, with temperatures of 46–48°C.

Methodology

This study was carried out in Iqbalgadh area, and the whole area was surveyed during the year 2019. To understand the habitat use, the whole area was divided into grids of 5 x 5 km. In each grid, strip transects of 100 m were laid to cover the minimum distance of 2 km (Cottom & Curtis, 1956; Akhtar et al., 2007). A total of 47 strip transects were covered. Sloth bear presence signs like scat, pug marks, claw marks, and digging marks were recorded. Scats were collected whenever animals were encountered during transect survey, at water points, and at feeding sites.

For determining the availability of food resources for the sloth bear, a minimum of five quadrates of 10 x 10 m in size were laid in each grid depending on the terrain. A total of 97 quadrates were laid in whole study area to study vegetation. Information on habitat variables, including terrain, vegetation type, tree, shrub spp., and termite mounds were also recorded. The density of available fruiting species was estimated to correlate with the content that appeared in the scats.

The second objective of the study required interaction with locals residing in the neighbouring villages of the sanctuary. A pre-designed questionnaire was prepared including different questions related to socio-economic conditions, occupation, knowledge of sloth bears, and resources available in the forest and their use (Riley, 2007; Dharmorikar et al., 2017). Interviews were recorded with the permission of the interviewee and were taken in their local dialect in Gujarati language. Respondents were selected on a random basis in seven villages located in close vicinity of the study area. Interviews were conducted in morning and evening, as locals were usually working or inside the forest during the day time. Recorded interviews were translated and converted into data sheets. Later, data was coded for ease in analysing and deducing the result.

The data collected from the field was scrutinised and statistically analysed to deduce the results on resource sharing. Spearman rank correlation was used to determine the relative importance of the different variables likely to affect fruit selection (Riley, 2007; Obbard et al., 2014). For this, human bear conflict data of last 17 years was used.

To identify overlapping in usage of fruiting tree species by humans and sloth bears, a Venn diagram was prepared (Rodgers et al., 2013). Maps were prepared using QGIS 3.6 software to show the habitat and resources used by sloth bears and humans.

Results and Discussion

The study was carried out to understand the extent of resource sharing between sloth bears and humans residing close to the forest area in Jessore Sloth Bear Sanctuary. The range surveyed around Iqbalgadh was approximately 125 sq km and was surveyed through 47 strip transects of 100 m, during which a total of 101 sloth bear signs were recorded. All 35 scats collected during the field work were analysed to find out food composition using the undigested matter in the scats. Analysis of scat content revealed four different food items including three plant species and ant heads. In all scats, a major component was plant matter, accounting for 66%, while animal matter was found up to 34%. Similar studies have been conducted in the past showing fruits as a major part of the sloth bear's diet during the winter season (Sukhadiya et al., 2013; Mewada et al., 2010).

In the plant matter, fruits were more frequently observed in winter due to availability of fruiting species compared to animal matter like ants and termites. Sloth bears were also observed to consume leftover harvested crops, which might have contributed to the higher content of plant matter in their diet during winter (Bargali et al., 2004).

Among the plants, *Zizyphus* spp. (n=19) accounted as highest, followed by *Cassia fistula* and *Grewia flavescens* (n=2). The rest of the content was mainly red and black ants. With the frequent occurrence of Zizyphus seeds in scat, it is assumed that *Ziziphus* was the most available food to the sloth bear in the study area, especially during winter. It was also observed that the forest department have been planting *Ziziphus*, replacing *Prosopis spp.* as a part of sloth bear habitat restoration. In another study conducted in Central India, insects were found more in scats during monsoon and winter, whereas plant matter dominated in summer (Bargali et al., 2004). On the other hand, Gopal (1991) reported that sloth bears consume fruits throughout the year mainly in February to June, but termites, ants, and honey form the major composition of their diet during other months. It can be concluded that fruits, ants, termites, and honey were the principle food of sloth bears in their range; however, the occurrence of these items in their diets may vary with the time of year, geographic location, and availability in the forest.

A total of 97 quadrates were traversed, covering an area of 125 sq km. Fifty-three tree species were recorded through all the quadrates laid on the strip transects. Out of these recorded plants, 12 tree species were identified as sloth bears' preferred food based on the literature, and only six were in the fruiting stage during winter, viz. *Aegle marmelos*, *C. fistula*, *Ziziphus* spp., *Ficus*. spp., *G. flavescens* spp., and *Diospora melonoxylon*. The abundance of identified tree species showed that the most abundant species was *D. melonoxylon* (53.22%), locally known as timru, followed by *Ziziphus* (24.95%), *G. flavescens* (18.91%), and *A. marmelos* (14.23%). *Cassia fistula* and *Ficus* spp. were the ones that were found throughout the year and form an important food source for sloth bears in the area.

Figure 12.1 shows the density of fruiting tree species recorded in the study area during the field survey in winter. *D. melanoxylon* was found with the highest density as compared to other tree species, followed by *Ziziphus* spp., *A. marmelos*, and *G. flavescens*. Many of these species are not in fruiting state during the winters (November-February) of 2019 in the study area. Although *D. melanoxylon* was found in high density during the survey, it was not recorded in the scats, as its fruits generally ripen in the late winter, whereas *Ziziphus* fruits are available in the early winter through to the end of February.

The overall results of the vegetation survey indicated that the density of fruiting plant species is quite low and patchily distributed in the study area (Figure 12.3). Similar findings have been reported in studies conducted earlier (Mewada et al., 2010; Dharaiya et al., 2009). At many places these food plant species are present, but they have not reached the fruiting stage to provide food to the bears,

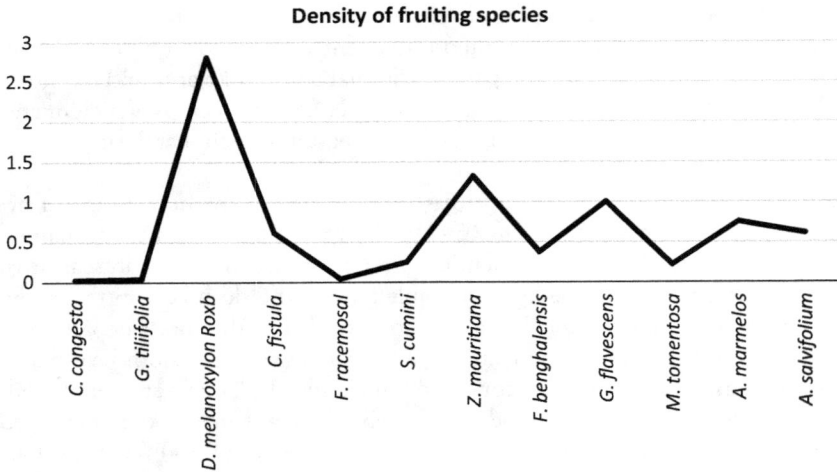

Figure 12.1 Density of fruit plants as sloth bear food in Jessore WLS.

however in future they can be a good source of food in the sanctuary and can also be exploited by forest dwellers.

Human Use of Forest Resources

The questionnaire-based survey revealed that the majority of the respondents practice agriculture and tend cattle as their main occupation. However, agriculture and cattle rearing are not fulfilling requirements. Therefore, people also depend on forest resources for their livelihoods. This results in vast exploitation of those forest resources which are largely shared with the sloth bear (Bargali et al., 2005). The main forest produce utilised by humans are firewood, timber, fodder, and fruits. The majority of the villagers use wood-fuel collected from the forest for domestic purposes (Figure 12.2). Increases in human population in the villages may directly affect the dependency of locals on forest produce. Moreover, their frequent visits to the forest area make them vulnerable to sloth bear encounters. In North Bilaspur forest, local people, during the non-growing season, turned towards non-timber forest products (Bargali et al., 2005), which increased sloth bear attacks due to accidental encounters in the forest.

As already discussed, sloth bear movement in and outside the sanctuary is mainly influenced by the availability of food and water. Studies by Gondaliya (2012) and Malik et al. (2018) established a positive correlation between bear movement and food availability in the forest, which indicates that food and water could be the main limiting factors in Jessore Wildlife Sanctuary. Probably due to scarcity in these resources, the bear is forced to leave their original habitat. The increased use of forest resources by humans also leads to over-exploitation of the

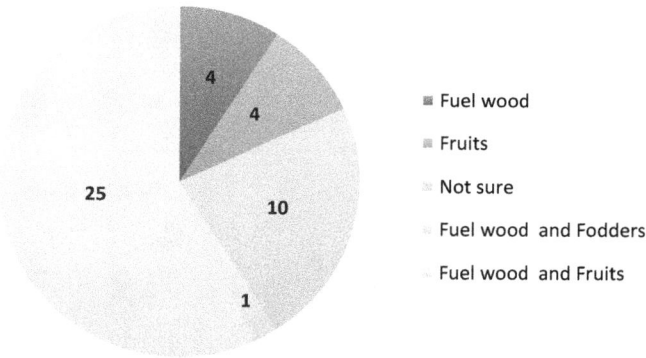

Figure 12.2 Villagers visiting forest for different purposes.

forest, resulting in disturbances as well as degradation of the sloth bear habitat. A study conducted by Dhamorikar et al. (2017) also suggested that continuous degradation of habitat has driven bears towards villages. Interestingly, out of 12 tree species which the sloth bear uses as food, three species were utilised by both humans and sloth bears, indicating the possibility of resource sharing.

Food resource sharing between humans and wildlife has been studied in many countries, and it has been suggested that this could be one of the significant reasons for human-wildlife interactions. In Indonesia, cacao and other fruits are eaten by macaque and also by humans, which was reported as a prime reason for conflicts (Riley, 2007). During the harvesting periods, the American black bear comes into the farm, and the cases of bear attacks have increased (Obbard et al., 2014). Very few studies on food resource sharing between humans and sloth bears are available; however, many researchers have suggested that resource sharing may be one of the reasons of sloth bear attacks and conflicts. Similarly, banana crop raiding by sloth bears was significantly reduced in Uganda when the availability of the fruiting tree Mimusops bagshawei was increased in forest areas (Naughton et al., 1998).

The data of the tree species identified was overlaid on the map of Jessore Sanctuary to understand the proximity of sloth bear signs around those species of trees in relation to human settlement areas. It was clearly depicted that none of the recent sloth bear attacks took place around the areas of high density of fruiting tree species (Figure 12.3). This indicates that there may be many other factors responsible for the encounter between humans and sloth bears in the study area, such as water, sloth bear movement close to villages, and sudden encounters with the sloth bear (Dharaiya et al., 2009; Ratnayeke et al., 2014; Koike, 2009).

People living in and around the sanctuary depend on forest resources. The food resources for the bears may have reduced due to habitat degradation and changes in land use over a period of time (Rajpurohit & Chauhan, 1996). The common practice observed during this study was the growing of fruiting trees

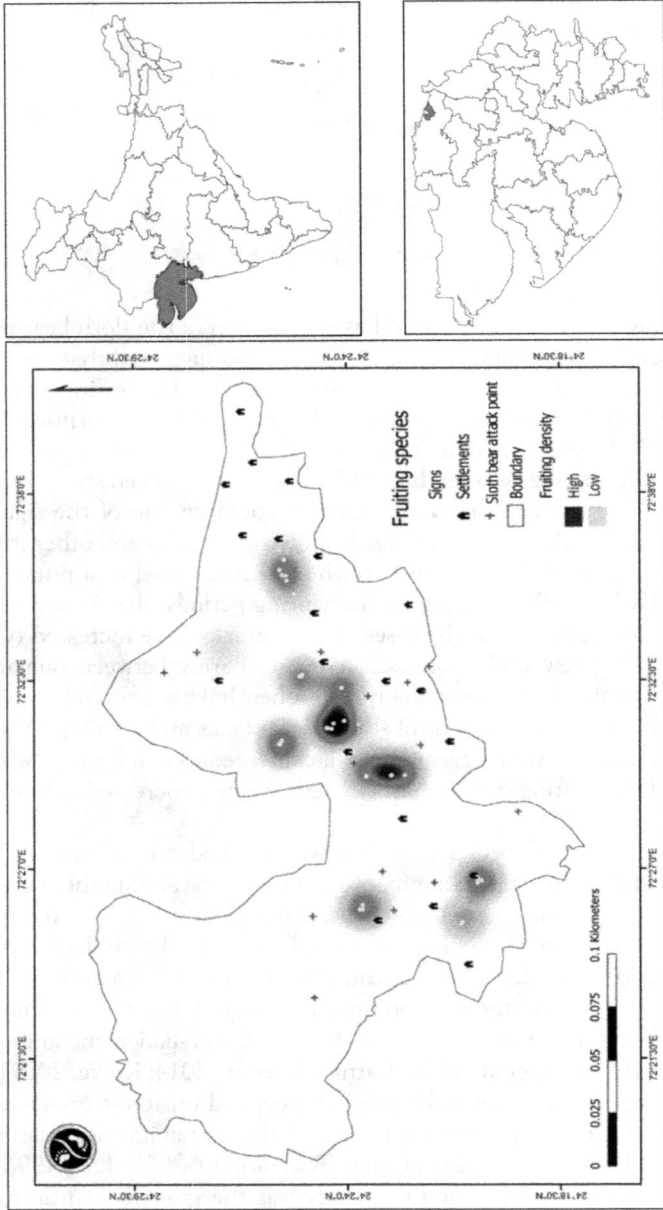

Figure 12.3 Fruiting tree species in the study area along with sloth bear signs, human settlements, and sloth bear attack locations.

such as *Ziziphus* spp., *S. cumini*, *C. fistula*, and *A. marmelos* in and around villages or on the border of agricultural fields. This may attract the bears to frequently come towards human habitation and agricultural fields in search of food.

The result suggests that current patterns of overlapping resource use between sloth bears and villagers in Jessore Wildlife Sanctuary cannot be the only reason behind the conflicts. Human use of the forests in Jessore Wildlife Sanctuary can include both the clearing of forests for agriculture, which results in a lower density of key food trees and a less productive environment (Riley, 2007; Honda, 2009), and the selective removal of particular tree species. Continuous overlapping of resources may negatively affect the sloth bear in the near future. Due to over-exploitation and degradation of habitat, bears are forced to move out of their areas and towards human settlements, increasing the chances of encounters.

The most important tree species exploited by villagers are *Ziziphus* spp., *S.cumini*, and *C. dichotoma*. *Ziziphus* was also found in high levels in the diet of the sloth bear (Figure 12.3). This might be due to the high density of *Ziziphus* recorded in the area. Also, *Ziziphus* was observed to be in fruiting stage in winter, unlike the other species recorded. According to respondents, they go into the forest to collect *Ziziphus* fruits. This is the only plant species found in sloth bear scat during the winter, revealing that people and sloth bears use the same food resources. To further clarify resource sharing in the area, a Venn diagram was prepared to analyse the extent of sharing of food resources between humans and sloth bears (Figure 12.4) (Rodgers et al., 2013). In interviews, 19 people mentioned collection of *Zizyphus* fruits from the forest area. Scat analysis shows a 19% occurrence of *Zizyphus* in the sloth bear's diet. Conclusively, in winter, only one fruiting tree species is shared between sloth bears and humans. *Ziziphus* is typically grown close to human settlements as well as in secondary lowland to hill forests. *Ziziphus* are widely recognised as keystone resources for many frugivorous

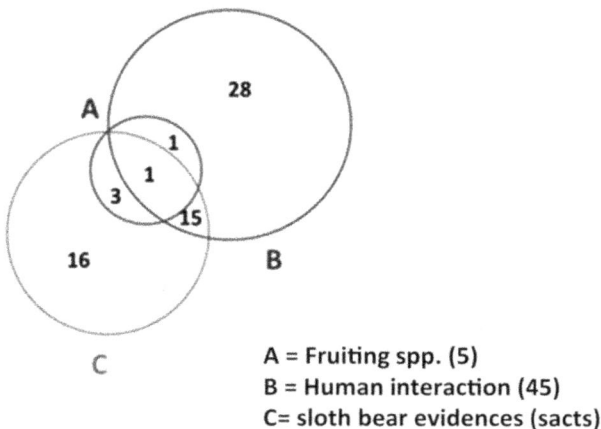

Figure 12.4 Venn diagram showing the sharing of food resources between human and bears.

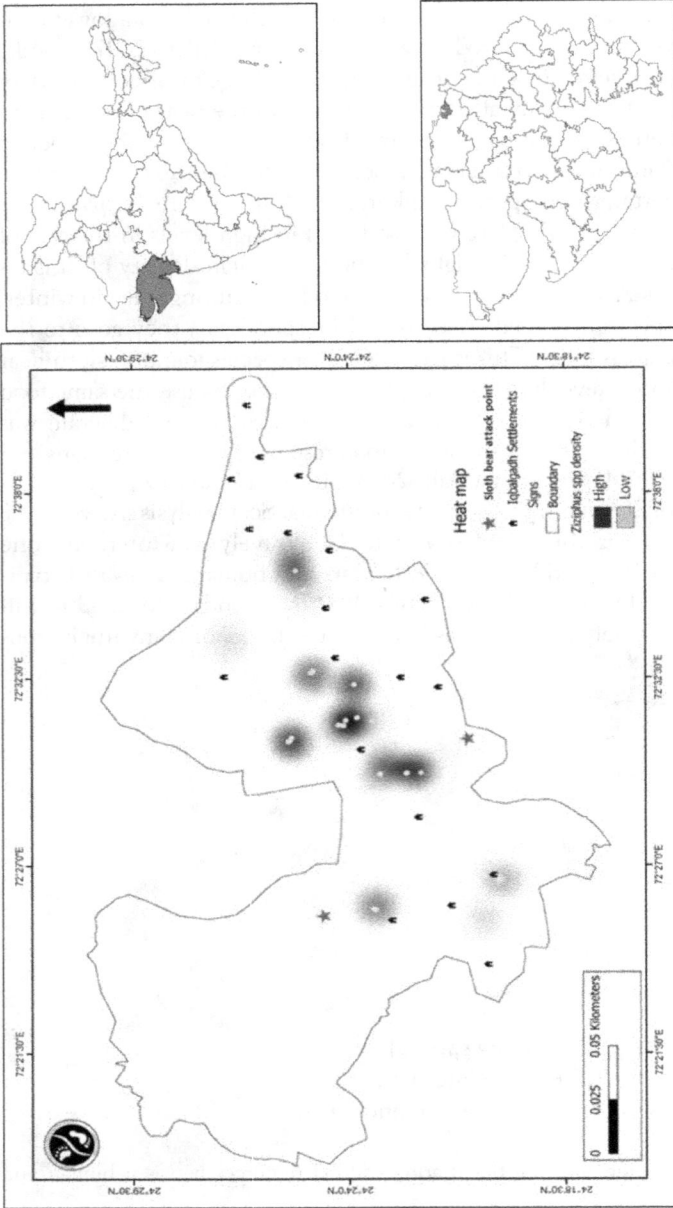

Figure 12.5 Density of Ziziphus along with sloth bear signs, human settlements, and conflict locations in Iqbalgadh range.

animals, because fruiting patterns result in a dependable food resource during times of general fruit shortage (Koike, 2009; Debata et al., 2016; Sreekumar & Balakrishana, 2002).

The distribution of *Ziziphus* spp. was further overlaid on the map to show the high-density areas in relation to locations of conflicts that happened in the past and in the human settlements. It can be easily interpreted from the map that sloth bear attacks happened at a long distance from the areas representing high density of *Ziziphus*. Hence, this rejects the hypothesis that sharing of food resources is a reason behind human–bear conflict, as chances of encounter decreases especially during winter. However, it cannot be entirely denied, because data shows partial sharing of food resources (Figure 12.5). A study conducted on the Asiatic black bear recognised an overlap in resource sharing between humans and bears in the area of Kashmir (India), leading to an increase in conflicts (Charoo et al., 2011). However, there were other factors like crop and livestock depredation for bear attacks on humans which is not observed in case of sloth bears. Due to a dearth of data on resource sharing between humans and bears, it is imperative to understand the pattern of resource sharing in other seasons throughout the year.

Spearman correlation index was calculated to find out if there is any relation between fruiting species in the sanctuary and sloth bear attack over a period of time (Spearman correlation coefficient, fruiting species rs = 776, n=17, ρ = -0.473). The Spearmen correlation value (ρ = -0.473) reveals that there is a negative correlation between availability of fruiting trees and number of sloth bear attacks in the study area; indicating that food resources might not be the only factor responsible for human sloth bear interactions (Bargali et al., 2004). However, dominance of *Ziziphus* seeds in the sloth bear's scats was in correlation with fruiting *Ziziphus* trees recorded in high density. Moreover, the social survey revealed the villagers' dependency on *Ziziphus* in winter, suggesting the possibility of resource sharing. Therefore, no significant correlation can be explained by the sparse and patchy distribution of vegetation reducing the chances of encounter. However, it is important to study food availability and food consumption on a seasonal basis for longer durations to understand resource sharing between humans and sloth bears.

Conclusion

This study found that sloth bears consume more fruits than animal matter such as ants, termites, insects, or honey in the winter season. Higher availability of fruits like *Ziziphus* in winter might be a preliminary reason behind the high fruit consumption by the sloth bear. The findings of this study may provide important information that can help in formulating habitat management strategies. The interviews conducted with villagers revealed that people use the forest resources throughout the year mostly to collect fire wood and fruits, and to graze cattle. While wandering in the forest, people encounter the sloth bear, and sometimes the accidental encounter results in conflict. Food resources were not found as a significant factor responsible for sloth bear attacks. It can be concluded that, during the dry season, when food and water fall short within the forest, sloth bears visit the villages in search of food and water. It can be also concluded that because of human use of forest areas and changes in land use patterns in the vicinity, the habitat of the sloth bear may be affected, and the chances of sloth bear encounters may be increased. This study can be used as a base for long-term research in western India, the western most distribution range of the sloth bear, and can help understand resource use patterns and the drivers through which human life and wildlife can co-exist. Further, a study covering all the seasons and with a larger sample size may provide a much clearer picture and answer our questions about resource sharing.

Acknowledgements: We thank the International Association for Bear Research and Management and Wildlife Conservation and Biology Research Lab, Hemchandracharya North Gujarat University, Patan (Gujarat) for the financial assistance for field visits. We are also thankful to the Chief Wildlife Warden, Gujarat state for providing permission to carry out the field work. We would like to express our gratitude towards field staff of Jessore and the local villagers for assisting us in our field work.

References

Akhtar, N., Bargali, H. S., & Chauhan, N. P. S. (2007). Characteristics of sloth bear day dens & use in disturbed and unprotected habitat of North Bilaspur Forest division, Chhattisgarh, Central India. *Ursus, 18*(2) 203–208.

Bargali, H. S., Akhtar, N., & Chauhan, N. P. S. (2004). Feeding ecology of sloth bears in a disturbed area in central India. *Ursus, 15*(2), 212–217.

Bargali, H. S., Akhtar, N., & Chauhan, N. P. S. (2005). Characteristics of sloth bear attacks and human casualties in North Bilaspur Forest division, Chhattisgarh, India. *Ursus, 16*(2), 263–267.

Baskaran, N. (1990). An ecological investigation on the dietary composition and habitat utilization of sloth bear at Madhumalai wildlife sanctuary, Tamil Nadu (South India) [Ph.D. Thesis]. Bharathidasan University, p. 57.

Basakarn, N., Sivagiganesan, N., & Krishnamoorthy, J. (1997). Food habits of sloth bear in Mudumalai wildlife sanctuary, Tamil Nadu, Southern India. *Journal of Bombay Natural History Society, 94*(1), 1–9.

Charoo, S. A. L., Sharma, K., & Sathyakumar, S. (2011). Asiatic Black bear-human interactions around Dachigam National Park, Kashmir, India. *Ursus, 22*(2), 106–113.

Cottm, G., & Curtis, J. T. (1956). The use of distance measures in phytosociological sampling. *Journal of Ecology, 37*(3), 451–460.

Debata, S., Swain, K. K., Sahu, H., & Palei, H. S. (2016). Human-sloth bear conflict in human-dominated landscape of Northern Odisha, India. *Ursus, 27*(2), 90–98.

Desai, A. A., Bhaskaran, N., & Venkatesh, S. (1997). Behavioural ecology of the sloth bears in Madumalai wildife sanctuary and national park. Report Tamil Nadu & BNHS collaborative project, Mumbai.

Dhamorikar, A. H., Mehta, P., Bargaliand, H., & Gore, K. (2017). Characteristics of human sloth bear (*Melursus ursinus*) encounters and the resulting human casualties in the Kanha-Pench corridor, Madhya Pradesh, India. *PLOS ONE, 12*(4), e0176612.

Dharaiya, N. (2009). *Evaluating habitat & human sloth bear conflicts in North Gujarat, India, to seek solutions for human – Bear coexistence.* Research project report I- submitted to the small grant's division. Rufford Foundation.

Dharaiya, N., Bargali, H. S., & Sharp, T. (2016). *Melursus ursinus.* The IUCN red list of threatened species 2016.

Dharaiya, N., & Ratnayeke, S. (2009). Escalating human-sloth bear conflict in North Gujarat: A though time to encourage support for bear conservation. *International Bear News, IBA, 18,* 12–14.

Garshelis, D. L., Joshi, A. R., Smith, J. L. D., & Rice, C. G. (1999). Sloth bear conservation action plan. In C. Servheen, S. Herrero, & B. Peyton (Eds.), *Bear: Status survey & conservation action plan.* IUCN/SSC Bear & Polar Bear Specialist Groups.

Garshelis, D. L., & Pelton, M. R. (1981). Movements of black bears in the Great Smoky Mountains national park. *The Journal of Wildlife Management,* 912–925.

Gokula, V. (1991). *Some aspects on the feeding habits of the sloth bear (Melursus ursinus) at Mundanthurai wildlife sanctuary, TamilNadu (South India)* [Ph.D. Thesis]. AVC College.

Gondaliya, H. (2012). *Evaluation of sloth bear habitat in Jessore* [M.Phil. Thesis]. Wildlife Sanctuary with Special Emphasis on Feeding Resources.

Gopal, R. (1991). Ethological observation on the sloth bear (*Melursusursinus*). *Indian Forester, 117,* 915–920.

Honda, T. (2009). Environmental factors affecting the distribution of the wild boar, sika deer, Asiatic Black bear and Japanese macaque in central Japan, with implication for the human-wildlife conflicts. *Mammal Study, 34*(2), 107–116.

Iswariah, V. (1984). *Status survey report and recommendation for conservation of the sloth bears in Ramnagaram taluk, Karnataka.* WWF-India [Unpublished report]. Bangalore, 34.

Johnsingh, A. J. T. (1981). *Ecology and behaviour of the dhole with special reference to prey-predator relation in Bandipur* [Ph.D. Dissertation]. Madurai University, pp. 306.

Joshi, A. R., Garshelis, D. L., & Smith, J. L. D. (1995). Home range of sloth bears in Nepal: Implication for conservation. *Journal of Wildlife Management, 59*(2), 204–214.

Joshi, A. R., Garshelis, D. L., & Smith, J. L. D. (1997). Seasonal and habitat-related diets of sloth bears in Nepal. *Journal of Mammalogy, 78*(2), 584–597.

Koike, S. (2009). Fruiting phenology & its effect on fruit feeding behaviour of Asiatic Black bears. *Mammal Study, 34*(1), 47–52.

Laurie, A., & Seidensticker, J. (1997). Behavioural ecology of the sloth bear (Melursus ursinus). *Journal of Zoological Society, 182*(2), 187–204.

Malik, A., Dharaiya, N., & Proctor, M. (2018). Use of geospatial techniques to target water sources for sloth bears, aimed at alleviating conflicts with people. *International Bear News, 27*(3), 12–14

Mattson, D. J. (2001). Myrmecophagy by Yellowstone grizzly bears. *Canadian Journal of Zoology, 79*(5), 779–793.

Mewada, T., & Dharaiya, N. (2010). Seasonal dietary composition of sloth bear (Melursus ursinus) in the reserve forest of Vijayanagar, North Gujarat, India. *TIGERPAPER, 37*(2), 8–13

Naughton Treve, L., Treves, A., Chapman, C., & Wrangham, R. (1998). Temporal patterns of crop-raiding by primates: Linking food availability in croplands and adjacent forest. *Journal for Applied Ecology, 35*(4), 596–606.

Norris, T. (1969). Ceylon sloth bear. *Animal, 12,* 300–303.

Obbard, M. E., Howel, E. J., Wall, L., Allison, B., Black, R., Davis, P., Dix-Gibson, L., Gatt, M., & Hall, M. (2014). Relationships among food availability, harvest, and human–bear conflict at landscape scales in Ontario, Canada. *Ursus, 25*(2), 98–110.

Peek, J. M., Pelton, M. R., Picton, H. D., Schoen, J. W., & Zager, P. (1987). Grizzly bear conservation and management: A review. *Wildlife Society Bulletin, 15*, 160–169.

Rajpurohit, K. S., & Chauhan, N. P. S. (1996). *Study of animal damage problems in and around protected areas and managed forest in India. Phase-I: Madhya Pradesh, Bihar and Orissa* [Report]. Wildlife Institute of India.

Rajpurohit, K. S., & Krausman, P. R. (2000). Human-sloth-bear conflicts in Madhya Pradesh, India. *Wildlife Society Bulletin, 28*(2), 393–399.

Ratnayeke, S., Manen, F., Pieris, R., & Pragash, V. (2014). Challenges of large carnivore conservation: Sloth bear attacks in Sri Lanka. *Human Ecology an Interdisciplinary Journal, 42*, 467–479.

Riley, E. (2007). The human – Macaque interfaces conservation implication of current and future overlap and conflicts in lore Lindu national park, Sulawesi, Indonesia. *American Anthropologist, 109*(3), 473.

Rodgers, P., Stapleton, G., Flower, J.,&Howse, J. (2013). Drawing area-proportional Euler diagrams representing up to three sets. *IEEE Transactions on Visualization and Computer Graphics, 20*(1), 1–1.

Schaller, G. B. (1967). *The deer and the tiger - A study of wildlife in India.* University of Chicago Press, pp. 370.

Seidensticker, J. (1976). Ungulate populations in Chitawan Valley, Nepal. *Biological Conservation, 10*(3), 183–210.

Sreekumar, P. G., & Balakrishnan, M. (2002). Seed dispersal by the sloth bears (Melursus ursinus) in South India. *Biotropica, 34*(3), 474–477.

Sukhadiya, D., Joshi, J. U., & Dharaiya, N. (2013). Feeding ecology and habitat use of sloth bear (Melurses Ursinus) in Jessore wildlife sanctuary, Gujarat, India. *Indian Journal of Ecology, 40*(1), 14–18.

Swenson, J. E., Jansson, A., Riig, G. R., & Sandegren, F. (1999). Bears & ants: Myrmecophagy by brown bears in central Scandinavia. *Canadian Journal of Zoology, 77*(4), 551–561.

Yoganand, K., Rice, C. G., Johnsingh, A. J. T., & Seidensticker, J. (2006). Is the sloth bear in India secure? A preliminary report on distribution, threats & conservation requirements. *Journal of Bombay Natural History Society, 103*(2–3), 172–181.

13 Bird Community Structure in Restored and Unrestored Areas in Delhi, India

Huda Tarannum, Aisha Sultana, Orus Ilyas, and Mohd Shah Hussain

Introduction

The Aravalli Biodiversity Park, Delhi, India had many mining pits operating during the 1980s. It is a part of the Aravalli hill range continuing into the Delhi ridge. The area was over mined for sandstone, mica and clay. The mining activities and other disturbances had resulted in barren hill slopes, a deep water table, and poor soil cover. The remnant forests on the site were highly degraded and invaded by *Prosopis juliflora*, a Mexican species. The government took initiative and restored the area as a biodiversity park. A study was conducted to assess the components of biodiversity. A biotic community is a naturally occurring and interacting assemblage of plants and animals living in the same environment and fixing, utilising and transferring energy in some manner. Some ecologists use this term to describe groups of similar organisms occurring in the same habitat, such as a community of grassland birds. Birds are one of the best indicators of environmental quality of any ecosystem (Ali & Ripley, 1978). Ecological communities are the characteristics and properties of the assemblages of species population (Koromondy, 1989) or a group of various kinds of species act as better predictors of the quality of habitat and health of the ecosystem as compared to single species. Over the years there has been a noticeable decline in the bird population all over the world for various reasons. Urbanisation-induced environmental changes favours the spread of invasive species, thereby reducing accessible area for native species (Akram et al., 2015, Barhadiya et al., 2021). Birds, along with other organisms, both micro and macro, constitute a significant component of our planet's biodiversity.

Study Area

Aravalli Biodiversity Park (ABP) is a part of Delhi ridge which was over mined and completely infected with an exotic plant species *Prosopis juliflora* (Vilayati kikar) and has lost its indigenous flora and fauna (Maheshwari, 1965). The park is spread over an area of 692 acres and lies between 77°7'52" to 77°10'49"E longitude and 28°34'27" to 28°34'25"N latitude (Figure 13.1). The ecological restoration work of the park was started in October 2004, and to date about 2,000 plant species native to Aravalli have been ecologically assembled which have

DOI: 10.4324/9781003321422-15

become home for many insects, birds, reptiles, and mammals. Delhi Development Authority in collaboration with the Centre for Environmental Management of Degraded Ecosystems, University of Delhi (CEMDE) started the restoration work to establish the biodiversity parks and bring back the lost biodiversity of Delhi ridge and the Yamuna River plain. The main flora consists of various developed plant communities such as *Anogeissus* dominant community, *Holoptelea* community, *Acacia* community, *Anogeissus* and *Boswellia* community, *Butea* community, grassland, bush forest, mixed dry deciduous, and thorny forest. More than 200 bird species have been recorded so far in ABP. The habitat created provides a food base and nesting sites to many bird species. Aravalli Biodiversity Park is a paradise for bird watchers. Common birds of the area include munias, kingfisher, water hen, partridge, woodshrikes, parakeets, starlings, and barbets. This is also an ideal habitat for passage migrants such as the red-throated flycatcher. Among the reptiles are the Indian cobra, monitor lizard, and common rat snake. More than 100 species of butterflies and moths have been spotted, and among them the notable species are red pierrot, pansies, plain tiger, striped tiger, common rose, and pioneer.

Material and Methods

Data Collection

Data were collected between the months of February and May 2018. The study area was stratified into restored and unrestored areas based on the status of habitat. The point count method (Bibby et al., 1992) and species richness counting method (MacKinnon & Philipps, 1993) was used for bird sampling. For point count, a total of 14 points were laid in the study area and five monitorings done on each point. Using the species richness counting method, a total of 85 lists were prepared to have ten species in each checklist, and no species was included more than once in each list. For the point count method, overall 14 points were laid out, with eight points in restored and six points in unrestored habitats, with five monitorings of each point for 20 minutes in an open width circular plot. At each point, bird species, number of individuals, radial and vertical distance, strata of tree used, activity, and height of the tree were used. Vegetation sampling was conducted to collect data on habitat parameters and to correlate the habitat parameters with bird density, diversity, and richness of the birds.

Data Analysis

Density

The densities were estimated using DISTANCE 7.1 program. The densities were calculated overall as well as in restored and unrestored areas. The densities for morning and evening monitorings were also calculated. T-test was conducted in SPSS version 20 to check the significant differences in densities of the two habitats.

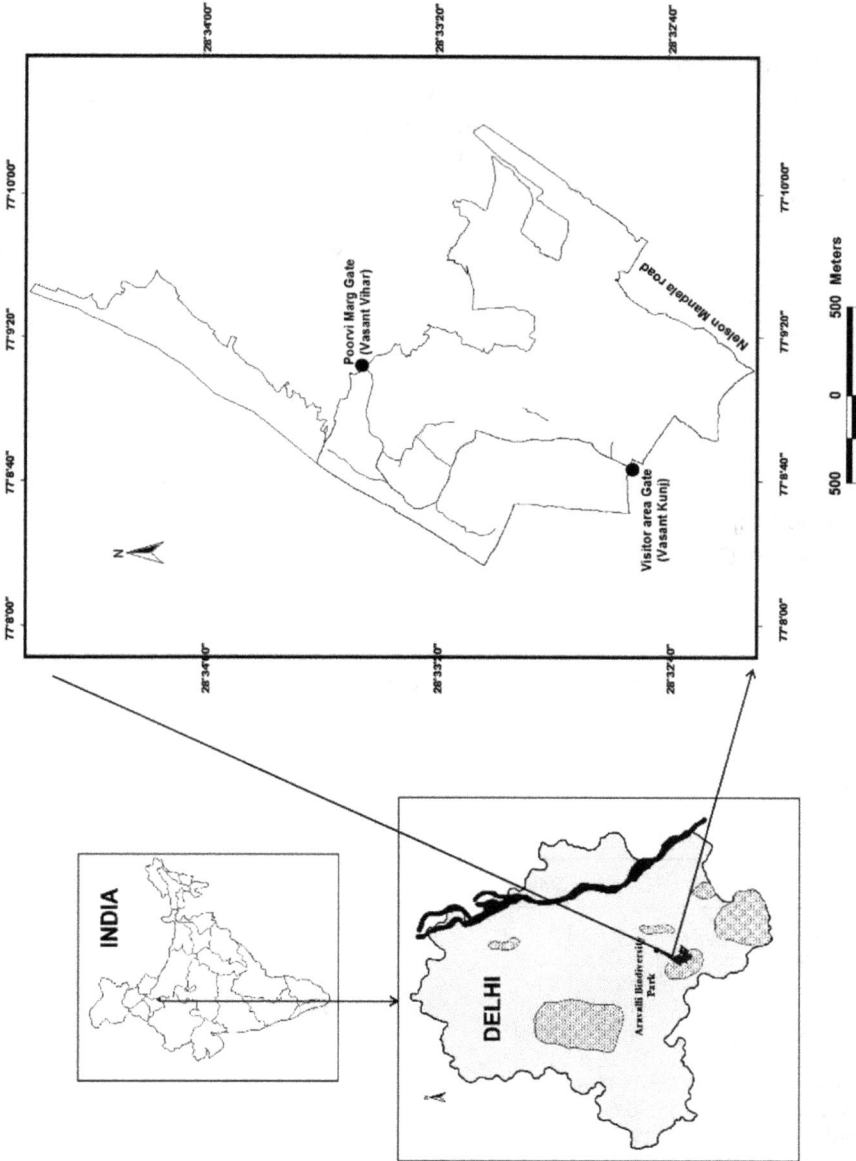

Figure 13.1 Map of the Aravalli Biodiversity Park.

Diversity and Richness

The bird species diversity was calculated in Microsoft Excel 2007 using Shannon-Weiner diversity index: $H' = -\Sigma Pi * \ln Pi$

Where, Pi is the proportion of the ith species.

The bird species richness was calculated using Margalef index: $R = S - 1/\ln(n)$ where

S = total number of species in area sampled.

n = total number of individuals observed.

Spearman Rank Correlation Analysis was done in SPSS version 20 to check the correlation between bird species density, diversity, and richness and the vegetation parameters. The bird species were also categorised into their feeding guilds following Ali and Ripley (1978) and abundance categories based on their number of sightings during the study period.

An abundance status of each species was assessed based on the number of sightings, and the bird species were ranked according to the following abundance categories:

1 = rare (0–5 sightings)
2 = common (6–25 sightings)
3 = abundant (26–50 sightings)
4 = very abundant (>50 sightings)

Results

The overall bird density was 9.295 individuals per hectare (SE = 1.262, 95% CI = 11.412, % CV = 10.48). The densities of birds in the restored and unrestored habitats were 10.973 and 9.087 individuals per hectare respectively. The densities of birds in the morning and evening were 8.606 and 11.247 individuals per hectare respectively. Restored habitat had a high diversity of birds as compared to the unrestored habitat (Table 13.1). The bird species diversity and richness were also observed to be higher in restored habitat in comparison to unrestored habitat. The bird densities in both the habitats did not differ significantly (t=1.281,

Table 13.1 Bird densities in restored and unrestored habitats

Habitat	Density (individuals/hectare)±SE	95% CI
Restored habitat	10.97 ± 1.54	14.40
Unrestored habitat	9.08 ± 1.26	11.92
Morning	8.60 ± 1.01	10.84
Evening	11.24 ± 1.50	14.61
Restored morning	11.89 ± 1.58	15.45
Restored evening	11.21 ± 2.18	16.41
Unrestored morning	8.35 ± 1.30	11.32
Unrestored evening	8.64 ± 1.67	12.62
Overall	**9.29 ± 1.26**	**11.41**

df=12, p=0.05). The Spearman rank correlation analysis reveals that the bird density positively correlated with shrub diversity (r=0.57, p=0.05) and foliage cover (r=0.64, p=0.05) among all the vegetation parameters. As the shrub diversity and foliage cover increased, the bird density also increased. The tree density, diversity, or richness did not affect the bird densities.

The feeding guild structure revealed that the dominant guild was insectivorous (50 %) with 55 species followed by omnivores, granivores, frugivorous, carnivores, and nectarivores representing 23.25%, 13.95%, 8.13%, 3.48%, and 1.16% respectively. Out of 103 bird species, 76 species were residents, 20 were winter migrants, and seven species were passage migrants. The analysis of bird species diversity and bird species richness in the restored and unrestored habitat revealed that it was higher in restored habitat. The bird species richness was higher by using the MacKinnons species richness method (R1 = 9.845) then the point count (R1 = 5.609). The analysis of bird abundance in their respective abundance categories shows that a higher number of bird species had fallen in the rare category (47 species). Among these 47 species of birds some were winter migrants (e.g. yellow wagtail, clamorous reed warbler, orange headed thrush, etc.) and some were passage migrants (e.g. ashy drongo, asian paradise flycatcher etc.), and among residents, the rare birds are mostly small birds. The house sparrow, the state bird of Delhi, also falls in the rare category and only had three sightings during the whole study period. The maximum number of species belonged to the order Passeriformes (58 species) followed by Accipitriformes (five species), while Columbiformes and Cuculiformes had four species in each order and the order Pssitaciformes and Piciformes were represented by three species each; Pelicaniformes, Galliformes, and Bucerotiformes had two species each. Gruiformes, Strigiformes, and Coraciformes had a single species each (Tables 13.2 and 13.3).

Table 13.2 Percentage and number of bird species in various abundance categories

Category	Number	Percent
Rare	47	55.29
Common	28	32.94
Abundant	6	7.05
Very abundant	3	3.52
Total	84	100

Table 13.3 Number of birds and total percent in their respective feeding guilds

Habitat	Insectivore	Omnivore	Frugivore	Granivore	Carnivore	Nectarivore
Restored	25	12	3	8	0	120
Unrestored	20	11	3	5	1	1
Overall	43	20	7	12	3	1
Total %	50	23.25	8.13	13.95	3.48	1.16

Appendix I: Checklist of bird species recorded at Aravalli Biodiversity Park. Feeding guild: 1=Insectivore, 2=omnivore, 3=granivore, 4=frugivore, 5=carnivore, 6=nectarivores. Habitat: 1=Restored habitat, 2=Unrestored habitat.

Abundance status: 1=Rare, 2=common, 3=abundant, 4=very abundant.

IUCN status: LC=Least concerned, NT=near threatened, V=vulnerable, E=endangered, NA=not assessed.

Common name	Species name	Feeding guild	Habitat	Abundance status	IUCN Status
Alexandrine parakeet	*Psittacula eupatria*	4	1	1	NT
Ashy crowned sparrow lark	*Eremopterix griseus*	1	2	1	LC
Ashy drongo	*Dicrurus leucophaeus*	1	1	3	LC
Ashy prinia	*Prinia socialis*	1	1, 2	1	LC
Asian koel	*Eudynamyss colopaceus*	4	1, 2	2	LC
Asian paradise flycatcher	*Terpsiphone paradise*	1	1	1	LC
Bank myna	*Acridotheres ginginianus*	2	1, 2	1	LC
Baya weaver	*Ploceus philippinus*	4	1	1	LC
Black-breasted weaver	*Ploceus benghalensis*	4	1	1	LC
Black drongo	*Dicrurus macrocerus*	1	1, 2	2	LC
Black kite	*Milvus migrans*	5	1, 2	2	LC
Black redstart	*Phoenicuru sochruros*	1	1, 2	2	LC
Black-winged kite	*Elanus caeruleus*	5	2	1	LC
Black-winged stilt	*Himantopus himantopus*	1	2	1	LC
Blue-tailed bee-eater	*Merops philippinus*	1	1	1	LC
Blue throat	*Luscinias vecica*	1	1	1	LC
Booted warbler	*Iduna caligata*	1	1	2	LC
Brahminy starling	*Sturnus pagodarum*	2	1	1	LC
Brown-headed barbet	*Megalaima zeylanica*	4	1, 2	2	LC
Brown rockchat	*Cercomela fusca*	1	1, 2	1	LC
Cattle egret	*Bubulcus ibis*	1	2	2	LC
Clamorous reed warbler	*Acrocephalus stentoreus*	1	1	1	LC
Common babbler	*Turdoides caudatus*	1	2	1	LC
Common chiffchaff	*Phylloscopus collybita*	1	1,2	1	LC
Common hawk cuckoo	*Hierococcyx varius*	2	1	1	LC
Common myna	*Acridotheres tristis*	2	1, 2	2	LC
Common rosefinch	*Carpodacus pulcherrimmus*	4	1	1	LC
Common stonechat	*Saxicola torquatus*	1	1, 2	2	LC
Common tailorbird	*Orthotomuss utorius*	4	1	3	LC
Common woodshrike	*Tephrodornis pondicerianus*	1	1	1	LC
Coppersmith barbet	*Megalaima haemacephala*	4	1	2	LC
Dusky warbler	*Phylloscopus fuscatus*	1	1	1	LC
Eurasian collered dove	*Streptopelia decaoto*	4	1, 2	1	LC

(*Continued*)

Common name	Species name	Feeding guild	Habitat	Abundance status	IUCN Status
Eurasian eagle owl	*Bubo bubo*	5	2	1	LC
Graceful prinia	*Prinia gracilis*	1	1	1	LC
Greater caucal	*Centropus sinensis*	2	1, 2	2	LC
Green bee eater	*Merops orientalis*	1	1, 2	2	LC
Greenish warbler	*Phylloscopus trochiloides*	1	1	1	LC
Grey breasted prinia	*Prinia hodgsonii*	1	1, 2	1	LC
Grey francolin	*Francolinus pondicerianus*	2	1, 2	2	LC
Grey headed canary flycatcher	*Culicicapa ceylonensis*	1	1	1	LC
Grey hornbill	*Ocyceros birostris*	4	1	1	LC
Hoopoe	*Upupa epops*	1	1	1	LC
House crow	*Corvus splendens*	2	1, 2	2	LC
House sparrow	*Passer domesticus*	3	2	1	LC
Hume's leaf warbler	*Phylloscopus humei*	1	1	1	LC
Indian robin	*Saxicoloides fulicatus*	1	1, 2	3	LC
Indian roller	*Coracias benghalensis*	1	2	1	LC
Indian silverbill	*Euodicema labarica*	4	1, 2	2	LC
Jungle babbler	*Turdoides striatus*	1	1, 2	4	LC
Jungle bush quail	*Perdicula asiatica*	4	2	1	LC
Large-billed crow	*Corvus macrorynchus*	2	1	2	LC
Large grey babbler	*Turdoides malcolmi*	1	1, 2	2	LC
Laughing dove	*Streptopelia seneghalensis*	4	1, 2	2	LC
Lesser goldenback	*Dinopium benghalensis*	1	1	2	NA
Lesser white throat	*Sylvia curruca*	1	1, 2	2	NA
Long-tailed shrike	*Lanius schach*	1	1, 2	1	LC
Moustached warbler	*Acrocephalus melanopogon*	1	1	1	LC
Orange-headed thrush	*Zoothera citrina*	1	1	1	LC
Oriental honey buzzard	*Pernisptilo rhynchus*	5	1	1	LC
Oriental magpie robin	*Copsychus saularis*	1	1, 2	1	LC
Oriental skylark	*Alauda gulgula*	1	2	1	LC
Oriental white-eye	*Zosterops palpebrosus*	1	1, 2	2	LC
Orphean warbler	*Sylvia hortensis*	1	1	1	LC
Paddyfield pipit	*Anthus rufulus*	1	2	1	LC
Pallid harrier	*Circus macrourus*	5	2	1	NT
Indian peafowl	*Pavo cristatus*	2	1, 2	3	LC
Pied bushchat	*Saxicola caprata*	1	1, 2	1	LC
Pied myna	*Gracupica contra*	2	1, 2	1	NA
Plain prinia	*Prinia inornata*	1	1, 2	2	LC
Plum-headed parakeet	*Psittacula cyanocephala*	4	1	1	LC
Purple sunbird	*Cinnyris asiaticus*	6	1, 2	4	LC
Red avadavat	*Amandava amandava*	4	1	1	LC
Red-naped ibis	*Pseudibis papillosa*	2	2	1	LC
Red-throated flycatcher	*Ficedula parva*	1	1	2	LC
Red-vented bulbul	*Pycnonotus cafer*	2	1, 2	4	LC

(*Continued*)

Common name	Species name	Feeding guild	Habitat	Abundance status	IUCN Status
Red-wattled lapwing	*Vallanus indicus*	1	1, 2	1	LC
Red-whiskered bulbul	*Pycnonotus joscosus*	2	1, 2	3	LC
Rock pigeon	*Columba livia*	4	1, 2	2	LC
Rose-ringed parakeet	*Psittacula krameri*	4	1, 2	2	LC
Rosy starling	*Pastor roseus*	2	1, 2	1	LC
Rufous-fronted prinia	*Prinia buchanani*	1	1	1	LC
Rufous-tailed lark	*Ammomane sphoenicura*	1	2	1	LC
Rufous treepie	*Dendrocitta vagabunda*	2	1, 2	3	LC
Scaly breasted munia	*Lonchura punctulata*	4	1, 2	1	LC
Shikra	*Accipiter badius*	2	1	1	LC
Sirkeer malkoha	*Taccocuales chenaultii*	1	1, 2	1	LC
Southern grey shrike	*Lanius meridionalis*	1	1	1	V
Spotted owlet	*Athene brama*	2	1	1	LC
Steppe eagle	*Aquila nipalensis*	5	2	1	E
Sulphur-bellied warbler	*Phylloscopus griseolus*	1	1, 2	1	LC
Tawny eagle	*Aquila rapax*	5	2	1	LC
White-throated kingfisher	*Halcyon smyrnensis*	2	1, 2	2	LC
White-breasted waterhen	*Amaurornis phoenicurus*	2	1	1	LC
White-browed wagtail	*Motacilla maderaspatensis*	1	2	1	LC
White-capped bunting	*Emberiza stewarti*	1	1	1	LC
White-eared bulbul	*Pycnonotus leucotis*	2	1, 2	2	LC
Wire-tailed swallow	*Hirundo smithii*	1	2	1	LC
Yellow-bellied prinia	*Prinia flaviventris*	1	1	1	LC
Yellow-eyed babbler	*Chrysom masinense*	1	1, 2	2	LC
Yellow-footed green pigeon	*Treron phoenicoptera*	4	1, 2	1	LC
Yellow wagtail	*Motacilla flava*	1	2	1	LC
Zitting cisticola	*Cisticola juncidis*	1	1	1	LC

Discussion

Aravalli Biodiversity Park in the art of Delhi is one of the best examples of habitat restoration. It used to be a mining site and was highly degraded with barren land and some exotic species such as *Prosopis juliflora*. The restoration work started in 2004, and up to now about 1,000 plant species native to Aravalli have been ecologically assembled which have become home for many insects, birds, reptiles, and mammals. More than 200 bird species have been recorded from the ABP. This study is an attempt to understand the ecological aspects of the ABP. The restored as well as unrestored areas were compared to assess significant differences, using various sampling strategies. The point count method and MacKinnon species richness methods have been used by in other studies to assess the density and richness (Bibby et al., 1992; MacKinnon & Phillips, 1993; Javed, 1996; Kidwai et al., 2013; Singh et al., 2014).

The bird densities in the restored and unrestored habitats did not differ significantly. In the restored habitat there are mature trees, and the density of trees is higher, while the unrestored habitat is under a restoration process, and tree density and diversity is much lower than the restored habitat. The density of birds was higher in the evening than the morning, as birds were comparatively more active than in the morning. The bird densities in restored and unrestored habitat did not differ very much, but the diversity of birds in both the habitats differed, because the unrestored habitat supported the bird species which are commonly seen near human habitations in large numbers. An analysis of bird species diversity and bird species richness in the restored and unrestored habitat showed that it was higher in restored than in unrestored habitats. The bird species richness was higher by using the MacKinnon richness method (R1 = 9.845) than the point count method (R1= 5.609).

The feeding guild structure revealed that the insectivore guild is dominant, which is a common trend in bird species (Sultana & Khan, 2000; Sultana et al., 2007; Ahmed et al., 2015) both in restored and unrestored habitats, followed by omnivore guild. Most of the birds are dependent on insect populations to feed their young (Dickinson, 1999) were more abundant in restored forest (Barhadiya et al., 2021).

Several rare birds have been sighted in Aravalli Biodiversity Park earlier (Sultana et al., 2013, Arafat et al. 2015). The Aravalli Biodiversity Park is a very good model for developing degraded land into natural forest. This model can be used in other degraded areas, especially in urban areas, to create a viable lung for crowded cities and a refuge for birds.

Acknowledgments

Authors are thankful to Prof. C.R. Babu, Project Incharge, Biodiversity Parks Programme, for his continous support and guidance. Thanks are due to Delhi Development Authority and staff of Aravalli Biodiversity Park also for their all help.

References

Ahmad, T., Khan, A., & Chandan, P. (2015). A pilot survey of the avifauna of Rangdum valley, Kargil, Ladakh (Indian trans-Himalayas). *Journal of Threatened Taxa, 7*(6), 7274–7281.

Akram, F., Ilyas, O., & Prusty, B. A. (2015). Impact of urbanization on bird community structure in and around Aligarh City, Uttar Pradesh, India. *International Journal of Engineering Technology Science and Research, 2*(10), 1–11.

Ali, S., & Ripley, S. D. (1978). *Handbook of the birds of India and Pakistan* (Vol. 1–10). Oxford University Press.

Arafat, Y., Hussain, M. S., & Sultana, A. (2015). Oriental Pied Hornbill *Anthracocerus albirostris*, and Indian Pitta *Pitta brachyura* in Delhi, India. *Indian Birds, 10*(5), 132–133.

Barhadiya, G., Sultana, A., Khudsar, F. A., & Hussain, M. S. (2021). Is there any impact of non-native vegetation on bird communities in Delhi, India? *Tropical Ecology.* https://doi.org/10.1007/s42965-021-00181-2.

Bibby, C. J., Hill, D. A., Burgess, N. D., & Mustoe, S. (1992). *Bird census techniques.* Academic Press.

Dickinson, M. B. (1999). *Field guide to the birds of North America.* 3rd ed. National Geographic Society.

Javed, S. (1996). *Study on bird community structure of Tarai forest of Dudhwa National Park, India* [Ph.D. Thesis]. Department of Wildlife Sciences, Aligarh Muslim University.

Kidwai, Z., Matwal, M., Kumar, U., Shrotriya, S., Masood, F., Moheb, Z., Ansari, N. A., & Singh, K. (2013). Comparative study of bird's community structure and function in two different forest types of Corbett National Park, Uttarakhand, India. *Asian Journal of Conservation Biology, 2*(2), 151–163.

Kormondy, E. J. (1989). *International handbook of pollution control.* Greenwood Press.

MacKinnon, J., & Phillipps, K. (1993). *The birds of Borneo, Sumatra, Java and Bali.* Oxford University Press.

Maheshwari, J. K. (1965). *Illustrations to flora of Delhi.* NISC, CSIR, pp. 282.

Singh, R., Dev, K., Kaur, D. N., & Sahi, D. N. (2014). Bird community structure along with species diversity, relative abundance and habitat use of tehsil Udhampur, Jammu & Kashmir, India. *International Journal for Pure and Applied Zoology, 2*(1), 26–40.

Sultana, A., Hussain, M. S., & Khan, J. A. (2007). Bird communities of proposed Naina and Pindari wildlife sanctuaries, Kumaon Himalayas, Uttarakhand, India. *Journal of the Bombay Natural History Society, 104*(1), 19–29.

Sultana, A., & Khan, J. A. (2000). Birds of oak forests in the Kumaon Himalayas, Uttar Pradesh, India. *Forktail, 16,* 131–146.

Sultana, A., Hussain, M. S., & Arafat, Y. (2013). Nesting of Indian Eagle owl *Bubo bengalensis* in Aravalli Biodiversity Park, a restored site in Delhi. *Indian Birds, 18*(3): 84.

14 Impact of Grazing on Bird Community in Binsar Wildlife Sanctuary

Orus Ilyas and Junaid Nazeer Shah

Introduction

The unprecedented rise in the human population, along with large-scale clearance of forests for extensive agriculture and rapid industrialisation, has resulted in fragmentation and a reduction in forest cover. This has not only led to a serious decline in the abundance of a number of faunal and floral species but also excessive dependence of local people on the resources of protected areas (PAs). A majority of the PAs in India are facing the problem of varying levels of livestock grazing pressure. Livestock grazing has been linked to a number of negative impacts on PAs and their biodiversity values. Conservationists in general perceive grazing to be responsible for habitat degradation and the decline in populations of a number of wildlife species. It is generally believed that severe over-grazing often leads to loss of ground cover, resulting in soil erosion, decrease in grass biomass, increase in unpalatable grass species, and weed abundance.

Binsar Wildlife Sanctuary in the Almorah district of Uttarakhand is one of the areas affected by livestock grazing and local dependency problems. These problems often result in conflict between park management and local communities. It is also the most neglected aspect from the point of view of wildlife research and management. To consider this, efforts were made to assess the bird species' density, diversity, and richness in different grazing zones of the Binsar Wildlife Sanctuary in 2002.

Study Area

The Binsar Wildlife Sanctuary (BWLS) lies at 29° 42' 3.2"N latitude and 79° 45' E longitude, at a distance of 30 km in north-east of Almora town (Figure 14.1). It was declared a wildlife sanctuary in May 1988. It covers an area of 46 km². The sanctuary is hilly with steep slopes. The highest elevation is 2,440 m, whereas the elevation of the surrounding villages inside the sanctuary barely reaches 1,500 m. Four different seasons (pre-monsoon, monsoon, post-monsoon, and winter) have been recognised in Binsar. BWLS represents characteristic floral elements of the moist temperate type of forest described by Saxena and Singh (1982). The oak forest is mainly in the centre of Binsar, and the surrounding

DOI: 10.4324/9781003321422-16

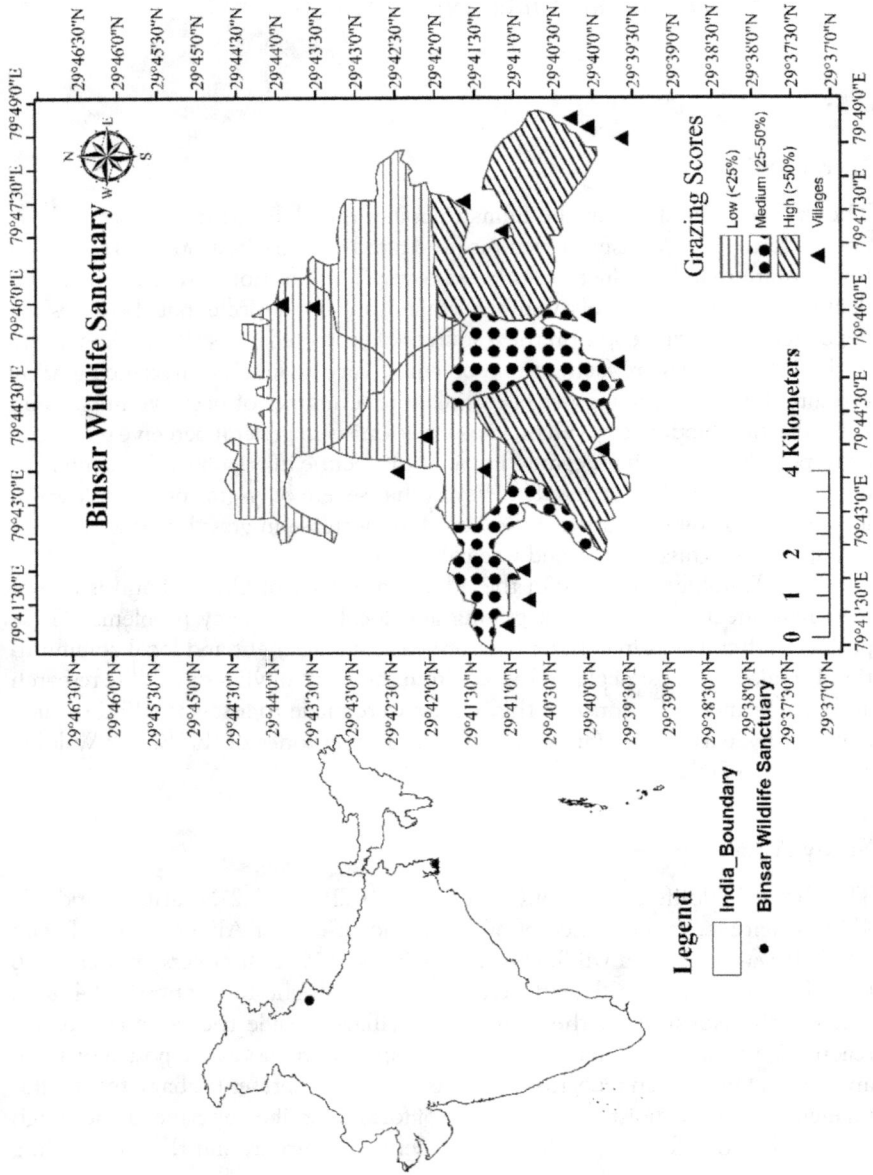

Figure 14.1 Forest beat-wise grazing pressure and location of villages in Binsar Wildlife Sanctuary.

ranges are mostly covered with chir pine forests interspersed with agricultural land. The oak forest on lower altitudes consists of tree species such as *Alnus nepalensis* and *Quercus leucotrichophora* and shrub species such as *Berberis aristata* and *Rubus ellypticus*. On higher altitudes (around 2,400 m), the oak forest comprises of *Quercus floribunda*, *Quercus leucotrichophora*, *Quercus glauca*, *Rhododendron arboreum*, and *Lyonia ovalifolia*, with fern species such as *Aetherium sp.*, *Polistichum sp.*, and *Pierkiium sp.*, and shrub species such as *Myrsine africana*, *Rubus biflorus*, *Daphne papyracea*. The chir pine forest comprises *Pinus roxburghii* along with *Vibumum cotinifolium*, among shrubs like *Pyracantha crenulata*, *Myrsine africana*, and *Desmodium sp.*

Methodology

Data Collection

Stratified random sampling was carried out in all the seven beats of the Binsar Wildlife Sanctuary, covering all the three habitats for the investigation of the grazing pressure in the Binsar Wildlife Sanctuary during the pre-monsoon season of 2000.

For birds, the open width point count method was used. Data was collected on 85 sampling points established in different beats, habitats, and different categories of grazed areas in the Binsar Wildlife Sanctuary. Data on bird species, number of individuals, and radial distance were collected on each bird point. Along with each bird data, data on habitat parameters were also collected, such as grazed category, percent tree cover, percent grass cover, percent bare ground, percent herb cover, nearest village, and broad habitat type.

Data on socio-economic aspects was collected for 17 villages situated within a 5 km radius of the sanctuary. Data on the following parameters was collected: Name of the village, distance from the sanctuary, total number of houses, total number of humans, cattle population, and goat population. Along with percentage of different religion/cast, percentage literacy rate, mulching animals, fuelwood consumption/day/house hold, fodder consumption/day/cattle, grass consumption/day/cattle, data on environmental awareness, conservation problems and their alternatives, and the number of the guns in each village were also collected.

Data on various disturbance attributes was collected at each sampling plot. The number of cut and lopped trees, number of cattle dung piles, and goat pellets were counted within a 12.62 m radius circular plot. The grazing was assessed in three different categories, i.e., low grazing pressure (35%), medium grazing pressure (35–70%), and high grazing pressure (70–100%). The distance from the nearest human settlement was also recorded.

For data analysis, bird species density was calculated using the following formula:

D = No of individuals/Area

Bird species numbers for each plot were added together to calculate mean densities and standard deviation for different beats, habitats, and for different grazed areas. The Shannon-Weiner Index (H') species diversity and Margalef index (RI) for species richness were used to calculate the species diversity and richness of birds for each plot using the following formula:

$$H' = \sum pi \times \log pi \text{ and } RI = S - 1 / \ln N$$

where pi is the proportion of ith species in the sample, S is the number of species in the sample, and N is the number of total individuals. These values were used for the calculation of mean diversity and richness for different sites.

One-way ANOVA and Scheffe's post hoc test were used to test for significant differences in various parameters, viz. grazing pressure across different beats and habitats of BWLS. The test was performed using the computer program SPSS, following Zar (1984).

Results

Assessment of Impact of Differential Grazing on Bird Species Diversity and Richness

Bird species density, diversity, and richness were calculated in different grazing pressure zones, and it was found that the bird species density was maximum in the no grazing zone (316.8±253.17) followed by the low grazing zone (200.7±141.6), and the result was found to be significant (F3 81= 7.4 p>0.006). Bird species diversity, however, was maximum in medium grazed area while it was minimum in high grazed areas. Bird species richness was maximum in no grazing area (2.12±0.8) and minimum in the high grazing are (1.33±0.57) The result was found to be significant both for bird species diversity as well as for richness (F3 81=4.47 p>.006 and F3 81=3.3 p>.02 respectively) (Table 14.1).

Bird species' density, diversity, and richness were examined across different habitats (Table 14.2). Bird species density was maximum in oak habitat (253.34±204.6), and minimum in pine (62.83±61.7). Bird species diversity was, maximum in mixed forest (0.67±0.18) and minimum in pine (0.41±0.3).

Table 14.1 Bird species density, diversity, and richness in different grazing zones of BWLS

Bird indices	Controlled plot	Low grazing	Medium grazing	High grazing
Bird species density	316.8±253.17	200.7±141.6	130.3±106.2	86.64±85.7
Bird species diversity	0.63±0.32	0.51±0.3	0.64±0.2	0.32±0.27
Bird species richness	2.12±0.8	1.67±0.7	1.9±0.8	1.33±0.57

Table 14.2 Bird species density, diversity, and richness in different habitats of BWLS

Bird indices	Mixed	Oak	Pine
Bird species density	189.79±150.7	253.34±204.6	62.83±61.7
Bird species diversity	0.67±0.18	0.59±0.3	0.41±0.3
Bird species richness	1.9±0.7	1.94±0.84	1.55±0.7

As far as bird species richness is concerned, it was seen maximum in oak forest (1.94±0.84) and a minimum in pine (1.55±0.7). However, the results were found to be significant only for bird species density and diversity (F2 82=9.1 p>.000 and F2 82=4.03 p>.031 respectively).

Bird species density, diversity, and richness was studies across different beats of BWLS (Table 14.3). While bird species density was highest in Gaunaph beat (346±240.6) followed by Satri beat (239.7±159.5) and Binsar (236±201.7), it was lowest in Kaparkhan beat (61.46±38.1). Bird species diversity was, highest in Dhaulchina beat (0.76±0.15) followed by Gaunaph beat (0.7±0.32), and it was lowest in Kaparkhan beat (0.29±0.28). Bird species richness was however highest in Gaunaph beat (2.4±0.7) followed by Dhaulchina beat (2.3±0.6), and the lowest species richness was in Kaparkhan beat (1.36±0.69). The results were found to be significant for bird species density, diversity, and richness (F6 78=3.7 p>.000, F6 78=3.7 p>.003 and F6 78=4.3 p>. 00 1 respectively).

Assessment of Impact of Grazing on Disturbance Factors

Among different grazing zones the abundance of cattle dung was maximum in high grazing area (5.9±4.7) and minimum in controlled plot (0.34±1.1), and the result was also found significant (F3 146=28.08 p>0.000). Goat pellets were also found highest in high grazing zone (1.13±3.7; F3 146=3.19 p>0.025). Tree cutting pressure was, however, highest in controlled plot (2.3±2.6; F3 146=9.08 p>0.025), while tree lopping was highest in controlled plot (2.3±2.6; F3 146=9.08 p>0.000) (Table 14.4).

Table 14.5 shows that among different habitats, abundance of cattle dung was highest in pine forest (4.7±4.7; F2 147=3.5 p>0.03). Tree cutting abundance were found highest in oak (2.6±2.8; F2 147=3.18 p>0.04). Tree lopping pressure was found to be highest in mixed forest (6.3±6.3; F2 147=p>0.000).

Among different beats (Table 14.6) abundance of cattle dung was highest in Kaparkhan beat (6.65±5.58; F6 143=14.3 p>0.000). Goat pellets were highest in Badaur beat (2.2±5.2; F6 143=3.8 p>0.001). Tree cutting as well as lopping pressure were highest in Badaur beat (3.9±4.06 and 10.4±9.5 respectively; F6 143 =5.7 p>0.003 and F6 143 =18.8 p>0.000 respectively).

Table 14.3 Bird species density, diversity, and richness in different beats of BWLS

Bird indices	Badaur	Binsar	Dhaulchina	Gaunaph	Kaparkhan	Pataniyanail	Satri
Bird species density	139±87.2	236±201.7	148.9±112.1	346.4±240.6	61.46±38.10	76.9±81.6	239.7±159.5
Bird species diversity	0.51±0.1	0.5±0.3	0.76±0.15	0.7±0.32	0.29±0.28	0.57±0.25	0.56±0.3
Bird species richness	1.4±0.39	1.7±0.7	2.3±0.6	2.4±0.7	1.36±0.69	1.7±0.8	1.6±0.8

Table 14.4 Mean values of disturbance in different grazing zones in BWLS

Disturbance factor	No grazing	Low grazing	Medium grazing	High grazing
Cattle dung	0.34±1.1	1.6±2.7	3.3±3.4	5.9±4.7
Goat pellets	0.13±0.9	0.41±0.74	0.77±1.45	1.13±3.7
Cut trees	2.34±2.6	2.04±2.08	2.2±2.4	1.3±1.39
Lopped trees	1.38±3.4	4.34±4.8	4.9±6.67	3.1±6.1

Table 14.5 Mean values of disturbance in different habitats of BWLS

Disturbance factor	Mixed	Oak	Pine
Cattle dung	2.8±3.5	0.68±1.58	4.7±4.77
Goat pellets	1±2.7	0.12±0.84	0.26±0.66
Cut trees	1.62±1.9	2.6±2.8	1.8±1.7
Lopped trees	6.3±6.3	1.7±4.2	0.6±1.2

Grazing Pressure due to Dependency of Local Community

There were 803 households in the 17 villages surveyed. A total human population of about 5,189 and a cattle population of about 2,555 as well as 1,357 goats was recorded from these surveyed villages. More than 80% villagers were unemployed and were totally dependent on the forest resources of BWLS. The large number of cattle mostly depends on the forest of BWLS for fodder and grazing. The fodder collection is high especially during snowfall when grass is not available for grazing. The villagers also collect grass to feed their cattle.

Discussion

Problems and Management Constraints

The management plan is silent on the problem of grazing within the sanctuary from the peripheral villages; however, the staff during informal discussions did consider grazing a major problem, which needs to be urgently addressed. Moreover, there is the problem related to feral cattle within the sanctuary. Apart from these two problems, a major management constraint is concerned with tackling the grazing of livestock belonging to villages lying within the sanctuary, as they have no nearby forest to graze their cattle. These facts have been collaborated by the results of the study, which show medium to high grazing pressure across the sanctuary. The southern portion of the sanctuary is under high grazing pressure, as this area also has pressure from 11 adjacent villages lying within a 5 km radius of the sanctuary boundary. Unfortunately, no records of grazing offences are being maintained in BWLS with which the findings of the study could be related.

Although eco-development work has been initiated in all the dependent villages in and around BWLS, unfortunately the villagers lack awareness of the

Table 14.6 Mean values of disturbance in different beats of BWLS

Disturbance factor	Badaur	Binsar	Dhaulchina	Gaunaph	Kaparkhan	Pataniyanail	Satri
Cattle dung	5.4±5.4	0.7±1.04	3±0.82	0.4±0.75	6.65±5.58	3.52±2.31	0.85±2.06
Goat pellets	2.2±5.2	2±1.33	0.0±0.0	0.45±0.89	0.08±0.28	0.35±1.18	0.35±1.18
Cut trees	3.9±4.06	1.65±2.06	2±1.76	2.2±2.2	2.7±1.72	1±1.19	1.75±1.8
Lopped trees	10.4±9.5	0.55±1.19	7.3±6.3	4.05±5.2	0.85±1.3	0.64±1.07	8.15±3.6

scheme. Moreover, none of the activities taken under eco-development planning target the grazing pressure on the sanctuary, which needs urgent attention. Although plantations under this scheme were raised around the forest a few years ago, unfortunately they were only of exotic species and not of indigenous fodder species. In addition, these plantations were burnt down in fire.

Due to unavailability of alternatives, villages lying within the sanctuary were totally dependent on the forests of BWLS for grazing their livestock and also for meeting their subsistence needs. Thus, cattle grazing in BWLS poses a major problem given the fragile ecosystem of the region, compounded by various management constraints and heavy biotic pressure on the small patch of forest of the PA and therefore needs immediate attention.

Recommendations

BWLS is rich in bird species diversity (Table 14.7) and supports the overall preference for low to medium grazing intensity areas by birds. It would be an ideal situation to completely remove all biotic disturbances from BWLS as a large portion was under high grazing pressure. However, it is an unrealistic aspiration considering the high level of biotic dependence. The major lacuna, which has emerged from the findings of the study, is the absence of a grazing policy. This can be seen that grazing is a continuous process, and it is necessary to evolve long-term site-specific grazing policies.

BWLS has villages within the sanctuary. It is imperative that for long-term conservation policy, relocating these villages to suitable sites outside the sanctuary should be worked out. As long as there is no grazing policy, the PA management should look for ways to implement controlled and rotational grazing in areas under medium-high grazing intensity. To help implement such major changes it would be necessary to involve the local dependent people.

Green fodder could also be made available to the villagers at the beginning of the monsoon season by hiring irrigated fields, during the gestation period, to enable people to adopt rotational grazing flopping and stall feeding.

Micro-planning activities should focus on viable eco-development activities, which would target reduction of grazing pressure on the forests of the PA. Although the southern portion of the sanctuary is under medium to high grazing pressure, the forest in the north is comparatively under low grazing pressure. Moreover, the northern portion of the sanctuary is also more rugged and has fewer habitations, apart from being at a considerable distance from the adjacent villages. Therefore, based on the spatial distribution of grazing pressure, it is suggested that the forest under the following compartments should be developed as a core zone: (1) All compartments of Satri beat, except compartment 7 and 25; (2) compartment numbers 20 and 21 of Gunapah beat; and (3) compartment numbers 3 and 30 of Binsar beat. The suggested core zone should be provided with protection, and local villagers should be deterred from grazing their cattle in this area, so that at least a small portion of this fragile ecosystem is protected. Villages falling on the boundary of the suggested core should be immediately relocated, as they have no alternatives available.

Table 14.7 Checklist of birds in BWLS

	Common name	Scientific name
Family: Accipitridae		
1	Sparrow hawk	*Accipiter nisus melaschistos*
2	Black eagle	*Ichtinaetus malayensis*
3	Himalayan griffon	*Gyps himalayensis*
4	Bearded vulture	*Gypaetus barbatus*
Family: Falconidae		
5	Kestrel	*Falco tinnunculus*
Family: Phasianidae		
6	Black partridge	*Francolinus francolinus*
7	Common hill partridge	*Arborophila torqueola*
8	Kaleej pheasant	*Lophura leucomelana*
9	Koklas pheasant	*Pucrasia macrolopha*
Family: Columbidae		
10	Yellow legged green pigeon	*Treron phoenicoptera*
11	Rufous turtle dove	*Streptopelia orientalis*
12	Indian ring dove	*Streptopelia decaocta*
13	Spotted dove	*Streplopelia chinensis*
Family: Psittacidae		
14	Blossom headed parakeet	*Psittacula cyanocephala*
Family: Cuculidae		
15	Large hawk-cuckoo	*Cuculus sparverioides*
Family: Strigidae		
16	Brown wood owl	*Strix leptogrammica*
Family: Caprimulgidae		
17	Indian jungle nighijar	*Caprimulgus indicus*
Family: Apopidae		
18	White-throated spintail swift	*Chaetura caudacuta*
19	Large white-rumped swift	*Apus pacificus*
Family: Alcedinidae		
20	White-breasted kingfisher	*Halcyon smymensis*
Family: Upupidae		
21	Hoopoe	*Upupa epops*
Family: Capitonidae		
22	Great hill barbet	*Megalaima virens*
Family: Picidae		
23	Scaly-bellied green woodpecker	*Picus squamatus*
24	Black-naped green woodpecker	*Picus canus*
25	Rufous-bellied woodpecker	*Hypopicus hyperythrus*
26	Himalayan pied woodpecker	*Picoides himalayensis*
27	Brown-fronted pied woodpecker	*Picoides auriceps*
Family: Hirundinidae		
28	Nepal house marten	*Delichon nipalensis*
Family: Dicruridae		
29	Black drongo	*Dicrurus adsimilis*
30	Ashy drongo	*Dicrurus leucophaeus*
Family: Sturnidae		
31	Common myna	*Acridotheres tristis*
32	Jungle myna	*Acridotheres fuscus*

(*Continued*)

Table 14.7 (Continued)

	Common name	Scientific name
Family: Corvidae		
33	Jay	*Garrulus glandarius*
34	Black-throated jay	*Garrulus lanceolatus*
35	Red-billed blue magpie	*Cissa erythrorhyncha*
36	Himalayan tree pie	*Dendrocitta formosae*
37	House crow	*Corvus splendens*
38	Jungle crow	*Corvus macrorhynchos*
Family: Campephagidae		
39	Pied flycatcher shrike	*Hemipus picatus*
40	Smaller grey cuckoo-shrike	*Coracina melaschistos*
41	Long-tailed minivet	*Pericrocotus ethgus*
Family: Pycnonotidae		
42	Red-vented bulbul	*Pycnonotus cafer*
43	Black bulbul	*Hypsipetes medagascariensis*
Family: Muscicapidae		
Sub Family: Timaliinae		
44	White-throated laughing thrush	*Getrulsx albogularis*
45	Striated laughing thrush	*Garrolax striatus*
46	Streaked laughing thrush	*Garrolax lineatus*
47	Redheaded laughing thrush	*Garrolax erythrocephalus*
48	Red-winged shrike babbler	*Pleruthius flaviscapis*
49	Yellow-naped yuhina	*Yuhina flavicollis*
50	Black-capped sibia	*Heterophasia capistrata*
Sub family: Muscicapinae		
51	Sooty flycatcher	*Muscicapa sibirica*
52	Rufous tailed flycatcher	*Muscicapa ruficauda*
53	Little pied flycatcher	*Muscicapa westermanni*
54	Rufous-bellied Niltava	*Muscicapa sundara*
55	White-browed blue flycatcher	*Muscicapa superciliaris*
56	Verditer flycatcher	*Muscicapa thalassina*
57	Grey-headed flycatcher	*Culicicapa ceylonensis*
Sub family: Sylviinae		
58	Aberrant bush warbler	*Cettia flavolivacea*
59	Plain leaf warbler	*Phylloscopus negiectus*
60	Tickell's leaf warbler	*Phylloscopus affinis*
61	Yellow-browed leaf warbler	*Phylloscopus pulcher*
62	Grey-headed flycatcher	*Seicercus xanthoschistos*
Sub family: Turdinae		
63	Orange-flanked bush robin	*Erithacus cyanurus*
64	Black redstart	*P. ochruros phoenicuroides*
65	Spotted forktail	*Enicurus maculatus*
66	Pied bush chat	*Saxicola capraata*
67	Chestnut-bellied rock thrush	*Monticola rufiventris*
68	Blue rock thrush	*Monticola salitarius*
69	Blue whistling thrush	*Myiophonus caeruleus*
70	Plain-backed mountain thrush	*Zoothera mollissima*
71	Golden mountain thrush	*Zoothera dauma*
72	Tickell's thrush	*Turdus unicolor*
73	Grey-winged black bird	*Turdus boulboul*

(*Continued*)

Table 14.7 (Continued)

	Common name	Scientific name
74	Black-throated thrush	*Turdus ruficollis atrogularis*
75	Mistle thrush	*Turdus viscivorus*
Family: Paridae		
76	Green-backed tit	*Parus monticolus*
77	Crested black tit	*Paras melanolophus*
78	Yellow-cheeked tit	*Parus xanthogenys*
79	Redheaded tit	*Aegithalos coucinnus*
80	White-throated tit	*Aegithalos niveogularis*
Family: Sittidae		
81	Chestnut-bellied nuthatch	*Sitta castanea*
Family: Certhiidae		
82	Himalayan tree creeper	*Certhia himalayana*
Family: Motacillidae		
83	Grey wagtail	*Motacilla cinerea*
Family: Zesteropidae		
84	White eye	*Zosterops palpebrosa*
Family: Ploceidae		
Sub family: Passerinae		
85	House sparrow	*Paser domesticus*
Family: Fringillidae		
Sub family: Fringillinae		
86	Himalayan green finch	*Carduclis spinoides*
Family: Emberizidae		
87	Crested bunting	*Melophus lat*

References

Saxena, A. K., & Singh, J. S. (1982). A phyto-sociological analysis of woody plant species in forest communities of a part of Kumaon Himalayas. *Vegetatio, 50*(3), 22.

Zar, J. H. (1984). *Biostatistical analysis*. Prentice Hall.

15 Insect Diversity in a Semi-Natural Environment

Anam Shakil, Afifullah Khan, and Ayesha Qamar

Introduction

Insects make up about 75% of all the world's known biodiversity and play a very important role in ecosystem functioning. Insects are found in all types of habitats and show a variety in shape, size, and form. According to Gullan and Cranston (2014), they pollinate more than half a quarter of a million species of flowering plants and even provide us with various materials such as silk, honey, wax, dye, etc. While working as decomposers they remove dead and decayed waste materials from our earth. India documents nearly 7% of global faunal diversity, and insects here comprise 6.13% of the total species recorded.

Key ecosystem functions that insects fulfil relate to interaction with vegetation. This includes various types of herbivorous links but also many mutuality relationships like pollination, seed dispersal, or predator defence in exchange for shelter (Qin & Wang, 2001). Plants provide the key habitat parameters for many insect species ranging from shelter to breeding sites. Plant-insect interactions have direct effects, for example on the storage and cycling of carbon and nutrients, as well as strongly influencing succession and competition patterns in plant communities and food web interactions (Swank et al., 1981; Weisser & Siemann, 2004)

Some of the taxa are even used as bio-indicators (Kati et al., 2004; Choi, 2006, Zou Yi et al., 2011). According to Chen et al., butterflies and grasshoppers have been successfully used as bio-indicators for environmental pollution and heavy metal contamination near industrial areas and even urban areas, as they are sensitive to environmental changes and quality. Butterfly population dynamics have been suggested as indicators of species richness for pollinators overall and of the structural and floristic diversity of habitats, as indicators of climate change and further ecological parameters, and of landscape distinctiveness (Pyle, 1976; Heath, 1981; Pe'er & Settele, 2008; and Zou Yi et al., 2011). Ground beetles are also commonly used as bio-indicators for changes in environmental conditions due to their sensitivity to habitat change and because carabid studies are highly cost efficient (Rainio & Niemelä, 2003). Insects are vital ecosystem components. Many of the key ecosystem functions that insects fulfil relate to interactions with vegetation (Zou Yi et al., 2011). Plant-insect

DOI: 10.4324/9781003321422-17

interactions have direct effects, for example on the storage and cycling of carbon and nutrients, as well as strongly influencing succession and competition patterns in plant communities and food web interactions (Swank et al., 1981; Weisser & Siemann, 2004).

Study Area

Aligarh district lies in the saline-alkaline belt of Uttar Pradesh. The region falls in upper Gangetic Plains of the Indo Gangetic Plains of India and comprises the northernmost part of the Agra division in the upper Ganga-Yamuna doab. The area of the district spread between 27°29' and 28°11' north altitude and 77°29' and 78°38' east longitude (Figure 15.1). The relative humidity during the winter (November to February) ranges from 63–71% at 8.30 a.m. and 28–45% at 5.30 p.m. In summer it varies between 30–37% at 8.30 a.m. and 16–21% at 5.30 p.m. in the month of May. During monsoon the relative humidity increases up to 87% at 8.30 a.m. and 78% at 5.30 p.m. in the month of August.

Aligarh is classified as the babul Savanna of Saline/alkaline scrub savanna type of the region. The region has been placed in the sub-group Northern Tropical Thorn Forest of which Acacias and their allies are characteristic plant species. The dominant tree species vary in height from 4.5 m to 10 m and tend to occur in clumps leaving bare ground in between. Aligarh comprises usar plains vegetation (Champion & Seth, 1968), but since the district is well drained by the Ganga canal, it is not as dry as its contemporary areas in the desert. Earlier in Aligarh there were natural patches of dhak (*Butea monosperma*) and Jhan (*Tamrix* sp.) forest, but due to the conversion of forest land into agriculture land these forest patches have now disappeared from the district.

Families

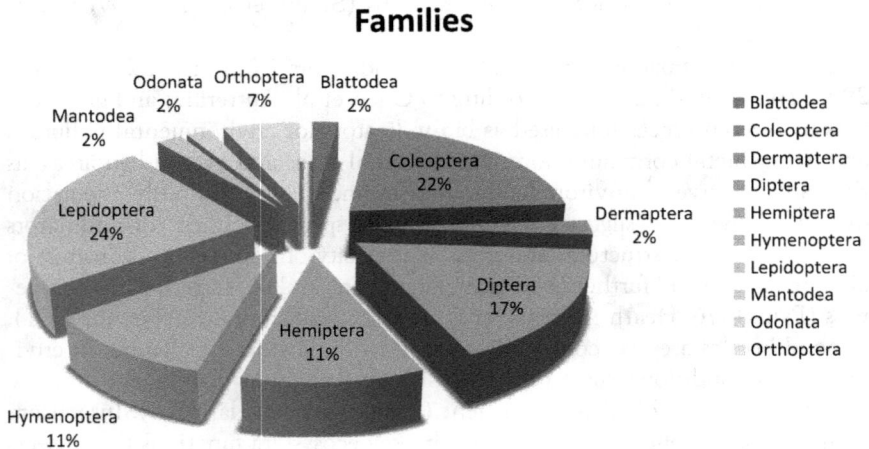

Figure 15.1 Map of the study area.

Methodology

The study was carried out between March and April. Four main sampling methods were employed in the collection of insects. Quadrate method was used for ground dwelling insects (Zaller et al., 2015). A quadrate was made using a plastic strip and by marking off each side of a square of 0.25 x 0.25 m. The quadrates were laid randomly in the study area at a distance of about 100 m and for a time period of ten minutes. The quadrates were laid one after the other with the gap of ten minutes and each of them were searched manually in each quadrate. Data on insects and numbers of individuals were collected The number of quadrates laid varied according to the diversity of insects occurring. Insects were captured daily by using the method during the whole survey.

Sweep net method was used for flying and fast moving insects (Malaise, 1937; Mazon & Bordera, 2008; Aguiar & Santos, 2010). In this method, samples were taken by just walking slowly and passing the open end of the linen mesh net (Yi et al., 2012) randomly through the vegetation (Laroca & Orth, 2002). The net is moved back and forth while moving and insects were captured by pressing the open end of the net to the ground at an angle of about 180°. This method was used to collect insects every day, but the time period for carrying out the method varied according to the species sampled during the field.

Light traps were used to sample nocturnal insects like moths and mosquitoes which are active during the night (Szentkiralyi, 2002). A light trap is constructed by using a 100 watt fluorescent bulb placed on the bulb hanger in front of a white sheet of size of about 2 x 2 m. The trap was placed in the evening and was monitored the next morning before sunrise. The insects sitting on the sheet were sprayed with the Hit insect killer and were collected in the sample box. Each trap was laid after three days from the previous date, and insects captured were placed in a separate jar with label of the trap date.

The analysis of the data included the determination of alpha-diversity, by applying various statistical methodologies.

In quadrate method, α diversity of insects was estimated by means of the Shannon-Wiener diversity index (H') while species richness was determined by using Margalef's species richness method.

Shannon-Weiner diversity index

$$H' = -\sum pi \ln pi$$

where pi is the proportion of ith species occurring in sample

Margalef's species richness formula

$$M = \frac{(S-1)}{\log N}$$

where S = total number of species present in an area, and N = total number of individuals.

And density is calculated by formula

$$D = \frac{\text{Number of individuals occuring in a quadrate}}{\text{Area of the quadrate}}$$

Density, diversity, and richness were calculated in each quadrate. Habitat-wise grouping of densities, diversities, and richness was calculated, and Kruskal Wallis One Way Anova (H) was used to determine any significant differences among the various habitats. $H = \left\{ \dfrac{12}{N(N+1)} * \sum \dfrac{(Rj)^2}{N} \right\} - 3(N+1)$

In the sweep net method, density is calculated by using the density formula

$$D = \frac{\text{Number of individuals in one sweep}}{\text{Area of the sweep net}}$$

The area of the sweep net was calculated by taking out the radius from the circumference of the open end. The graph is being prepared based on the densities of the insects captured during the method. Generally, in most of the sweeps only one species occurred so diversity and richness values couldn't be calculated.

mbu

Another method used in the present study to analyse the ground dwelling animals is the pitfall trapping method. From the data obtained by the method, relative abundance of each species is calculated by using the formula

$$\text{Relative abundance} = \frac{\text{sum of ith species individuals}}{\text{Total sum of all the species individuals}}$$

The relative abundance of the species was then used to relate the abundance of species found in each trap and a graph was made from data obtained.

Results and Discussion

A total of 80 species were recorded, representing 10 orders and 47 families (Figure 15.2). The overall density of insects in various habitats was found not significantly different and proving the null hypothesis of no difference in densities of insects in all four habitats (Table 15.1). Density per habitat ranges from 56.09–64.01 individuals/unit.

The highest diversity of insects was found in herbaceous and plantation habitats followed by grassy patches and least is found in shrubs. The highest richness of insects was found in herbaceous habitats followed by grassy patches and

Families

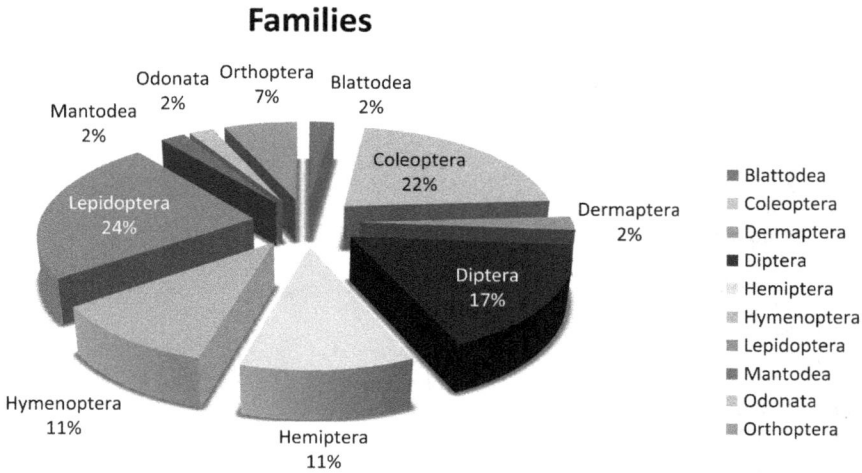

Mantodea 2%
Odonata 2%
Orthoptera 7%
Blattodea 2%
Coleoptera 22%
Lepidoptera 24%
Dermaptera 2%
Diptera 17%
Hymenoptera 11%
Hemiptera 11%

- Blattodea
- Coleoptera
- Dermaptera
- Diptera
- Hemiptera
- Hymenoptera
- Lepidoptera
- Mantodea
- Odonata
- Orthoptera

Figure 15.2 Percentage distribution of all the orders found during the study.

Table 15.1 Richness, diversity, and density of insects in various habitat types in the study area (quadrate method)

Habitat category	Richness	Diversity	Density (unit)	Relative abundance
Grassy patches	0.46	0.21	56.09	0.17
Herbaceous	0.56	0.22	64.51	0.43
Plantation	0.46	0.22	62.34	0.12
Shrubs	0.44	0.20	62.40	0.27

plantation and least in shrubs (Table 15.2). The highest relative abundance of insects was found in the herbaceous vegetation (0.43) while lowest abundance was recorded in the plantations (0.12). Among the insect species, the most abundant species is Formica species, showing its presence in nearly all quadrates.

The mean density of all the species after calculation comes out to be 179.59 with standard error of ± 58.45 in the sweep net method. Musca domestica comes out to be the most abundant among all, with a density of 2828.5714, followed by Anopheles with a density of 1614.2857. Among the butterfly species *Pieris brassica* (814.2857) followed by *Catocrysopes strabo* (342.8571) show the highest densities (Table 15.3).

In the light trap method Anopheles shows the highest relative abundance (0.2255) (Table 15.3).

There is no scientific information on the insect's diversity of the study area, so the present study provides the preliminary information on the common insects of the study area. The study was carried out over a short period of time. So some species may have been missed out. The selection of the sampling sites within the

Table 15.2 Densities of insects captured in sweep net

Species	Density
Anopheles	1614.2857
Apis dorsata	157.1428
Apis florae	28.5714
Canton simplex	85.7142
Catochrysopes strabo	342.8571
Catopsila crocale	100
Chilades laius	14.2857
Chilades trochylus	128.5714
Chinavia hilaris	142.8571
Coccinella septempuctata	257.1428
Coccinella transversalis	242.8571
Danaus chrysippus	871.4285
Danaus genutia	71.4285
Eurema hecabe	42.8571
Gastrophysa cyanea	14.2857
Graphium doson	71.4285
Graphium sarpedon	14.2857
Ixais pyrene	57.1428
Ixias marianne	71.4285
Junonia almanac	42.8571
Junonia hierta	14.2587
Junonia orithya	14.2587
Lucilia sericata	185.7142
Musca domestica	2828.5714
Musca nebulo	42.8571
Muscina prolapsa	42.8571
Mycalensis perseus	71.4285
Orussus terminalis	28.5714
Pachliopta aristolochiae	42.8571
Papilio demoleus	42.8571
Pieris brassica	814.2857
Pieris canidia	42.8571
Plecia amplipennis	114.2587
Plecia neoartica	28.5714
Prodasineura verticalis	28.5714
Sarcophaga nodosa	14.2857
Tirumala limniace	57.1428
Zygogramma	42.8571
Utethesia pulchella	14.2857
Xylocopa violacea	28.5714
Ypthima huebneri	257.142
Ypthima incia	14.2857
Trichogramma species	42.8571

campus was done randomly by the authors, as no established information was available from the AMU campus on the type of habitat used by the insects.

Different methods had given different values of the analysed data. Several studies compared the efficiency of the methods assessing ground dwelling arthropods (Henderson, 2003; Schmidt et al., 2006; Sabu et al., 2011; Corti et al., 2013). In

Table 15.3 Species and their relative abundance in light trap method

Name of species	Relative abundance
Anopheles	0.2255
Archerontia styx	0.0044
Coccinella transversalis	0.0014
Creatonotus gangis	0.0028
Eudocima maternal	0.0018
Formica species	0.0063
Grammodes geometrica	0.0031
Gryllus assimilis	0.0009
Musca domestica	0.0044
Scirpophaga novella	0.0037
Spodoptera litura	0.0047
Thysanoplusia orichalcea	0.004
Uthethesia pulchella	0.0037

general, the habitat of the study area is the main variable that affects the diversity and abundance of insects present (Missa et al., 2009). But in the quadrate analysis for the ground dwelling insects in our study, habitat structure doesn't show a significant difference in the densities of the insects sampled. However, they show a big difference in their abundance. According to the result, the most common species occurring in the quadrate survey is ant belongs to the Hymenoptera order. Ants have shown their presence in nearly all the quadrates sampled making them the most abundant species present on the ground. However according to the diversity estimates, coleopterans are the most diversified, showing their presence in nine families during the survey. The next common genera are the Hemiptera, having species such as bugs and aphids, showing their presence in five families, revealing that species of this order are also present in abundance on the ground. Using a sweep net is a very common sampling method (Arun & Vijayan, 2004), as they are easy to use, cost efficient, and light (Buffington & Redak, 1998; Southwood & Henderson, 2000; Yi et al., 2012). The sweep net methods results show that the most abundant species is the Musca genus, *Musca domestica*, which shows that it is the most abundant species in all the campus and that the habitat is preferable to the species. Likewise, *Musca nebulo*, *Muscina prolapse*, and other flies also show their presence in high numbers. Along with flies, the mosquito species Anopheles also shows its presence. The presence of mosquitoes reveals that the area does not have a clean environment. Butterfly *Pieris brassicae* shows its presence in large numbers. The presence of these species in high numbers shows that these are the species which are common in the study area and were abundant at the time when the study took place. Some butterfly species like *Chilades laius*, *Junonia hierta*, and *J. orithya* are present in somewhat smaller numbers and don't show their presence in most of the areas sampled, showing that they are the least abundant butterfly species in the campus. This could be due to it being an unfavourable habitat or an unfavourable time period of their capturing. From this study, the most abundant

Table 15.4 Orders, families, and number of species of total insects captured during the study

Order	Family	Species identified
Blattodea	Blattidae	*Periplanata americana*
Coleoptera	Buprestidae	*Agrilus acutus*
	Carabidae	*Proterostichus* species
	Chrysomelidae	*Altica chalybea*
		Gastrophysa cyanea
		Zygogramma species
	Coccinellidae	*Coccinella septumpunctata*
		Coccinella transversalis
		Epilachna vigintiopunctata
	Curculionidae	*Sitophilus oryzae*
	Geotrupidae	*Geotrupus stercorarius*
	Hydrophilladae	*Hydrophilla* species
	Pyrrhochoridae	*Dysdercus cingulatus*
	Scarabaeidae	*Scarabaeus* species
	Tenebrionidae	*Gonocephalum civicum*
Dermaptera	Forficulidae	*Forficula auricularia*
Diptera	Bibionidae	*Plecia nearctica*
	Calliphoridae	*Calliphora vomitoria*
		Lucilia seicata
	Culicidae	*Anopheles stephensi*
	Dolichopodidae	*Chrysosoma* species
	Muscidae	*Musca domestica nebulo*
		Muscina prolapsa
	Sarchophagidae	*Sarchophaga nodosa*
	Tipulidae	*Tipula* species
	Stratiomyidae	*Hermetia illucens*
Hemiptera	Coccidae	*Pulvinaria indica*
	Pentatomidae	*Chinavia hilaris*
	Scutelleridae	*Chrysocoris stolii*
	Lygaeidae	*Lygaeus equestris*
	Aphididae	*Aphis gossypii*
Hymenoptera	Apidae	*Apis dorsata*
		Apis cerana
		Xylocopa violacea
	Formicidae	*Formica ligniperda*
		Lasius niger
	Orussidae	*Orussus terminalis*
	Trichogrammatidae	*Trichogramma* species
	Torymidae	*Torymus dalman* species
Lepidoptera	Arctiidae	*Grammia parthenice*
	Crambidae	*Scirpophaga nivella*
	Erebidae	*Creatonotus gangis*
		Eudocima phalonia
		Eudocima hecabe
		Grammodes geometrica
		Uthesia pulchella
	Lasiocampidae	*Trabala vishnou*

(Continued)

Table 15.4 (Continued)

Order	Family	Species identified
	Lycaenidae	*Catochrysops strabo*
		Chilades trochylus
		Chilades 8aius
	Noctuidae	*Spodoptera litura*
		Thysanoplusia orichalcea
	Nymphalidae	*Danaus chrysippus*
		Danauas genutia
		Jumonia almanac
		Jumonia hierta
		Jumonia orithya
		Mycalesis perseus
		Tirumala limniace
		Ypthima heubneri
		Ypthima incia
	Papilionidae	*Graphium serpedon*
		Graphium dorson
		Pachliopta aristolochiae
		Papilio demolus
		Papilio polytes
	Pieridae	*Catopsilia pomana*
		Catopsilia crocale
		Ixias Marianne
		Ixias pyrene
		Pieris brassicae
		Pieris canidia
	Pyralidae	*Galleria mellonella*
	Sphingidae	*Archerontia styx*
Mantodea	Mantidae	*Amantis indica*
Odonata	Protoneuridae	*Prodasineura verticallis*
Orthoptera	Acrididae	*Acrida exaltata*
	Gryllidae	*Gryllus assimils*
	Pyrgomorphidae	*Chrotogonus trachypterus*

order is shown to be Coleopteran, so beetles are the most abundant insects present on the ground. This method could be used to capture the beetles easily, as they may fall down while walking on the ground in the pits, and the agent present inside beetles preserves the insects for future studies. The data reveals that among the beetles or Coleopterans, the Coccinellidae family is the most diverse among the ground dwellers, as four different species of the ground-dwelling coccinellid beetles were captured with this method. Coleopteran, Diptera, and Hemiptera, having bugs, and Heteropteran, having ants, also show their presence with 2 families each. The simple light trap gave good results for this study, because even this simple light trap could work to capture nocturnal insects. In this study, four different families of Lepidoptera were found, which were only moth species found. Among the lepidopterans, moths are the nocturnal species while butterflies are the diurnal species. The most abundant family was Noctuidae, showing that this family has the most common family members or that the area selected for the study is favourable habitat or favourable environment of Noctuidae. Diptera, Coleopterans, Hymenopterns, and Orthoptera also show their presence by having some nocturnal families. From the data, it was deducted that mostly Lepidopterans are dominant among the nocturnal insects while other insects are generally diurnals, having few families which are mainly nocturnal. Most moth species of the Lepidopterans are nocturnal. Climate change is predicted to be a key factor affecting future developments in biodiversity (Zou Yi et al., 2011; Hawkins et al., 2003). Insects species richness and species composition are known to be particularly strongly affected by environmental factors such as temperature and moisture (Brehm et al., 2003; Axmacher et al., 2009). On the basis of this study, we suggest long term investigation of the insects in and around AMU campus, also to see the impact of climate change on insect diversity.

References

Aguiar, A. P., & Santos, B. F. (2010). Discovery of potent, unsuspected sampling disparities for Malaise and Möricke traps, as shown for Neotropical Cryptini (Hymenoptera, Ichneumonidae). *Journal of Insect Conservation, 14*(2), 199–206.

Arun, P. R., & Vijayan, V. S. (2004). Patterns in abundance and seasonality of insects in the Siruvani Forest of Western Ghats, Nilgiri hills biosphere reserve, South India. *Scientific World Journal, 4*, 381–392.

Axmacher, J. C., Brehm, G., & Hemp, A. (2009). Determinants of diversity in afrotropical herbivorous insects (Lepidoptera: Geometridae): Plant diversity, vegetation structure or abiotic factors? *Journal of Biogeography, 36*(2), 337–349.

Brehm, G., & Fiedler, K. (2003). Faunal composition of geometrid moths changes with altitude in an Andean montane rain forest. *Journal of Biogeography, 30*(3), 431–440.

Buffington, M. L., & Redak, R. A. (1998). A comparision of vaccume sampling versus sweep netting for arthropods biodiversity measurement in California coastal sage scrub. *Journal of Insect Conservation, 11*, 399–408.

Champion, H. G., & Seth, S. K. (1968). *A revised survey of the forest types of India.* Government of India Publication.

Choi, S. W. (2006). Patterns of species description and species richness of geometrid moths (Lepidoptera: Geometridae) on the Korean Peninsula. *Zoological Science, 23*(2), 155–160.

Corti, R., Larned, S., & Datry, T. (2013). A comparison of pitfall trap and quadrate method for sampling ground dwelling invertebrates in dry riverbeds. *Hydrobiologia.* DOI 10.1007/s10750-013-1563-0

Gullan, P. J., & Cranston, P. (2014). *The insects: An outline of entomology.* (5th ed.). Wiley.

Hawkins, B. A., Field, R., Cornell, H., Hawkins, B. A., Currie, D. J., Guegan, J., Kaufman, D. M., Kerr, J., Mittelbach, G. G., Oberdorff, T., O'Brien, E. M., Porter, E. E., & Turner, J. R. G. (2003). Energy, water, and broad-scale geographic patterns of species richness. *Ecology, 84*(12), 3105–3117.

Heath, J. (1981). *Threatened Rhopalocera (butterflies) in Europe* (Vol. 23). Council of Europe.

Henderson, P. A. (2003). *Practical methods in ecology* (1st ed.). Blackwell Publishing.

Kati, V., Devillers, P., Dufrene, M., Legakis, A., Vokou, D., & Lebrun, P. (2004). Testing the value of six taxonomic groups as biodiversity indicators at a local scale. *Conservation Biology, 18*(3), 667–675.

Laroca, S., & Orth, A. (2002). *Melissocoenology: Historical perspective, method of sampling and recommendations to the "program of conversation and sustainable use of pollinators, with emphasis on Bees" (ONU)*. Pollinating Bees; the conversation link between Agriculture and Nature (pp. 217–225). Ministry of Environment.

Malaise, R. (1937). A new insect trap. *Entomologisk Tidskrift, 58,* 148–116.

Mazon, M., & Bordera, S. (2008). Effectiveness of two sampling methods used for collecting Ichneumonidae (Hymenoptera) I. Cabaneros national park (Spain). *European Journal of Entomology, 105*(5), 879–888.

Missa, O., Basset, Y., Alonso, A., Miller, S. E., Curletti, G., & De Meyer, M. (2009). Monitoring arthropods in a tropical landscape: Relative effects of sampling methods and habitat types on trap catches. *Journal of Insect Conservation, 13*(1), 103–118. https://doi.org/10.1007/s10841-007-9130-5.

Pe'er, G., & Settele, J. (2008). The rare butterfly *Tomares Nesimachus* (Lycaenidae) as a bioindicator for pollination services and ecosystem functioning in Northern Israel. *Israel Journal of Ecology and Evolution, 54*(1), 111–113.

Pyle, R. M. (1976). The eco-geographic basis for Lepidoptera conversation. Ph.D Thesis. Yele University, New Heaven. In R. Pyle, M. Bentzin, & P. Opler (Eds.). (1981) Insect conversation, *Annual Review of Entomolgy, 1926,* 1233–1258.

Qin, J. D., & Wang, C. Z. (2001). The relation of interaction between insects and plants to evolution. *Acta Ecological Sinica, 44*(3), 360–365. (In chinese).

Rainio, J., & Neimela, J. (2003). Ground beetles (Coleoptera: Carabidae) as bio-indicators. *Biodiversity and Conversation, 12*(3), 487–506.

Sabu, T. K., Shiju, R. T., & Vinod, K. (2011). A comparision of the pitfall trap, Winkler extractor and Berlese funnel for sampling ground dwelling arthropods in tropical montane. *Journal of Insect Science, 11*(28), 28.

Schmidt, M. H., Clough, Y., Schulz, W., Westphalen, A., & Tscharntke, T. (2006). Capture efficiency and preservation attributes of different fluids in pitfall traps. *Journal of Arachnology, 34*(1), 159–162.

Southwood, T. R. E., & Henderson, P. A. (2000). *Ecological methods* (3rd ed.). Blackwell Science.

Swank, W. T., Waide, J. B., Crossley, D. A., & Todd, R. L. (1981). Insects defoliation enhances nitrate export from forest ecosystems. *Oecologia, 51*(3), 297–299.

Szentkiralyi, F. (2002). Fifty year long insect survey in hungry: T. Jermys Contribution to Light Trapping. *Acta Zoologica Acadamiae Scientarum Huneraicae, 48,* 85–105.

Weisser, W. W., & Seimann, E. (2004). *Insects and ecosystem function.* Springer.

Yi, Z., Jinchao, F., Dayuan, X., Weiguo, S., & Axmacher, J. C. (2012). A comparison of terrestrial arthopod sampling methods. *Journal of Resources and Ecology, 2*(3), 174–182.

Zaller, J. G., Kerschbaumer, G., Rizzoli, R., Tiefenbacher, A., Gruber, E., & Sched, H. (2015). Monitoring arthropods in protected grasslands: Comparing pitfall trapping, quadrat sampling and video monitoring. *Web Ecology, 15*(1), 15–23

Zou Yi, J. F., Dayuan, X., & Sang, W. (2011). Insect diversity, addressing an important but strongly neglected research topic in China. *Journal of Resources and Ecology, 2*(4), 380–384.

16 Dipteran Flies Associated with Abuse and Neglect

A Forensic Indicator

Swaima Sharif, Tabassum Choudhary, Munawwar Husain, and Ayesha Qamar

Introduction

Forensic entomology is an important determinant of the abuse and neglect of elderly people, children, or animals (Benecke & Lessig, 2001). Insect evidence may represent how long a person or an animal was abused or neglected. In its classic form, forensic entomology is based on the timetable of arrival of the different species of Sarco-saprophagous insects, i.e., in the sequence in which they make their appearance in the carrion and on the recognition of the developmental stage of offspring of different species. In principle, this kind of evidence can be provided by every species of Sarco-saprophagous insects, but nevertheless, forensic applications are to be found mainly in the life histories of flies. Insects are one of the most diverse forms of life on earth and they can be found almost everywhere, viz. in soil, water, ground, concealed spaces, debris, faeces, garbage, among others. The most important insects are Sarco-saprophagous dipteran flies, whose larvae feed on the dead or decaying flesh of a corpse, as well as on wounds and bedraggled natural orifices of a living person (Beneckea, 2001). Dipterans can be found almost anywhere, and some species of them are human commensals all over the world. Regarding medical and veterinary aspects, flies are well adapted for picking up pathogens and transmitting them to humans and other animals, infesting live vertebrates with their larval stages and causing annoyances. They have medical, veterinary, agricultural, economic, social, and forensic importance (Akbarzadeh et al., 2017, 2018).

Maggots crawling on a dead body are considered a disgusting element that need to be washed off on the autopsy table. When a body is infested with insects, they must be collected in congruency with other biological evidence. These creatures can prove to be an important parameter to many unanswered questions about a case. This field is not only confined to death investigations but also includes wider aspects like child or elderly neglect cases, badly stored food products (Defilippo et al., 2017), poor hygienic conditions of restaurants, and wildlife poaching cases (Anderson, 1999). Flies have attracted the attention of entomologists because of their importance as transmitters of disease and as causes of livestock myiasis (Greenberg, 1965) and human myiasis (Greenberg, 1965).

DOI: 10.4324/9781003321422-18

Human myiasis may be the result of self-negligence in maintaining proper hygienic conditions of the body or the deliberate negligence by the caregiver of any child or maimed, paralyzed, or debilitated elderly person (Amendt, 2011). The odour, wound, any injury to the tissue, or injured surroundings attract different flies to oviposit within minutes. The eggs hatch, live, feed, and complete their life cycle in these conditions (Byrd & Castner, 2001). Once the life cycle starts repeating itself, a fresh army of insects is produced each time, thereby causing tremendous havoc (Beneckea et al., 2009). In some cases, the victim may develop bed sores and gangrene, which also increases the chances of flies being attracted and ovipositing. The eggs hatch into these places, and the larvae start feeding on the flesh of the body (Iqbal et al., 2011). If proper treatment is not given on time, death may occur.

Insect evidence can also be used to determine in what circumstances a death has occurred. Entomological shreds of evidence not only involve carcasses but also living victims. In the cases of abuse, neglect, and rape, blowflies are worthwhile forensic indicators. In the cases of sexual assault preceding death, blowflies are likely to oviposit in the injured or vaginal region. As a result, entomological evidence appears to be a valuable tool for assisting investigators in providing important clues and indications ante-mortem and post-mortem (Bonacci et al., 2017).

In cases where the body of an old person or animal that is incapacitated or has any impairments is found to be infested with maggots, the victim or his/her proxy need not produce testimony from any expert to establish the cause or effect of the neglect, as it will be proven by *res ipsa loquitor* ("the thing or fact speaks for itself").

Forensic Indicators of Neglect

Some insects, for example, *Chrysomya megacephala* (oriental latrine fly), are attracted to odours such as ammonia emanating from faecal and urine contamination. The incidence of this condition will be in cases of individuals or animals who lack voluntary control or are bound, helpless, and cannot maintain their routine hygiene. The author's collections of forensically significant and widely distributed dipteran associated with such cases include the following:

Chrysomya megacephala (Diptera: Calliphoridae): Oriental latrine fly
Chrysomya rufifacies (Diptera: Calliphoridae): Hairy maggot blow fly
Calliphora vicina (Diptera: Calliphoridae): Bluebottle fly
Cochliomya hominivorax (Diptera: Calliphoridae): Screw worm fly
Synthesiomya nudiseta (Muscidae Diptera)
Megaselia scalaris (Diptera: Phoridae): Humpbacked fly
Sarcophaga dux (Diptera: Sarcophagidae): Flesh fly

The life cycles and images of these insects have been depicted in Table 16.1 and Figure 16.1, respectively.

Table 16.1 Life cycle of some of the Dipteran associated with neglect*

Diptera	Family	Egg (hours)	First instar (hours)	Second instar (hours)	Third instar (hours)	Pupa (hours)	Temperature; Relative humidity %
Calliphora vicina (Meenakshi, 2009)	Calliphoridae	23	34	22	40	216	28 °C 75%
Chrysomya megacephala (Siddiki & Zambare, 2017)	Calliphoridae	18.37	25.45	26.35	29	22.40	24.1 °C 49.6%
Chrysomya ruffifacies (Siddiki & Zambare, 2017)	Calliphoridae	22.38	25.06	27.35	51.05	41.50	24 °C 42.1%
Cochliomya hominivorax (Laake et al., 1936)	Calliphoride	12–20	24	24	96–120	168–192	28 °C 75%
Synthesiomyia nudiseta (Kruger et al., 2002)	Muscidae	18–22	23–25	42–48	48	228–240	25 °C 75%
Megaselia scalaris (Chakraborty et al., 2016)	Phoridae	10	48	60	118	132	27 °C 35%
Sarcophaga dux (Babu et al., 2017)	Sarcophagidae	-	48	48	72	384	24 °C 65%

* Note: Life cycle of the flies depends on the ambient temperature and season

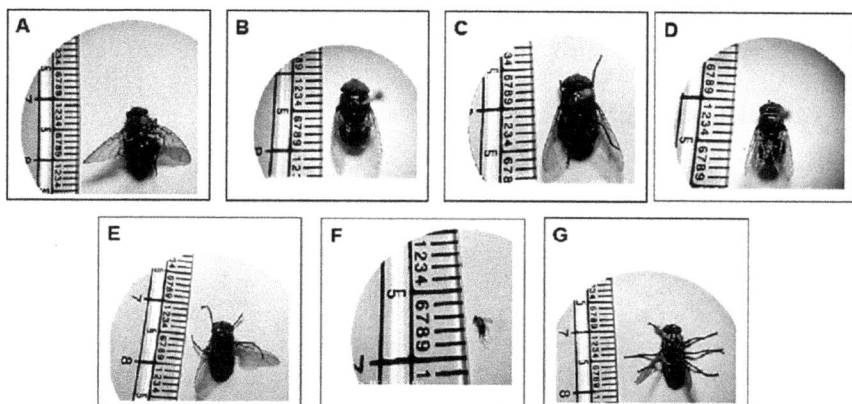

Figure 16.1 Images of some of the Dipteran associated with neglect.

The Doctrine of Res Ipsa Loquitor

The term literally means "the thing or fact speaks for itself."

1) As per the rules of the court, if the victim alleges that the caregiver was negligent, the onus is on the victim to prove that the caregiver was negligent, and it is not for the caregiver to prove that he was not negligent.
2) However, in certain cases, the negligence is so obvious that the victim does not have to prove anything and the fact speaks for itself (Doctrine of Res Ipsa Loquitor). Some obvious examples are:
   a. the presence of maggots in and around clothing, open wounds, ulcers, bed sores, or natural orifices of people who are unable to control themselves.
   b. when swabs, scissors, artery forceps, or other surgical instruments that are left in the abdomen following an operation are clearly visible on an X-ray.

Indian Law Contains Legal Provisions for the Elderly

The Maintenance and Welfare of Parents and Senior Citizens Act 2007
 The Act was enacted "to provide for more effective provisions for the maintenance and welfare of parents and senior citizens, guaranteed and recognized under the Constitution and for matters connected therewith or incidental thereto."
 The Code of Criminal Procedure, 1973

 Section 125. Order for maintenance of wives, children, and parents.

 1) If anyone with sufficient means neglects or refuses to maintain

A first-class Magistrate may, upon evidence of such neglect or refusal, order such person to make a monthly allowance for the maintenance of their father or mother, at such monthly rate not exceeding five hundred rupees in total, as such Magistrate thinks fit, and to pay the same to such person as the Magistrate may from time to time direct.

2)   Such an allowance is payable beginning on the date of the order or, if so ordered, beginning on the date of the application for maintenance.

3)   If any person so ordered fails without sufficient cause to comply with the order, any such magistrate may issue a warrant for levying the amount due in the manner provided for levying fines and may sentence such person, for the whole or any part of each month's allowances remaining unpaid after the execution of the warrant, to imprisonment for a term which may extend to one month or until payment is made; Provided that no warrant shall be issued for the recovery of any amount due under this section unless application be made to the Court to levy such amount within a period of one year from the date on which it became due.

Creatures

The Prevention of Cruelty to Animals Act, 1960

This is an Act to prevent the infliction of unnecessary pain or suffering on animals and to amend the law relating to the prevention of cruelty to animals for that purpose. Animals also have the ability to understand physical and emotional suffering, similar to human beings. They can also perceive the extent of any physical or emotional damage inflicted on them. This earth belongs to them as much as it belongs to us human beings. Thus, other living creatures must be allowed to live and thrive just like us.

With this aim, the Prevention of Cruelty to Animals Act, 1960 was passed to severely penalise those indulging in cruelty against animals by establishing an animal welfare board, identifying the acts amounting to cruelty against the animals, etc.

Salient features of the Prevention of Cruelty to Animals Act, 1960

The Act provides for punishment for causing unnecessary cruelty and suffering to animals.

The Act defines animals and different forms of animals.

This Act enshrines the provisions relating to the establishment of an animal welfare board, its constitution, powers, and functions.

This Act discusses different forms of cruelty, exceptions, and the killing of a suffering animal when any cruelty has been committed against it, so as to relieve it from further suffering.

This Act provides the guidelines relating to experimentation on animals for scientific purposes.

This Act enshrines the provisions relating to the exhibition of performing animals and offences committed against them.

This Act provides for a limitation period of 3 months, beyond which no prosecution shall lie for any offences under this Act.

Suggestive Measures to Prevent Neglect

In humans, myiasis is often the result of neglect and is most commonly found in the very old, the very young, or others who are unable or unwilling to ensure basic hygiene and wound cleanliness (Anderson et al., 2002). Myiasis in domestic animals has been studied more extensively, but continuous management is difficult and expensive. A key concern is the inadvertent introduction and global spread of agents of myiasis into non-endemic areas, facilitated by climate change and global transport.

In pet animals, myiasis can occur when a wound is left untreated or when neglect results in the accumulation of faeces or urine, which then attracts flies. Most cases of myiasis appeared to be the result of an untended wound or matted hair and the presence of excrement, which indicates that owner neglect plays a major role in the occurrence of myiasis. In many cases, this is probably the result of ignorance rather than deliberate neglect or abuse, and owner education may go a long way towards alleviating the problem. As a majority of people find maggots particularly repugnant, it is probable that explaining to the owner how the maggots came to be present on the animal and how such infestations can be prevented in the future may result in more diligence in care and grooming on the part of the owner.

Neglect is the harsh reality and is emerging as a severe problem, and tackling such issues individually is a difficult affair. Hence, the legal system can be a good alternative for handling such cases. At the same time, most prosecutions are time-consuming affairs that cause the victims more discomfort and agony than justice and relief. The problems get aggravated when many of the victims do not complain. One of the first reasons behind such is the further abuse and neglect and shame in the community, as reported by the Help Age India report 2015. The individuals should be made aware of their rights in the legal system so that they could come forward to register complaints against the neglect. The legal system should be made more accessible so that it becomes approachable and convenient. For this, some agents should be hired as "Social Welfare Officers" like in other countries to keep an eye on such homes/agencies and visit them at regular intervals. This will ensure that the disabled and neglected people, who are not in a position to move out, can register complaints regarding their well-being and welfare.

As mentioned (under the Maintenance and Welfare of Parents and Senior Citizens Act 2007 and Provisions), the penalty in the cases of elderly neglect is a monthly allowance of Rs. 10,000 or imprisonment of 3–6 months, or both. The disrespect, the trauma, the pain, the infestation of insects on the body, the eating away of flesh by maggots, the verbal abuse, the humiliation, the embarrassment, the abandonment, and other unaccountable tyrannies in the list only results in a penalty of 3–6 months of imprisonment. These penalties should be increased, and the accused caregiver (the child, relative, or son/daughter-in-law of the injured party) should be dismissed from the property. If the accused caregiver is a nurse or any other person who has been hired for the care of the elderly person, his or her license should be cancelled. Compensation should be given to the victim for suffering all the pain and trauma. A separate and reliable as well as responsible caregiver should be provided by the government/law to look after the elderly, children, and animals. Also, the elderly person should be made aware of the use of the helpline numbers provided by the various agencies by visiting homes or by advertising them on social media and mass media. The government can increase funding to provide training for people who work in aging-related care on the prevention and detection of elder abuse.

It may be concluded that the malfeasance of the animal or person is a serious issue. Forensic entomology can aid in such cases. If complaints get registered, then the presence of maggots in the anal region, wounds, and other uncleaned body parts of the maimed speak of the woe caused by neglect (*res ipsa loquitor*). Intentional neglect can progress to the point where it causes further infection in the individual's body. Thus, this neglect constitutes *actus reus* with associated *mens rea*.

References

Akbarzadeh, K., Mirzakhanlou, A. A., Lotfalizadeh, H., Malekian, A., Hazratian, H., & Talarposhti, K. R. (2017). Natural parasitism associated with species of Sarcophagidae Family of dipteran in Iran. *Annals of Tropical Medicine and Public Health*, *10*(1), 134–137.

Akbarzadeh, K., Saghafipour, A., Jesri, N., Karami-Jooshin, M., Arzamani, K., Hazratian, T., Kordshouli, R. S., & Afshar, A. A. (2018). Spatial distribution of necrophagous flies of infraorder Muscomorpha in Iran using geographical information system. *Journal of Medical Entomolology*, *55*(5), 1071–1085.

Amendt, J., Richards, C. S., Campobasso, C. P., Zehner, R., & Hall, M. J. R. (2011). Forensic entomology: Applications and limitations. *Forensic Science, Medicine and Pathology*, *7*(4), 379–392. https://doi.org/10.1007/s12024-010-9209-2.

Anderson, G. S. (1999). Wildlife forensic entomology: Determining time of death in two illegally killed black bear cubs. *Journal of Forensic Sciences*, *44*(4), 856–859.

Anderson, G. S., & Cervenka, V. J. (2002). Insects associated with the body: Their use and analyses. In W. D. Haglund & M. Sorg (Eds.), *Advances in forensic taphonomy method, theory and archeological perspectives* (pp. 174–200). CRC Press.

Babu, S., Sharma, S., & Upadhyay, S. (2017). Studies on the larval growth of forensically important flesh fly *Sarcophaga dux* Thompson 1869 (Diptera: Sarcophagidae) under outdoor ambient temperatures from central India. *International Journal of Applied Research*, *3*, 366–370.

Benecke, M., & Lessig, R. (2001). Child neglect and forensic entomology. *Forensic Science International*, *120*(1–2), 155–159. https://doi.org/10.1016/S0379-0738(01)00424-8.

Beneckea, M., Josephib, E., & Zweihoffb, R. (2009). Neglect of the elderly: Forensic entomology cases and consideration. *Forensic Science International*, *146*, 195–199. https://doi.org/10.1016/j.forsciint.2004.09.061.

Beneckea, M. A. (2001). Brief history of forensic entomology. *Forensic Science International*, *120*(1–2), 2–14. https://doi.org/10.1016/S0379-0738(01)00409-1.

Bonacci, T., Vercillo, V., & Benecke, M. (2017). A forensic entomological neglect case of an elderly man in Calabria, Southern Italy. *Romanian Journal of Legal Medicine, 25*(3), 283–286. https://doi.org/10.4323/rjlm.2017.283.

Byrd, J. H., & Castner, J. L. (2001). *Forensic entomology—The utility of arthropods in legal investigations* (1st ed.). CRC Press.

Chakraborty, A., Naskar, A., Parui, P., & Banerjee, D. (2016). Developmental variation of Indian thermophilic variety of scuttle fly *Megaselia scalaris* (Loew, 1866) (Diptera: Phoridae) on different substrates. *Scientifica, 4257081.* https://doi.org/10.1155/2016 /4257081

Defilippo, F., Pinna, M., Calzolari, M., Dottori, M., & Bonilauri, P. (2017). *Stored-product forensic entomology: Study of development of Plodia interpunctella (Lepidoptera: Pyralidae) and use of data in determining time intervals of food infestation.* https://doi.org/10.13140/ rg.2.2.24968.52488.

Greenberg, B. (1965). Flies and disease. *Scientific American, 213*(1), 92–99. https://doi.org /10.1038/scientificamerican0765-92.

Iqbal, J., Hira, P. R., & Marzouk, M. M. (2011). Pressure sores and myiasis: Flesh flies (Diptera: Sarcophagidae) complicating a decubitus ulcer. *Annals of Tropical Medicine and Parasitology, 105*(1), 91–94. https://doi.org/10.1179/136485910X12851868780469.

Kruger, R. F., Ribeiro, P. B., Carvalho, C. J. B. de, & Costa, P. R. P. (2002). Development of *Synthesiomyia Nudiseta* (Diptera, Muscidae) in laboratory. *Iheringia Série Zoologia, 92*(4), 25–30. http://doi.org/10.1590/S0073-47212002000400004.

Laake, E. W., Cushing, E. C., & Parish, H. E. (1936). Biology of the primary screw worm fly, *Cochliomyia Americana,* and a comparison of its stages with those of *C. macellaria.* Special Collections, USDA National Agricultural Library. Retrieved January 24, 2020, from http://www.nal.usda.gov/exhibits/speccoll/items/show/7338.

Meenakshi, B. (2009). Studies on life cycles of forensically important flies, calli-Phora vicina and Musca domestica Nebulo at different temperatures. *Journal of Entomological Research, 33*(3), 273–275.

Siddiki, S., & Zambare, S. P. (2017). Studies on time duration of life stages of *Chrysomya megacephala* and *Chrysomya rufifacies* (Diptera: Calliphoridae) during different seasons. *Journal of Forensic Research, 8,* 379. http://doi:10.4172/2157-7145.1000379.

Web References

The maintenance and welfare of parents and senior citizens act 2007. Retrieved November 22, 2019, from http://socialjustice.nic.in/writereaddata/UploadFile/Annexure -X635996104030434742.pdf.

The code of criminal procedure 1973 (Chapter IX, Section 125). Retrieved November 22, 2019, from https://indiacode.nic.in/bitstream/123456789/6796/1/ccp1973.pdf.

An overview of prevention of cruelty to animals act (1960). Retrieved January 16, 2021, from https://blog.ipleaders.in/overview-prevention-cruelty-animals-act-1960/.

Part III

Landscape Ecology and Conservation

17 Conservation of Pheasants in Jammu and Kashmir

A Review

Khursheed Ahmad

Wankukar is the general Kashmiri term used for pheasants. Pheasants have been associated with mankind since time immemorial. Earlier known as game birds, they were hunted for sport as well as kept as pets. Pheasants occur all along the Himalayas and seem to have originated in and around the Himalayas in the remote past, and from here have gradually radiated into Tibet, China, and other parts of South and Southeast Asia. Of the only 51 species of pheasants found worldwide, 50 species are from Asia. Pheasants constitute one of the most gorgeously beautiful, spectacular, and conspicuous family of birds in the world. The pheasants are characterised by conspicuous sexual dimorphism and the plumage of most male pheasants is striking while females are generally subtle. Facial adornments in the form of crests, wattles, ruffs, and hackles are present in the males.

The family *Phasianiadae*, comprising 38 genera, 155 species, and 399 taxa, is the largest and the most diverse family of the order Galliformes distributed throughout much of the old world with greater diversities in Southeast Asia and Africa (del Hoyo et al., 1994; McGowan et al., 1995; Fuller & Garson, 2000). The species have rather short, stout, sometimes strongly spurred tarsi (larger in males), a plump body, short bill, rounded wings, and a short to the very long tail (Johnsgard, 1986; del Hoyo et al., 1994; Fuller & Garson, 2000).

There have been earlier attempts made by taxonomists to classify this group. For example, Westmore (1960) placed pheasants under a separate family *Pheasianidae* under the subfamily *Phasianoidea*. Sibley and Ahlquist (1990) suggested that typical pheasants be included in a separate subfamily *Phasianinae* under the broad family *Phasianidae*. This subfamily was further classified into the tribe *Phasianini* which included pheasants exclusively. Wolters (1975–1982) considered the guinea fowl amongst one of the 15 subfamilies of the family *Phasianidae* and proposed that typical pheasants be divided into eight separate subfamilies. This classification was most widely followed. More recently, Sibley and Monroe (1990) based their classification on DNA-DNA hybridisation and classified pheasants and old-world partridges, quails, and francolins in the family *Phasianidae*.

Out of the 16 genera (Delacour, 1977) and 51 species (69 taxa) of pheasants known to us, 50 species are found in Asia (McGowan & Garson, 1995; Fuller & Garson, 2000) except one species the Congo pheasant (*Afropavao congensis*)

DOI: 10.4324/9781003321422-20

which is native to Zaire in Central Africa (Crowe et al., 1986). Several species of pheasants have been introduced outside their native home in Europe and the USA for sport-hunting (Bump, 1941; Pokorny & Pikula, 1987; Hill & Robertson, 1988a). The globally distributed domestic fowl is believed to have originated from red junglefowl (*Gallus gallus*) (Wood-Ghush, 1959).

In Asia, pheasants are distributed from Indonesia through to north-eastern China and from the Caucasus to Japan. All along their distribution range, pheasants are found in diverse habitats such as lowland tropical forests (mountain peacock pheasant *Polyplectron inopinatum*), temperate coniferous forests (western tragopan *Tragopan melanocephalus*), sub-alpine scrub (blood pheasant *Ithaginius cruentus*), alpine meadows (Chinese monal *Lophophorus lhuysii*), montane grass scrub (cheer pheasant *Catreus wallichii*), and broad-leaved evergreen forests (kalij pheasants *Lophura* spp. and koklass pheasant *Pucrasia macrolopha*).

Out of the total 50 species of pheasants found in Asia, out of which 17 species are present in India. Seven species of pheasants are found in remove the Jammu and Kashmir. These include three threatened species such as the western trago-pan, the rarest of the Himalayan pheasants, the Himalayan monal *Lophophorus impejanus* and the cheer pheasant. Other pheasant species found in Jammu and Kashmir include the koklass pheasant *Pucrasia macrolopha*, kalij pheasant *Lophura leucomelanos*, red junglefowl *Gallus gallus*, and Indian peafowl *Pavo cristatus*.

The western tragopan is one of the rarest pheasants of the world, with a narrow distribution range in the temperate forests of the northwest Himalaya (Rahmani, 2012). Although reported from swat catchment in North-West Frontier Province and Palas Valley of Pakistan, in India it is distributed from the north of Kashmir through Himachal Pradesh and up to west Uttarakhand between 2,400–3,600 m but as low as 2,000 m in winter (Rahmani, 2012). In Jammu and Kashmir, the species is found in significant numbers only in the Limber-Lachipora Wildlife Sanctuary (WLS) and the adjoining Ajas forests and Hirpora Wildlife sanctuary in the Pir Panjal mountainous region of Kashmir, Kishtawar High Altitude National Park, as well as being reported from the Sud Mahadev area of Jammu province. The bird has historically been distributed only along the Pir Panjal Range (which borders the valley with Himachal Pradesh) but not towards the inner Greater Himalayan Range (Rahmani et al., 2014). It has recently been recorded by Riyaz Ahmad in Badanari, Kalamund, and Hingli in Kalamund-Tatakuti, and Godinuk and Pathra in Khara Gali of Poonch along the southern slopes of Pir Panjal (Ahmad et al. 2017)). Abundance estimates available from the limber area only suggest a population of over 45 males in 26 km² of Limber Sanctuary. They appear to be fairly secure in the Limber-Lachipora area of the Kazinag range. The recent call count estimation survey confirmed the presence of western tragopan from Kazinag National Park, Limber, Lachipora, Khara Gali Conservation reserve and Tattakuti wildlife sanctuary (Ahmad et al. 2017). They also seem to enjoy a fairly good status in Kishtawar High Altitude National Park as is evident from various population census figures of the Jammu and Kashmir Wildlife Department (Anonymous, 2003).

The western tragopan occupies limited vegetation types in temperate and sub-alpine forests of western and north-western Himalayas in India and Pakistan. During the Important Bird Area (IBA) process, 12 sites have been reported where this species is known to occur in India. The Great Himalayan National Park in Himachal Pradesh is considered one of the best strongholds of the western tragopan with a reported population of 500 pairs reported (Sanjeeva Pandey, personal communication, 2010). Numbers of surveys by various workers (Pandey, 1993/1994; Narang, 1993; Sharma, 1993; Gaston, 1980, Gaston et al. (1981, 1983); Sangha & Naoroji, 2005; Sharma & Pandey 1989), and detailed studies by the Wildlife Institute of India in 1998 and Department of Wildlife Sciences, AMU, Aligarh in 2000 was carried out on the status and conservation of this species in Himachal Pradesh. However, except some brief site surveys by Knox and Walters (1994), Baker (1921–1930), Kaul (1989a,b), Qadri et al. (1990), Javed (1992), Akhtar et al. (1994), and Ward (1906–1908) in Wular, Lolab Valley, Limber Valley and Kishtwar, Kaul & Garson (1993) very little has been studied of western tragopan in Jammu and Kashmir. Recent new records confirming its occurrences and additional populations in Jammu and Kashmir suggests that further scientific studies are required to assess the current population status and distribution of this threatened species.

Among the three species of monal pheasants found in India, the Himalayan monal is distributed from the west to east of the Himalayan range. The Himalayan monal, or the monal pheasant as it is more commonly known, is among all the Himalayan pheasants the bird with undoubted Himalayan credentials. It is rightly called the Himalayan monal in view of its widespread distribution across the Himalayas from Afghanistan and Pakistan in the west to Arunachal Pradesh in the east within India, besides being found in Tibet, China, Bhutan, Nepal, and Myanmar (Ali & Ripley, 1987). No other pheasant species enjoys such a widespread distribution in the Himalayan region. In Jammu and Kashmir as well, the Himalayan monal is the most widely distributed pheasant. It is commonly known as *Sunal* or *Suna Murg*, which means the golden bird. Owing to the splendidly gorgeous plumage of the male bird, it is also called "the bird of nine colours." It is found at higher reaches at elevations ranging from 2,500 m to 5,000 m, but during winters may be found in the lower reaches as well. The Himalayan monal is common in almost all the protected areas within its distributional range.

In Jammu and Kashmir, the Himalayan monal can be sighted in good numbers in Limber WLS, Hirpora WLS, Dachigam National Pak, Overa-Aru WLS, Shikargah-Khiram Conservation Reserve, and Kishtawar High Altitude NP, besides other territorial and reserve forest areas of the state (author's personal observation). However, no population estimates of the species are available. The main threats were hunting in the past but much of that appears to have been controlled now, although they are still trapped in pit-fall traps.

The cheer pheasant is distributed mainly between 1,350 and 3,500 m in the Western and Central Himalayas, from north Pakistan, through Kashmir and Himachal Pradesh into Garhwal and Kumaon Hills of Uttarakhand in India to Central Nepal. It has, however, also been seen as low as 950 m at Birahi, Chamoli district (Bisht et al., 2007) and as high as 4,545 m in Uttarkashi district (Ghosh, 1997). In Jammu and Kashmir, however, the cheer pheasant has a limited distribution. There are confirmed reports of its presence only from Trikuta area in Jammu province, with one bird sighted and recorded near Mata Vaishno Devi temple (Hillaluddin Ahmad, personal communication, 2013). However, Riyaz Ahmed (personal communication, 2013) reported sighting four birds and call records of ten birds in Kalamund-Tatakuti and four call records from Khara Gali of Poonch along Pir Panjal. In Kashmir its occurrence is only confirmed from Limber/Lachipora WLS, although some records of its presence are also reported from Hirpora WLS (Islam & Rahmani, 2004).

The cheer pheasant is a well-studied species. Since the initial survey by Gaston (1983), extensive surveys and studies have been conducted on the species in Himachal Pradesh and Uttar Pradesh by Garson (1983), Singh et al. (1980), Young et al. (1987), Sathyakumar et al. (1993), Subedi (2003), Kalsi and Kaul (1997), Kalsi (1999), Bisht et al. 2003,. However, no such efforts have been made for surveying the cheer pheasant in Jammu and Kashmir. There are no recent records of its sighting from any area in its range in Jammu and Kashmir (Rahmani et al., 2014). Since this species appears to adapt to disturbed habitats and therefore may not be as threatened as thought to be, the southern slopes of the Pir Panjal mountainous range mainly in and around the areas of Ponch, Rajouri, and Hirpora WLS, as well as chir pine (*Pinus roxiburghii*) habitats along the Jammu-Srinagar National Highway needs to be surveyed to locate more sites.

Among the two species of koklas found in India, the Kashmir koklass *P. m. biddulphi* is distributed from northern Kashmir to Himachal Pradesh, while the common koklass *P. m. macrolopha* is found from southern Kashmir to Garhwal Himalayas. Koklas is also a relatively common pheasant (as in the other parts of the western Himalaya) and is found in most forested areas of the state, particularly in Overa-Aru WLS, Limber WLS, Lachipora WLS, Gulmarg WLS, Hirpora WLS, Shikargah-Khiram Conservation Reserve, Dachigam National Park, and Kishtawar National Park. Old records of the presence of this species also occur from the Boniyar area of the Kashmir valley (Lamba et al., 1982a,b). Population estimates for this pheasant are lacking. The pheasant survey of Kashmir by Lamba et al. in 1982 showed a very good population of Himalayan monal and koklas pheasants in Dachigam National Park (NP).

My personal observations suggest easy sighting of a significant population of Himalayan monal and koklas pheasants in Dachigam N.P., Overa-Aru WLS, Baltal-Thajwas WLS, Sindh Forest Division, and Shikargah Conservation Reserve (CR) in the Zanskar mountainous region of Kashmir, as well as Kishtwar High Altitude NP, Patnitop forests, and Chinab Valley.

Out of the two species of junglefowl, the red junglefowl *Gallus gallus* and grey junglefowl *Gallus sonneratii* are found in India The Indian red junglefowl *G. g. murhgi* is a highly versatile, widely distributed pheasant amongst the Indian pheasants. It is distributed along the Himalayan foot hills up to an altitude of 2,000 m, from Kashmir to Arunachal Pradesh and southwards to parts of Uttar

Pradesh, Madhya Pradesh, Bihar, Orissa, West Bengal, and Assam. The red junglefowl has proved extremely useful and economically very important to man, and by way of having a closer association with humans than any other bird in the world, has contributed significantly to the welfare of the human by being the progenitor of the world's domestic poultry. The red junglefowl is mainly found along the forested tracts in the Himalayan foothills (Shiwaliks) of Jammu region mainly in Ramnagar WLS, Nandni WLS, Sud Mahadev forests, reserve forests of Kathua, Chinab Valley, and in certain areas in the Pir Panjal Mountains in Jammu province. However, there are no reports of its presence from any part of the Kashmir Valley.

The kalij pheasant *Lophura leucomelanos* has the most extensive distribution of all the pheasant groups. It ranges from Pakistan in the west to Indonesia in the east. Out of the nine subspecies of the kalij pheasant found in the world, five subspecies are found in India. The white-crested kalij pheasant *L. l. hamiltoni* is distributed in the western Himalayas. Listed as least concern with a declining population trend, it inhabits secondary forests and scrubs, mostly between an altitudinal (Johnsgard 1986) range of 915 to 2135 m having direct contact with human habitations In Jammu and Kashmir, the kalij pheasant is largely a species confined to the Jammu region of the state and is not found in any significant numbers in the Valley although some reports indicate that they may be present in the lower parts of the Jhelum Valley which needs to be ascertained. However, the kalij pheasant is found in fearly good numbers along the Pir Panjal mountain range in the Kashmir Valley particularly in the Limber-Lachipora areas of Kajinag National Park. The distribution largely follows that of the red junglefowl within the state. The kalij pheasant has recently been declared as the Union Territory bird of Jammu and Kashmir is the least studies species in its natural habitats. The scanty scientific information available on the white-crested kalij pheasant necessitates the need for more in-depth research on the ecology and biology of the species in its distributional range areas including in the Jammu and Kashmir for effective conservation and management planning of the species and its habitats in the landscape.

Of the two subspecies of peafowl found in India, Indian peafowl *Pavo cristatus* and green peafowl *Pavo muticus*, the Indian peafowl are essentially birds of the plains but are also found in the lowlands and Himalayan foothills up to 1,800 m throughout the country including most parts of the Jammu region. This species occupies the areas adjacent to the Indian plains and is found in the Ramnagar, Nandini, and Nagrota WLS, as well as near agricultural farmland in the Sambha and Kathua areas of Jammu.

Despite their colourful plumage and diurnal habits, pheasants are poorly studied in the state of Jammu and Kashmir in particular and in the country in general. The pheasant census and survey carried out by the Zoological Survey of India (ZSI) (Lamba et al., 1982b) in Kashmir is the only effort made so far in this regard. Besides Kaul (1989), Qadri et al. (1990) did some surveys aimed to document the distribution and status of pheasants in the Kashmir valley. No surveys dedicated towards distribution of Galliformes in general and pheasants in particular in the Jammu region have been conducted except that in Kishtawar National Park (Scott et al., 1989). Distribution of Galliformes has been better documented in the Ladakh region by travellers and ornithologists.

Pheasants have been associated with mankind from the time immemorial. The domestic fowl, which we relish today and which has contributed significantly to the welfare of humans, has its origins in the red junglefowl since 5000 BC. However, despite their ecological and aesthetic values, the pheasants have been under severe threat of biotic interferences owing to habitat degradation, live-stock grazing, and fuel and firewood collection. Because of being a highly special-ised group of birds, the pheasants inhabit very fragile habitats with considerable human exploitive pressure. The human-induced land use changes and habitat modifications and other biotic pressures can cause large scale changes and conse-quently such habitat changes result in population decline of pheasants and pose threats for their long term conservation in these landscapes.

The pristine habitats of Jammu and Kashmir, as in other areas of the Himalayan region, have been under severe pressure ever since they were opened and exploited commercially in the colonial period, more than around 150 years ago.

Habitat loss or degradation, hunting or poaching, and other disturbances due to human activities are reported to have had cascading impacts on several species of birds and mammals (besides other environmental impacts), and this is shown either by population decline or by becoming locally extinct. With the increase in human and cattle population, the demands for land and fodder further led to a shrinkage of habitats available for pheasants, and it also led to the degradation of quality habitats in the remaining areas. These activities as well as persecu-tion by poachers have decreased the population of most of the pheasant species substantially, causing local extinction of some species and pushing others to the brink of extinction.

Apart from habitat degradation and fragmentation, which has been a com-mon feature throughout the Himalayas in the recent past, the main threats to the pheasants along their distribution range is the hunting for food, poaching for trade, genetic swamping, and pesticides. The Himalayan monal, despite its very wide distribution across the Himalayas, is a threatened pheasant because of often a victim of poachers and trappers for meat and plumes, especially during winter when the bird descends to lower altitudes near human habitations. Among all the pheasants, the Himalayan monal stands out for its royal crest and highly col-ourful and exquisite plumage that defies its description and makes it spectacular and indeed a sight to behold. As such it has been a popular bird in many public and private collections all over the world.

The cheer pheasant is confined to the Kazinag range where it survives in the Limber and Lachipora WLS. As is apparent from surveys in other areas, the cheer survives well outside protected areas as well as inside them. Thus, we do not per-ceive any major threats to their population in the state. However, the lack of sur-veys has meant that our knowledge about the cheer is restricted to only two sites in Kazinag (Kashmir) and Trikuta and adjoining hills in the Jammu areas. There is a need to survey south facing slopes of the Pir Panjal range and chir pine habitats in the Chinab valley along the Jammu-Srinagar highway to identify more sites with occurrence of cheer pheasant.

The western tragopan, on the other hand, is more dependent, perhaps because of the protection to its habitat, and is thus safe in Kazinag range (Limber and Lachipora areas). Their status is unknown in Kishtawar National Park though the potential habitat here is larger than in Limber and Lachipora. More surveys are required to establish the status of the western tragopan in Kishtwar NP and protected areas such as Hirpora WLS and Ajas forests in Pir Panjal ranges.

Hunting and poaching of pheasants for meat and trade, which have been under control in the state of Jammu and Kashmir for the last two decades, may possibly start happening again with changes in political scenario. Efforts are needed to be made to stop this, as any form of hunting is illegal under the Jammu and Kashmir Wildlife (Protection) Act 1978 (amended in 2002) as well as Indian Wildlife Protection Act 1972 (amended 2006).

The western tragopan, Himalayan monal, and cheer pheasant are among the six species of pheasants in the Himalayan region that are threatened (Fuller & Garson, 2000, IUCN/SSC/Red Data Book 2003) and considered vulnerable by Birdlife International (2001, 2013) because of their restricted distribution and small population (Rahmani et al., 2014). As such they require urgent protection and consideration as a flagship species for their long-term conservation and for conserving the unique ecosystems in the Indian Himalayas. There is a very urgent need to undertake status surveys aimed at updating information and enhancing our scientific knowledge on several aspects of the ecology and behaviour of Galliformes in general and pheasants in particular, which are a prerequisite for their effective management and long term conservation planning in Jammu and Kashmir. Besides this, science based conservation breeding programmes aimed at augmenting the existing populations of these threatened pheasant species in the wild are imperative for their long term survival and conservation.

References

Ahmad, R., Sharma, N., Pacchnanda, U., Suhail, I., Deb, K., Bhatnagar, Y. V., & Kaul, R. (2017). Distribution and conservation status of the western tragopan Tragopan melanocephalus in Jammu and Kashmir, India. *Current Science*, 112(9),1948–1953.

Akhtar, A., Vibhu, P., & Salim, J. (1994). The Western Tragopan-Bird of the Himalayas. *Sanctuary*, 14(2), 44–49.

Ali, S., & Ripley, S. D. (1987). *Compact handbook of the birds of India and Pakistan* (2nd ed.). Oxford University Press.

Baker, E. C. S. (1930). *Game birds of India, Burma and Ceylon*. John Bale and Son.

Bisht, M. S., Kathait, B. S., & Dobriyal, A. K. (2003). Some records of the endangered cheer pheasant in Garhwal, Central Himalaya. *Environmental Information System on Himalayan Ecology*, 10(1), 4–6.

Bump, G. (1941). The introduction and transplantation of game birds and mammals into the State of New York. *Transactions of the North American Wildlife Conference*, 5, 409–420.

Crowe, T. M., Keith, G. S., & Brown, L. H. (1986). Galliformes In E. K. Urban, C. H. Fry, & S. Keith (Eds.), *The birds of Africa* (Vol. II, pp. 1–75). Academic Press.

Delacour, J. (1977). *The pheasants of the world.* Spur Publications and The World Pheasant Association, pp. 395.

Delhoyo, J., Elliott, A., & Sargatal, J. (Eds.). (1994). *Handbook of the birds of the world. Vol. 2. New World vultures to guineafowl.* Lynx Edicions.

Fuller, R. A., & Garson, P. J. (Eds.). (2000). *Pheasants: Status survey and conservation action plan 2000–2004.* WPA/Birdlife/SSC Pheasant Specialist Group. IUCN, Gland. Switzerland and Cambridge, U.K. and the World Pheasant Association, Reading, U.K.

Garson, P. J. (1983). The cheer pheasant Catreus wallichii in Himachal Pradesh, western Himalayas: An update. *Journal of the World Pheasant Association, 8,* 29–39.

Gaston, A. J. (1980). Census techniques for Himalayan Pheasants, including notes on individual species. *Journal of the World Pheasant Association, 5,* 40–53.

Gaston, A. J., & Singh, J. (1980). The status of cheer pheasant *Catreus wallichii* in the Chail wildlife sanctuary, Himachal Pradesh. *Journal of the World Pheasant Association, 5,* 68–73.

Gaston, A. J., Garson, P. J., & Hunter, M. L. (1981). Present distribution and status of pheasants in Himachal Pradesh, Western Himalayas. *Journal of the World Pheasant Association, 6,* 10–30.

Gaston, A. J., Garson, P. J., & Hunter, M. L. (1983). The status and conservation of forest wildlife in Himachal Pradesh, Western Himalayas. *Biological Conservation, 27,* 291–314.

Gaston, A. J., Islam, K, & Crawford, J. A. (1983). The current status of the Western Tragopan Tragopan melanocephalus. *WPA-Journal,* VIII: 40–49.

Ghosh, S. (1997). Record of cheer pheasant, *Catreus wallichi* above 4545 m in the Western Himalayas. *Journal of the Bombay Natural History Society, 94,* 566–566.

Hill, D., & Robertson, P. (1988). *The pheasants: Ecology, management and conservation.* BSP Professional Books, pp. 281.

Islam, M. Z., & Rahmani, A. R. (2004). *Important bird areas in India: Priority sites for conservation. Indian bird conservation network, Bombay natural history society, and Birdlife International, U.K.* Oxford University Press, pp. xviii + 1133.

Javed, S. (1992). Birds of Limber valley forest (Jammu & Kashmir) Newsletter for. *Birdwatchers, 32*(5 & 6), 13–15.

Johnsgard, P. A. (1986). *The pheasants of the world.* Oxford University Press, pp. 300.

Kalsi, R. S. (1999). Status and habitat of the Cheer Pheasant in Himachal Pradesh. *Tragopan,* 13–14, 19–25.

Kaul, R. (1989a). *Ecology of the cheer pheasant Catreus wallichii in Kumaon Himalaya* [Ph.D. Thesis]. University of Kashmir

Kaul, R. (1989b). Western Tragopan surveys in Limber valley, Kashmir, India. *WPA News, 26,* 12–14.

Kaul, R., & Garson, P. J. (1993). Present distribution and status of pheasants in India. In D. Jenkins (Ed.), *Pheasants in Asia 1992* (pp. 157). World Pheasant Association.

Knox, A. G., & Walters, M. P. (1994). Extinct and endangered birds in the collections of the Natural History Museum. British Ornithologists' Club, Occasional Publications 1, Tring.

Lamba, B. S., Narang, M. L., Tyagi, A. K., Tak, P. C., Kumar, G., & Baccha, M. S. (1982a). Pheasant census survey of Kashmir valley – A report. *Second International Pheasant Symposium,* Srinagar.

Lamba, B. S., Narang, M. L., Tyagi, A. K., Tak, P. C., Kumar, G., & Bacha, M. S. (1982b). Pheasant census survey of Kashmir Valley – A report. In C. D. W. Savage & M. W. Ridley (Eds.), *Pheasants in Asia 1982* (pp. 52–54). World Pheasant Association.

McGowan, P. J. K., & Garson, P. J. (1995). Status survey and conservation action plan 1995–1999. *Pheasants*. IUCN and World Pheasant Association, Switzerland and U.K, pp. 116.

Narang, M. L. (1993). Searches for Western Tragopan in Himachal Pradesh, India, in 1987. In D. Jenkins (Ed.), *Pheasants in Asia 1992* (Vol. 89, pp. 55–57). World Pheasant Association.

Pandey, S. (1993). Pheasant surveys and conservation of protected areas in the upper Beas valley, Himachal Pradesh, India. In D. Jenkins (Ed.), *Pheasants in Asia 1992* (pp. 58–61). World Pheasant Association.

Pandey, S. (1994). A preliminary estimate of number of Western Tragopan in Daranghati sanctuary, Himachal Pradesh. *Annual Revision Wildlife Protection Act 1993/1994*, 49–56.

Pokorny, F., & Pikula, J. (1987). Artificial breeding, rearing and release of reeve pheasant (Syrmaticus reevesi Gray 1929) in Czechoslovakia. *Journal of the World Pheasant Association*, 12, 75–80.

Qadri, M. Y., Kaul, R., & Iqbal, M. (1990). Pheasant surveys in Jammu and Kashmir. In *Pheasants in Asia (1990)*. World Pheasant Association SARO.

Rahmani, A. R., Islam, M. Z., Ahmad, K., Intesar, S., Chandan, P., & Zarri, A. A. (2012). Important birds areas of Jammu & Kashmir. *Indian bird conservation network, Bombay natural history society, royal society for the protection of birds, and birdlife international.* Oxford University Press.

Rahmani, A. R., Suhail, I., Chandan, P., Ahmad, K., & Zarri, A. A. (2014). *Threatened birds of Jammu & Kashmir*. Oxford University Press, pp. 150.

Sanga, H., & Naoroji, R. (2005). Birds recorded during seven expeditions to Ladakh from 1997 to 2003. *Bombay Natural History Society*, 102(3), 290–304.

Sathyakumar, S., Prasad, S. N., Rawat, G. S., & Johnsingh, A. J. T. (1993). Ecology of kalij and monal pheasant in Kedarnath wildlife sanctuary, Western Himalaya. In D. Jenkins (Ed.), *Pheasants in Asia 1992* (pp. 83–90). World Pheasant Association.

Sharma, V. (1993). Ecological status of pheasants in Himachal Pradesh. In D. Jenkins (Ed.), *Pheasants in Asia 1992* (pp. 158). World Pheasant Association.

Sharma, V., & Pandey, S. (1989). Pheasant surveys in Shimla Hills of Himachal Pradesh, India. *Journal of the World Pheasant Association*, 14, 64–78.

Sibley, C. G., & Ahlquist, J. E. (1990). *Phylogeny and classification of birds: A study in molecular evolution*. Yale University Press.

Sibley, C. G., & Monroe, B. L. Jr. (1990). *Distribution and taxonomy of birds of the world*. Yale University Press.

Subedi, P. (2003). Status and distribution of cheer pheasant in Dhorpatan hunting Reserve (Doctoral dissertation) Instituted of Forestry Pokhara.

Wetmore, A. (1960). A classification of the birds of the world. *Smithson Miscellaneous Collection*, 139(11), 1–37.

Wood-Gush, D. G. M. (1959). A history of the domestic chicken from antiquity to the 19th century. *Poultry Science*, 38(2), 321–326.

18 Avifaunal Diversity at Baba Ghulam Shah Badshah University Campus

Kaleem Ahmed, Mohd Qasim, Abdul Hamid Malik, and Ali Asghar Shah

Introduction

In a sustainable ecological system, diversity is mostly used as an indicator, and avian diversity is among the critical components of the earth's biodiversity. It plays a crucial role in connecting the food chain within the environmental unit of nature. Birds, being the most important biotic component of any ecosystem (Dhindsa & Saini, 1994), also play a crucial role in maintaining ecological balance. The bird's activities develop relations within and between ecosystems and can have huge effects on other species (Faiz et al., 2015). Thus, the combined impact of climate change and urbanisation, leading to habitat loss, is one of the most dangerous conditions for the conservation of bird diversity. Under these conditions, remnants of wild vegetation and plantations on university campuses provide hope for bird conservation.

In India, there are a large number of universities with varying land sizes, and most of them have at least a few patches of natural vegetation and plantation. However, documentation of birds in such areas was not given the same importance as in the forests and other reserve areas. Thus, preparing a list of species is essential to the study of bird species in a place because a list of species stipulates diversity (Singh et al., 2018).

During the last few decades, the loss of species at a rapid pace, mainly due to the pressure of increasing human and livestock populations and developmental activities, has been noted with great concern. This depletion has led to the destruction or degradation of natural habitats or ecosystems. This has resulted in the extinction of many species, many even without being named. Besides this, a large number of species are being threatened and are on the verge of extinction. If the situation continues, an imbalance in the ecosystem is inevitable. As a result, the threatened ecosystems may become non-functional, and that may lead to the extinction of the human race. In order to chalk out any strategy for the conservation of biodiversity, the first step is to complete the inventory of species and to know the richness of species in a particular ecosystem or biogeographical zone. Besides this, a database on biodiversity is also considered a prerequisite before starting any development or industrial project at village or district level.

DOI: 10.4324/9781003321422-21

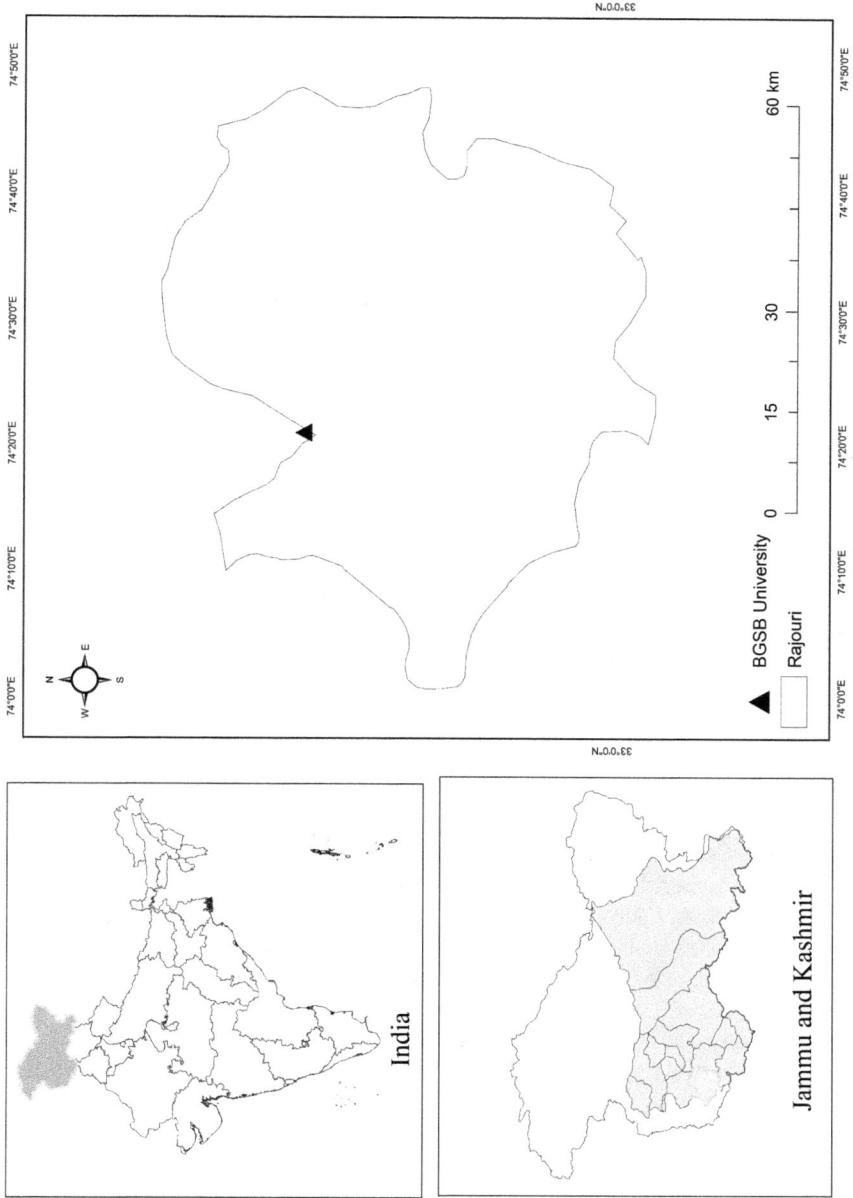

Figure 18.1 Map of study area showing location of BGSBU, Campus, Rajouri.

As many studies have been published on avian diversity in protected areas and other natural habitats, there is no information on any organised study from campuses of universities. Sporadic information about some of the university and college campuses is available. A comprehensive survey of bird diversity has not been conducted in the Pir-Panjal region of the district of Rajouri. However, there has never been a thorough survey of Pir-Panjal Biodiversity Park by Baba Ghulam Shah Badshah University (henceforth, BGSBU) for the study of avian fauna. The present study is the first study of its kind that was undertaken at the BGSBU campus. Hence, this study was conducted on the avifaunal diversity of the BGSBU campus, Rajouri, located in the most diverse ecosystem of the Pir-Panjal region of Jammu and Kashmir, and efforts have also been made to prepare a preliminary checklist of its birds.

Study Area

The present study was conducted from April 2019 to July 2019 at the BGSBU campus in Rajouri, Jammu and Kashmir, India. It is situated in the southern part of Jammu division, in the Pir-Panjal range, between 32°58′ and 33°-35′ latitude and 74°-81′ longitude, at an elevation of 370–6,000 mean sea level (Figure 18.1). This region contains a rich geographic, climatic, and biological diversity. It has a variety of trees, shrubs, grasses, berry plants, and flowering plants, which provide food for a large number of bird species. Small hills are also present in this region, which provide shelter for birds and other animals. Climatically, the area faces three main seasons, i.e., winter, summer, and rainy. Due to the typical topography of the area, it supports diverse vegetation, ranging from sub-tropical to temperate types. On the campus, the Pir-Panjal Biodiversity Park is being developed on more than ten acres. The water reservoir and vegetation planted have created a habitat that attracts many insects, reptiles, birds, and mammals to the campus.

Methodology

The point count method (Reynolds et al., 1980) and the species richness counting method (Mackinnon & Phillips, 1993) were used for sampling birds in the study area. During the survey, birds were sampled by monitoring permanently established points. Each point was monitored thrice, for a total sampling effort of 69. Efforts were made to monitor at least eight to ten points weekly in a rotational manner. The identification of species was done with the help of Grimmett et al. (2011) and, for others, family and scientific names (Parveen et al., 2020). The sampling points were taken randomly. A distance of approximately 250 m was maintained between the two sampling points. A 20-minute field time duration was spent at each point in the morning hours between 6.00 and 10.30 a.m. with a fixed radius of 20 m each. At each point station, data was recorded on the following variables: Species, number of individuals, perch height, stratum, height of the tree, and activity.

Foraging Behaviour of Birds

The foraging behaviour of birds was observed in the study area mostly throughout the day. Individual birds were observed through binoculars and the type of perch (electric power lines, trees, shrubs, and ground), perching height (height at which the bird was perched while feeding), foraging height, foraging substrate (the material from which the birds take food-classified into air, plants, and ground), and foraging method (classified into aerial feeding-a bird flew into the air to catch flying prey; gleaning-a stationary food item is picked from its substrate by a standing or hopping bird; and ground feeding-the bird picks prey from the ground) were used (Asokan, 1995). The heights were visually estimated.

Analysis

Shannon-Weiner Index (H') was used for diversity and Margalef's Index (RI) for richness and evenness computations. Each species list was treated as a separate sample for statistical analysis using the computer programme Biodiversity Pro (Neil, 1997). Non-parametric statistics from Chao 1 (abundance based model) were used to extrapolate species richness curves.

Bird densities per unit area were calculated for a given sampling point as the total number of groups of a species seen at a particular point divided by the total area of all points. First, mean group densities (Dg, number of groups /ha) and their standard error (SeDg) for all the sampling points were calculated. From this, the mean ecological densities (D, number of individuals /ha) and its standard error (SeD) were derived using the following equations (see Drummer, 1987; Karanth & Sunquist, 1992; Noor et al., 2016):

$$D = Dg \times Y$$
$$SeD2 = SeD2g \times Se\,Y2/n + SeD2g \times Y2 + SeY2/n \times D2g$$

Y = Mean cluster size, SeY = Standard error of mean cluster size, n = number of cluster of species detected

Results

Accuracy of Sample Size

A total of 883 individual birds belonging to 66 species were identified during the avifaunal survey on the BGSBU campus in Rajouri (Table 18.1). Chao 1 was used for stabilisation of the number of species observed in the study areas (Figure 18.2) and asymptote was reached after 72 birds using this model.

Table 18.1 Checklist of birds along with IUCN status from BGSBU, Campus, Rajouri

Order and family	Common and scientific name	No. of individuals	IUCN status
Galliformes	**Phasianidae**		
	Black francolin *Francolinus francolinus*	2	LC
	Indian peafowl *Pavo cristatus*	3	LC
	Red junglefowl *Gallus gallus*	5	LC
Columbiformes	**Columbidae**		
	Rock pigeon *Columba livia*	28	LC
	Eurasian collared dove *Streptopelia decaocto*	6	LC
	Oriental turtle dove *Streptopelia orientalis*	13	LC
	Spotted dove *Streptopelia chinensis*	14	LC
Caprimulgiformes	**Apodidae**		
	Common swift *Apus apus*	2	LC
Pelecaniformes	**Ardeidae**		
	Cattle egret *Bubulcus ibis*	9	LC
Accipitriformes	**Accipitridae**		
	Black eagle *Ictinaetus malaiensis*	3	LC
	Black-winged kite *Elanus caeruleus*	2	LC
	Himalayan vulture *Gyps himalayensis*	3	NT
Strigiformes	**Strigidae**		
	Spotted owlet *Athene brama*	11	LC
Bucerotoformes	**Upupidae**		
	Common Poopoe *Upupa epops*	4	LC
Coraciiformes	**Alcedinidae**		
	Common kingfisher *Alcedo atthis*	3	LC
	White-throated kingfisher *Halcyon smyrnensis*	4	LC
	Meropidae		
	Green bee-eater *Merops orientalis*	5	LC
Piciformes	**Picidae**		
	Himalayan woodpecker *Dendrocopos himalayensis*	3	LC
Falconiformes	**Falconidae**		
	Common kestrel *Falco tinnunculus*	1	LC
Psittaciformes	**Psittaculidae**		
	Rose-ringed parakeet *Psittacula krameri*	43	LC
	Slaty-headed parakeet *Psittacula himalayana*	15	LC
Passeriformes	**Campephagidae**		
	Scarlet minivet *Pericrocotus speciosus*		
	Dicruridae		
	Ashy drongo *Dicrurus leucophaeus*	5	LC
	Black drongo *Dicrurus macrocercus*	5	LC
	Monarchidae		
	Indian paradise-flycatcher *Terpsiphone paradisi*	3	LC
	Laniidae		
	Isabelline shrike *Lanius isabellinus*	2	LC
	Long-tailed shrike *Lanius schach*	7	LC
	Corvidae		
	Black-headed jay *Garrulus lanceolatus*	1	LC
	House crow *Corvuss plendens*	22	LC
	Large-billed crow *Corvus macrorhynchos*	27	LC
	Red-billed blue magpie *Urocissaery throryncha*	1	LC

(Continued)

Table 18.1 (Continued)

Order and family	Common and scientific name	No. of individuals	IUCN status
	Rufous Treepie *Dendrocitta vagabunda*	18	LC
	Yellow-billed blue magpie *Urocissa flavirostris*	3	LC
Paridae			
	Cinereous tit *Parus cinereus*	5	LC
Locustellidae			
	West Himalayan bush warbler *Locustella kashmirensis*	1	LC
Pycnonotidae			
	Black bulbul *Hypsipetes leucocephalus*	6	LC
	Himalayan bulbul *Pycnonotus leucogenis*	7	LC
	Red-ventedbBulbul *Pycnonotus cafer*	36	LC
Phylloscopidae			
	Grey-hooded warbler *Phylloscopus xanthoschistos*	35	LC
	Tytler's Leaf warbler *Phylloscopus tytleri*	5	NT
	Wood warbler *Phylloscopus sibilatrix*	7	LC
Cettiidae			
	Brownish-flanked bush warbler *Horornis fortipes*	6	LC
Aegithalidae			
	White-throated tit *Aegithalos niveogularis*	5	LC
Zosteropidae			
	Indian white-eye *Zosterops palpebrosus*	3	LC
Leiothrichidae			
	Common babbler *Argya caudata*	34	LC
	Jungle babbler *Argya striata*	134	LC
Regulidae			
	Goldcrest *Regulus regulus*	3	LC
Tichodromidae			
	Wallcreeper *Tichodroma muraria*	3	LC
Certhiidae			
	Bar-tailed treecreeper *Certhia himalayana*	4	LC
Sturnidae			
	Common myna *Acridotheres tristis*	33	LC
Muscicapidae			
	Blue-capped redstart *Phoenicurus coeruleocephala*	3	LC
	Blue whistling thrush *Myophonus caeruleus*	13	LC
	Grey bushchat *Saxicola ferreus*	6	LC
	Indian robin *Copsychus fulicatus*	24	LC
	Pied bushchat *Saxicola caprata*	13	LC
	Spotted forktail *Enicurus maculatus*	3	LC
	Variable wheatear *Oenanthe picata*	4	LC
	Verditer flycatcher *Eumyias thalassinus*	5	LC
	White-capped water redstart *Phoenicurus leucocephalus*	10	LC
Nectariniidae			
	Purple sunbird *Cinnyrisa siaticus*	4	LC
Estrildidae			
	Tricoloured munia *Lonchura malacca*	9	LC
Passeridae			

(*Continued*)

Table 18.1 (Continued)

Order and family	Common and scientific name	No. of individuals	IUCN status
	House sparrow *Passer domesticus*	64	LC
	Russet sparrow *Passer cinnamomeus*	85	LC
	Motacillidae		
	Paddyfield pipit *Anthus rufulus*	3	LC
	White wagtail *Motacilla alba*	19	LC
	Western yellow wagtail *Motacilla flava*	10	LC

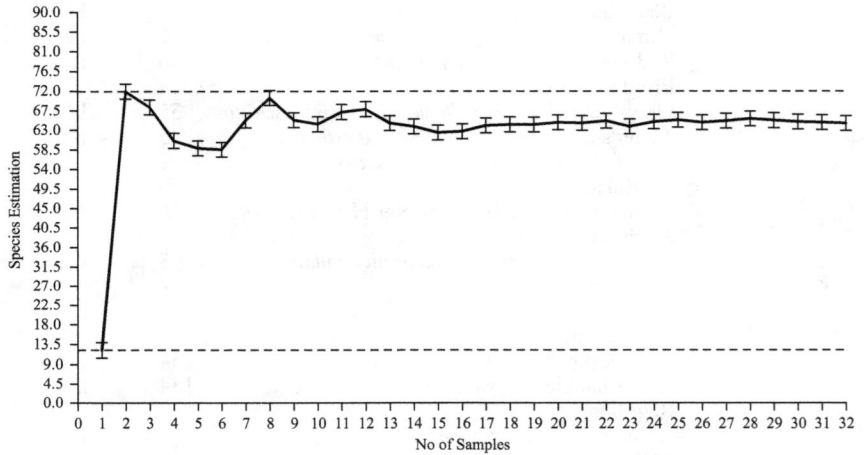

Figure 18.2 Performance of Chao 1 Model for estimating the bird species in BGSBU, Campus, Rajouri.

Bird Density

Overall bird density (D±SE) was 26.66 ± 1.67/ha, the density of clusters (Dg) was 14.57 ± 0.693/ha and the average cluster size (Avs) was 6.47 ± 0.31. In total, 12 orders and 34 families were recorded in the study area. Passeriformes order has the highest percentage (68.18%), and the lowest recorded for Caprimulgiformes, Pelecaniformes, Strigiformes, Bucerotoformes, Piciformes, and Falconiformes (Table 18.2). Similarly, the highest percentage (13.64%) was recorded for the Muscicapidae family and the lowest recorded for 20 other families ($z = 5.47$, $P > 0.05$) (Table 18.2).

Species Diversity, Richness, and Evenness

Overall species diversity in the study area was (3.45), richness (10.23) and evenness (0.432), ($t = 1.62$, $P < 0.05$). Species diversity was found highest for the

Table 18.2 Species diversity, richness, and evenness along with the percentage (%) of the orders and families found in the BGSBU, Campus, Rajouri

Order	%	Family	%	Diversity	Richness	Evenness
Galliformes	4.55	Phasianidae	4.55	0.9	0.962	0.82
Columbiformes	6.06	Columbidae	6.06	1.383	0.962	0.798
Caprimulgiformes	1.52	Apodidae	1.52	-	-	-
Pelecaniformes	1.52	Ardeidae	1.52	-	-	-
Accipitriformes	4.55	Accipitridae	4.55	1.424	1.61	0.831
Strigiformes	1.52	Strigidae	1.52	-	-	-
Bucerotoformes	1.52	Upupidae	1.52	-	-	-
Coraciiformes	4.55	Alcedinidae	3.03	0.683	0.514	0.99
		Meropidae	1.52	-	-	-
Piciformes	1.52	Picidae	1.52	-	-	-
Falconiformes	1.52	Falconidae	1.52	-	-	-
Psittaciformes	3.03	Psittaculidae	3.03	0.572	0.246	0.886
Passeriformes	68.18	Campephagidae	1.52	-	-	-
		Dicruridae	3.03	0.693	0.434	1
		Monarchidae	1.52	-	-	-
		Laniidae	3.03	0.53	0.459	0.849
		Corvidae	9.09	1.372	1.141	0.657
		Paridae	1.52	-	-	-
		Locustellidae	1.52	-	-	-
		Pycnonotidae	4.55	0.938	0.57	0.639
		Phylloscopidae	4.55	0.742	0.52	0.7
		Cettiidae	1.52	-	-	-
		Aegithalidae	1.52	-	-	-
		Zosteropidae	1.52	-	-	-
		Leiothrichidae	3.03	0.504	0.195	0.827
		Regulidae	1.52	-	-	-
		Tichodromidae	1.52	-	-	-
		Certhiidae	1.52	-	-	-
		Sturnidae	1.52	-	-	-
		Muscicapidae	13.64	1.93	1.791	0.766
		Nectariniidae	1.52	-	-	-
		Estrildidae	1.52	-	-	-
		Passeridae	3.03	1.094	0.364	0.995
		Motacillidae	4.55	0.769	0.588	0.719

Muscicapidae family (1.93) and lowest for the Leiothrichida family (0.5037), (t = 8.59, P < 0.05). The highest avifaunal species richness was found highest for the Muscicapidae family (1.791) and lowest for the Psittacidae family (0.246), (t = 25.72, P < 0.05). Species evenness was found highest for the Dicruridae family (1.00) and lowest for the Pycnonotidae family (0.638), (t = 5.64, P < 0.05) (Table 2).

Guild Structure

The percentage of birds based on different feeding guild were found highest for omnivorous (43%), insectivorous (34%), granivorous (8%), carnivores

(5%), frugivorous (5%), and nectarivorous (5%) ($z = 2.44$, $P = 0.01$). Guild 1 consists of species that usually forage on the ground. It includes francolin, myna, peafowl, sparrow, and pipit in the study area. Guild 2 consists of species that utilise top canopy and tree trunks like minivet, treecreeper, dove, and tit. Insectivore birds dominated this guild. Guild 3 consists of species that utilise bushes and a small tree for foraging like a crow, chat, babbler, russet sparrow, and bulbul. Guild 4 consists of species that occupy both top and middle canopy which includes species like a woodpecker, white eye, and egret in the study area (Figure 18.3).

Discussion

A total of 66 species of birds were recorded from the study area. Most of the observed species are breeding residents, mainly due to the occurrence of various types of microhabitats within the campus, nearby streams, and riverine habitats. The density values of the present study 26.66/ha were almost close to the other studies conducted in the Western Himalayas by Dar et al. (2008): (30.48/ha in Phakot) and Ahmed (2010): (29.97/ha in Dabka and 28.35/ha in Khulgarh) but lower than the values recorded in Sultana and Hussain (2012): (12.75/ha and 13.93/ha) from the Binsar WLS and Vinayak forest reserves respectively. These variations in density may be due to the reason that birds are sensitive to the local landscape, and changes in vegetation patterns can affect the bird population in their area (Sauvjo et al., 1998). Moreover, the present study was conducted in the summer season and generally, high bird densities were found in the summer (Osborne & Green, 1992).

Overall omnivorous and insectivorous birds are the most dominant species in the study area. This seems due to the availability of different prey items as well as diverse habitat in the study area. This type of pattern was also observed in other studies conducted by Singh et al., 2018 in Banda University of Agriculture and Technology, Uttar Pradesh, India; Dapke et al., 2015 in Laxminarayan Institute of Technology, Nagpur, India; Dey et al., 2014 in Maharaja Bir Bikram College, Tripura, India, and Wadatkar (2001) in Amravati University Campus, Amravati, Maharashtra, India. Moreover, domination of the bird community by insectivores is a common trend and observed in other studies in the Himalayan and the North-east regions (Acharya, 2010; Dutta, 2011). Due to their specialised diet and low availability of preferable food resources, carnivores, granivores, frugivorous, and others are generally less represented. Our results also indicate that all the bird species used different perch types in the study area. All species, except the common myna that preferred to use the ground, showed great use of electric power lines. The electric power lines are a common feature in the study area, especially around the agricultural habitats. These structures provided suitable perches for birds for detecting prey. Lammers and Collopy (2007) stated that avian predators are attracted to overhead utility power lines because they provide perches for various activities, including hunting prey. The other dominant perch sites used by the birds in our study area were trees and shrubs. With the absence of electric power lines in areas like riverine habitats, birds preferred and readily utilised trees and shrubs as perches.

Bray-Curtis Cluster Analysis (Single Link)

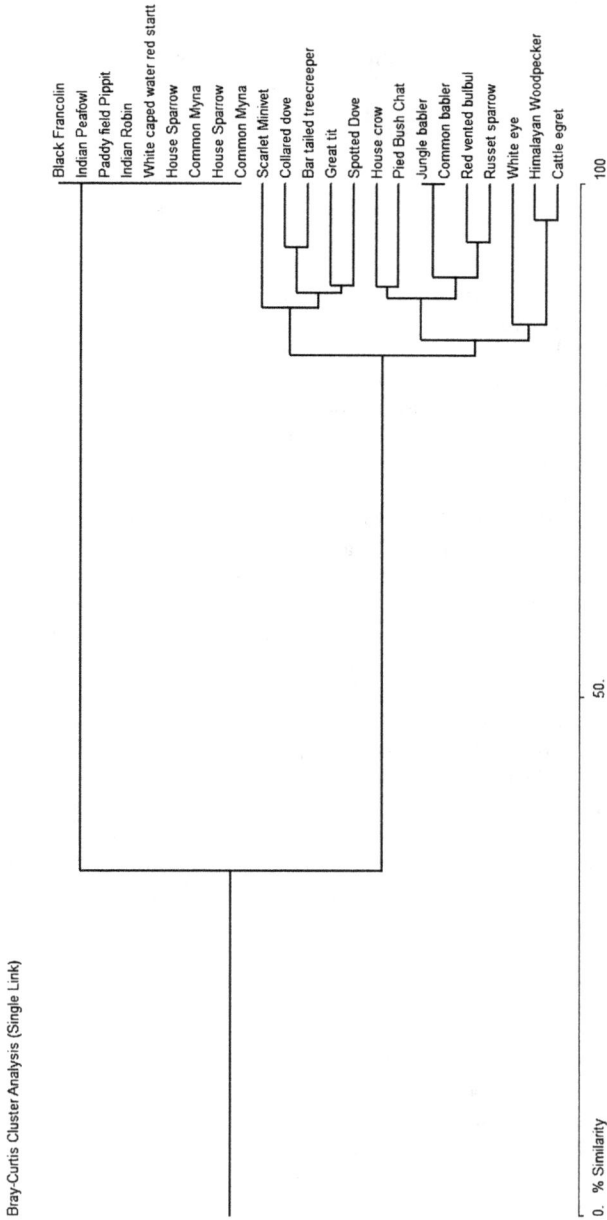

Figure 18.3 Dendrogram of bird community based on feeding niche in BGSBU, Campus, Rajouri.

The study highlights the importance of green space on University campuses for maintenance of the ecological balance and sustainable urban development. The present study on the avifauna of BGSB, campus preliminary in nature, indicated that the study area is supporting many species of birds. Further, this study provides baseline information for future studies to understand and compare the changes. To extract exhaustive information and assess the importance of avian fauna in the ecosystem, there is an urgent, in fact, compelling need to have comprehensive knowledge of the avifaunal diversity of the campus.

References

Acharya, B. K., Vijayan, L., & Chettri, B. (2010). The bird community of Shingba rhododendron wildlife sanctuary, Sikkim, Eastern Himalaya, India. *Tropical Ecology*, 51(2), 149–159.

Ahmed, K. (2010). *A faunal diversity of Dabka and Khulgarh watershed areas of Kumoan Himalayas, Uttarakhand, India* [Ph.D. Dissertation]. Department of Wildlife Sciences, Aligarh Muslim University.

Asokan, S. (1995). Ecology of the Small Green Bee-eater, *Merops orientalis* Latham 1801 with special reference to its population, feeding and breeding in Mayiladuthurai, Tamil Nadu, South India [Ph.D. Thesis]. Bharathidasan University.

Dapke, S., Didolkar, S., & Koushik, S. (2015). Studies on diversity and abundance of avifauna in and around Laxminarayan Institute of Technology, Nagpur. *Journal of Entomology and Zoology Studies*, 3(5), 141–146.

Dar, T. A., Habib, B., Khan, J. A., Kushwaha, S. P., & Mendiratta, N. (2008). Bird community structure in Phakot and Pathrirao watershed areas in Uttarakhand, India. *International Journal of Ecology and Environmental Sciences*, 34(2), 193–205.

Dey, A., Deb, D., Chaudhuri, S., & Chaudhuri, P. (2014). A preliminary study on avifaunal species diversity of Maharaja Birbikram college campus, Tripura, North East India. *International Multidisciplinary Research Journal*, 3(2), 36–43.

Dhindsa, M. S., & Saini, H. K. (1994). Agricultural ornithology: An Indian perspective. *Journal of Biosciences*, 19(4), 391–402.

Drummer, T. D. (1987). *Program documentation and user's guide for SIZETRAN*. Michigan Technological University.

Dutta, N. N., Baruah, D., & Borah, S. (2011). Avifaunal diversity in an IBA site Northeast India and their conservation. *Annals of Biological Research*, 2(5), 374–384.

Faiz, A. H., Abbas, F. I., Ali, Z., & Zahra, L. (2015). Avifaunal diversity of Tolipir National Park Jammu and Kashmir Pakistan. *Journal of Animal and Plant Sciences*, 25(2), 404–409.

Grimmett, R., Inskippand, L., & Inskipp, T. (2011). *Birds of the Indian subcontinent*. Christopher Helm, pp. 528.

Karanth, K. U., & Sunquist, M. E. (1992). Population structure, density and biomass of large herbivores in the tropical forests of Nagarahole, India. *Journal of Tropical Ecology*, 8(1), 21–35.

Lammers, W., & Collopy, M. W. (2007). Effectiveness of avian predator perch Deterrentson electric transmission lines. *Journal of Wildlife Management*, 71(8), 2752–2758.

MacKinnon, J., & Phillips, K. (1993). *A field guide to the birds of Borneo, Sumatra, Java and Bali, The Greater Sunda Islands*. Oxford University Press.

Neil, M. (1997). *Biodiversity professional beta* (version 2.0). The Natural History Museum and Scottish Association of Marine Sciences.

Noor, A., Ahmed, K., Mir, Z. R. (2016). Estimating abundance of some wild faunal elements of Jasrota Wildlife Sanctuary, India. *Journal of King Saud University-Science*, 28(3), 232–238.

Osborne, W. S., & Green, K. (1992). Seasonal changes in composition, abundance and foraging behaviour of birds in the snowy mountains. *Emu – Austral Ornithology*, 92(2), 93–105.

Praveen, J., Jayapal, R., & Pittie, A. (2020). Checklist of the birds of India (v4.1) [Website]. Retrieved July 25, 2020, from http://www.indianbirds.in/india/.

Reyonds, R. T., Scott, J. M., & Nussbaum, R. A. (1980). A variable circular plot method for estimating bird numbers. *Condor*, 82(3), 309–313.

Sauvjot, R. M., Buechner, M., Kamradt, D. A., & Schonerwald, C. M. (1998). Patterns of human disturbances and response by small mammals and birds, in chapparal near urban development. *Urban Ecosystem*, 2(4), 279–297.

Singh, K., Maheshwari, A., & Dwivedi, S. V. (2018). Studies on avian diversity of Banda University of agriculture and technology campus, Banda, Uttar Pradesh, India. *International Journal of Avian & Wildlife Biology*, 3(2), 177–180. https://doi.org/10.15406/ijawb.2018.03.00082.

Sultana, A., & Hussain, M. S. (2012). Bird density and abundance pattern in different forest types of the Kumaon Himalayas, India – Fitting lognormal distribution model. *Avian Ecological Behavior*, 22, 35–54.

Wadatkar, J. S. (2001). Checklist of birds from Amravati University campus, Maharashtra. *Zoos Print Journal*, 16(5), 497–499.

19 Distribution and Conservation of Kashmir Gray Langur

*Zaffar Rais Mir, Athar Noor, Riyaz Ahmad,
Khursheed Ahmad, Intesar Suhail, Bilal Habib,
Shahid Ahmad Dar, Aimon Bushra, Rita Rath,
Orus Ilyas, and Junid N. Shah*

Introduction

The Kashmir grey langur (*Semnopithecus ajax*) is an endangered (IUCN, 2019) leaf-eating primate which is distributed in parts of Pakistan, Nepal, and India (Minhas et al., 2010). The taxonomic nomenclature of the Himalayan langurs has been debated for a long time. Several classification schemes have been proposed to resolve the taxonomy of the Himalayan langurs, of which Hill's (1939) proposed classification has been recommended based on integrated information from several sources (Arekar et al., 2019). Based on morphological features, Brandon-Jones in 2004 suggested that langurs in the Himalayan region be classified as the dark-armed Himalayan langur, commonly called the Kashmir grey langur or Himalayan grey langur (*Semnopithecus ajax*). In India, *S. ajax* is distributed in Jammu and Kashmir and Himachal Pradesh (Roberts, 1997; Walker & Molur, 2004; Anandam et al., 2013; Mir et al., 2015). In Himachal Pradesh, the species is found in the Great Himalayan National Park, Kallatope National Park, Khajjiar, and Manali Wildlife Sanctuaries, while in Jammu and Kashmir, its presence has been recorded in Dachigam National Park (Mir et al., 2015), the Wadwan area of Kishtwar National Park (Brandon-Jones, 2004), and Bhaderwah forest division (Sharma & Ahmed, 2017). The species has also been reported from the Neelum and Jhelum Valleys (Ahmed et al., 1999; Iftikhar, 2006) and Machiara National Park (PoK), Pakistan (Groves, 2001; Minhas et al., 2012, 2013).

Like most of the other primate taxa in South Asia, the major and common threats faced by this primate across its distribution range include habitat loss and degradation through developmental activities, unavailability of food, and several viral and bacterial diseases (Nowak, 1999; Biswas & Sankar, 2002; Bagchi et al., 2003).

Despite its tremendous ecological importance, the species is understudied, especially in the Kashmir valley. A few studies have been conducted on Kashmir grey langur in areas other than Kashmir, which concentrated mainly on distribution, behaviour, and feeding aspects (e.g., Sugiyama, 1976; Sayers & Norconk, 2008; Minhas et al., 2010, 2012, 2013). Moreover, a long-term study is going on

DOI: 10.4324/9781003321422-22

in Chamba, Himachal Pradesh, focusing on various ecological and conservation aspects of the Kashmir grey langur. The International Union for Conservation of Nature (IUCN) distribution map of Kashmir grey langur shows its distribution restricted to Himachal Pradesh, India only. A recent study from the Kashmir valley has ascertained the presence of Kashmir grey langurs in the valley and has given an account of population estimates and diet of Kashmir grey langurs in Dachigam National Park (Mir et al., 2015). Later, Sharma & Ahmed, 2017, reported the presence of Kashmir grey langur in Jammu and Kashmir's Bhaderwah forest division. Keeping in mind the uncertainty in the distribution of this threatened primate, the present study was undertaken to investigate the distribution of this species in various protected and non-protected forest areas of Kashmir, especially in North Kashmir. This study will serve as baseline information for future studies on the species and will be helpful in planning effective management strategies for langur conservation in its distributional range.

Study Area

The Kashmir Valley

The union territory of Jammu and Kashmir has two distinct regions, comprising the Kashmir Valley and the Jammu division. Kashmir forms the northernmost part of India and lies between the western extremes of the Himalayan mountains along a valley between the Pir Panjal and the Greater Himalayas (Figure 19.1). It is located in the North-West Himalayan Biogeographic Zone (2A) (Rodgers & Panwar, 1988). Ecologically, this region is of huge importance due to its diverse habitats, which harbor many primitive as well as newly evolved taxa.

The region lies between 32° 20' to 34° 54' northern latitudes and 73° 55' to 75° 35' eastern longitudes, comprising an area of 15,948 sq. km. About 65% of its total area is covered by mountains. The valley is an asymmetrical fertile basin stretching from southeast to north-western directions, measuring 187 km from the southeast to northwest corners, while the breadth varies considerably, being 115.6 km along the latitude of Srinagar. The altitude ranges from 1,600 m above sea level at Srinagar to 5,420 m at the highest peak, Kolohai.

It's a drier region with a fauna that is distinctly Palearctic. Meher-Homji (1971) classified the climate of Kashmir as of Mediterranean type, concluding that Srinagar may have several types of climate, viz., uni-, bi-, tri-, and quadrixeric conditions in a year. The temperature of the valley ranges from an average daily maximum of 31 °C and a minimum of 15 °C in July to an average daily maximum of 4 °C and a minimum of -5 °C in January. The maximum daily humidity ranges from 40–90% throughout the year. The average rainfall in Srinagar is about 659 mm per annum, and most of the precipitation occurs in the form of snow during winter and early spring.

The present study was conducted in the northern part of the Kashmir valley. Surveys were conducted in the protected as well as non-protected forest areas.

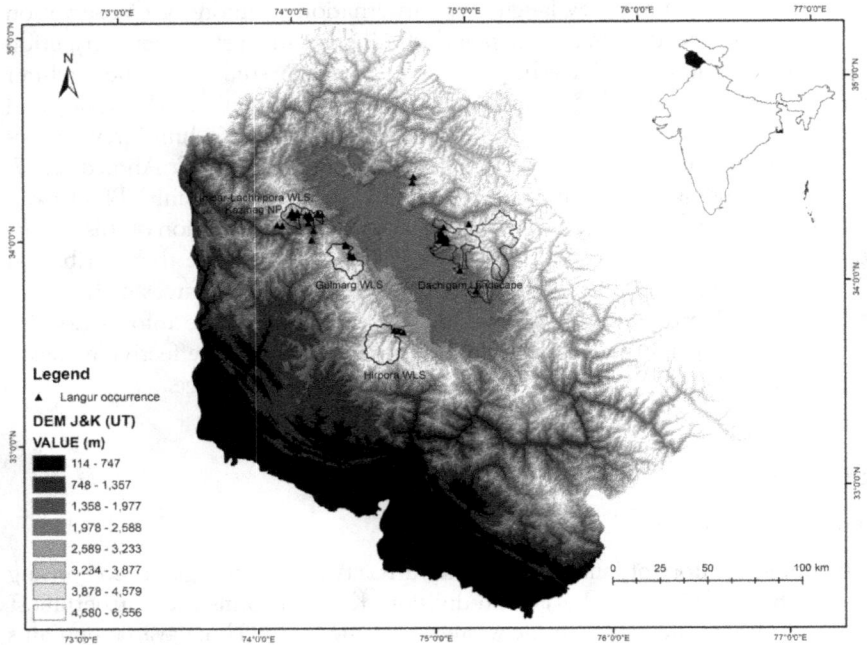

Figure 19.1 Digital elevation model of Jammu and Kashmir (Source: ASTER).

Limber and Lachhipora Wildlife Sanctuaries, as well as Naganari Conservation Reserve, were among the important protected areas surveyed. The northern parts of these three parts form the southern portion of the Kajinag National Park. Other protected areas surveyed include Hirpora Wildlife Sanctuary, Gulmarg Wildlife Sanctuary, and Dachigam Landscape. In addition to these areas, other non-protected forest areas lying in the districts of Uri, Kupwara, and Bandipora were also surveyed.

Results

A total of 349 km was travelled in surveying and trail monitoring across the protected and non-protected forest areas of the Kashmir valley. This initiative resulted in a total of 54 independent langur locations distributed throughout the surveyed area. The area clusters of Limber-Lachhipora-Kazinag, Uri, and areas clusters in the north of Kashmir valley had the highest concentration of locations (n = 22), closely followed (n = 21) by the Dachigam landscape in central Kashmir (n = 22) (Figure 19.2).

The encounter rates of langurs varied from a minimum of 0.21±0.15 indi-viduals/km to a maximum of 5.67 ± 2.42 individuals/km across the areas surveyed

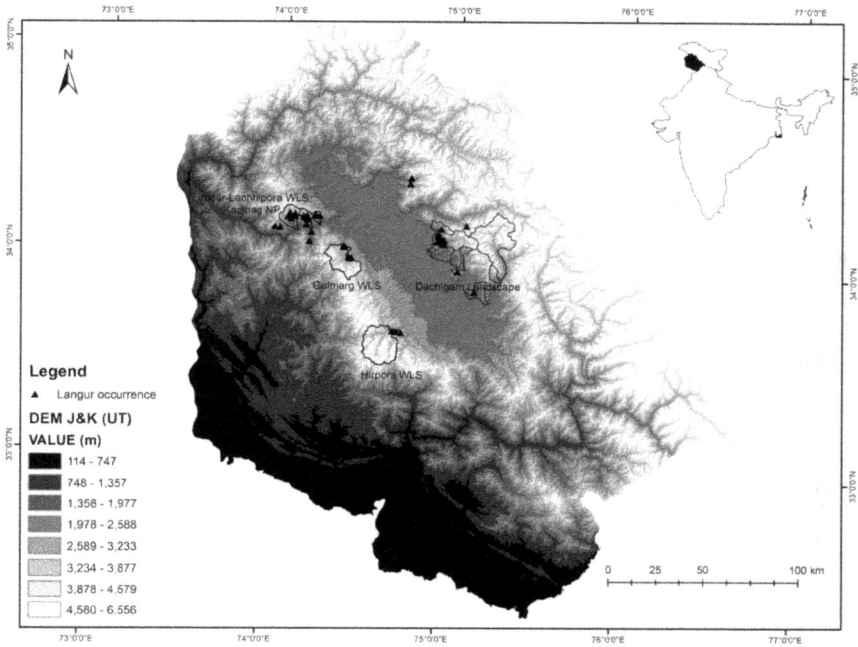

Figure 19.2 Locations of langur observed across the north, south and central areas of the Kashmir valley.

(Table 19.1). It was observed that the encounter rate of langurs was low in all the surveyed areas except in Dachigam National Park. The mean (±SE) group size of the langurs sighted across all the areas surveyed was 11.4 (±2.35) individuals. In this study, it was found that the adult langur has a rather bushy upper coat that tends to flow on either side of the body. Moreover, the forearms and thighs of adult langurs were dark grey in colour. On the basis of these features, the langur groups sighted during this study, were identified as *Semnopithecus ajax* (Brandon-Jones, 2004).

Questionnaire Surveys

A total of 68 respondents were interviewed across the protected and non-protected forest divisions of the Kashmir valley. The interviewees included 29 migratory herders, 12 shepherds, and 27 local villagers, aged between 22–80 years. The family monthly income of respondents ranged from 4,000 to 30,000 INR. Among the respondents, around 91.10% reported that they had sighted langur once or more in their life. Around 60.29% of respondents reported that the langur population is decreasing in the area, and different reasons were provided for

Table 19.1 Encounter rate of langur (per kilometre) across different protected and non-protected areas surveyed in Kashmir Valley

Area surveyed	Altitude range (m)		Protection status	Encounter rate of langur per kilometre
	Min.	Max.	Protected (P)/Non-protected (NP)	
Limber Wildlife Sanctuary	2036	3136	P	0.93±0.29
Lachhipora Wildlife Sanctuary	2111	2661	P	1.92±0.67
Nagnari Conservation Reserve	1681	2024	P	0.43±0.37
Kazinag National Park				1.52±0.91
Drang Forest Division	2589	2833	NP	0.34±0.17
Dachigam National Park	2601	2771	P	5.67±2.42
Surfraw Forest Division	2722	2871	NP	0.26±0.19
Dara Conservation Reserve	2053	2190	P	0.43±0.21
Sumlar Forest Division	2850	2952	NP	0.23±0.16
Uri Forest Division	2686	2771	NP	0.94±0.47
Shikargah Conservation Reserve	2674	2700	P	1.15±0.85
Hirpora Wildlife Sanctuary	2015	2850	P	0.21±0.15
Gulmarg Wildlife Sanctuary	2350	2740	P	0.34±0.14
Bandipur Forest Division	2750	2800	NP	0.22±0.17
Khangund Conservation Reserve	2365	2650	P	0.38±0.21

the decline (Figure 19.3). Predation by wild animals such as the common leopard (*Panthera pardus*), was reported by as many as 50% of respondents as a major cause of langur decline, followed by habitat loss (20%) and anthropogenic disturbance (18%) (Figure 19.3).

Approximately 88% of surveyed respondents were found to have a positive attitude towards langur conservation. The reason for this highly positive attitude

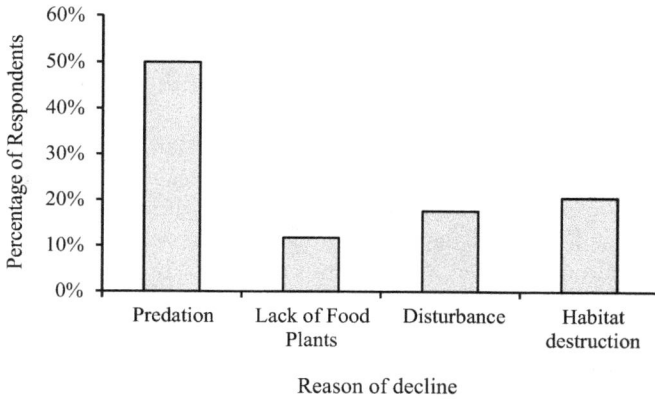

Figure 19.3 Reasons of population decline as mentioned by respondents.

was that, except in a few cases, langurs were not majorly involved in crop-raiding in the study area. None of the respondents reported any cases of langur hunting or predation by domestic dogs.

Discussion

This is the first study to report the distribution of Kashmir grey langur in most of the areas of the Kashmir valley. According to the physical characteristics provided by Brandon-Jones (2004), all the groups sighted during this study were identified as *Seminopethicus ajax*, although a recent study based on molecular phylogeny has suggested merging all three Himalayan langur species (*Semnopithecus ajax*, *Semnopithecus hector*, and *Semnopithecus chistaceus*) into one single species (Arekar et al., 2019). Through this study, which agrees with a previous study (Mir et al., 2015), we are confirming the presence of the Kashmir grey langur in different areas of the Kashmir valley and thus recommend that its range, which is restricted to the Chamba area of Himachal Pradesh, should be extended by the IUCN accordingly.

While the species is distributed in a few areas of Pakistan, Nepal, and in Jammu and Kashmir and Himachal Pradesh in India, the population of this endangered primate is low in the Kashmir valley, which is evident from low encounter rates recorded in all the protected areas except Dachigam National Park. The reasons for the low langur population in these areas need to be ascertained through a long-term ecological study in the valley. The healthy population of primates in Dachigam National Park is attributed to the good habitat conditions and protection provided inside the park.

The non-protected areas, which are under the administrative control of the Jammu and Kashmir Forest Department, form an important part of the langur distribution range. Some of these areas include Sindh Forest Division, Bandipora Forest Division, and Pir Panjal Forest Division. The current distribution map, as prepared from this study, meets the necessity of identifying potential areas that demand conservation concern. Ecological studies need to be taken up in these areas to the study population and behavioral aspects of this endangered primate.

It appears that there is no direct threat to the langur population in the region as no hunting was reported or directly observed. These langurs are threatened by habitat destruction and various anthropogenic activities. Political instability and the insurgency in the Kashmir valley have significantly affected the wildlife of the valley as conservation has taken a back seat in the area. The prey population in the Kashmir valley has shown a declining trend in the past few decades (Habib et al., 2014). This may be the cause of the huge predation pressure on langurs since other prey has become scarce. It is evident from the recent study of Dachigam National Park, which showed a high occurrence of langurs in the leopard diet (Habib et al., 2014; Mir, 2016).

The attitude of people towards langurs has been seen as positive throughout the study area, which is in line with a previous study in Chamba, Himachal Pradesh (Anandam et al., 2013). The reason for this positive attitude towards langurs can be the low human-langur conflict in the area. In the current study, a few crop-raiding incidents were only reported from the peripheral areas of Dachigam National Park which holds a good population of langurs. No conflict was reported from other protected as well as non-protected areas surveyed. One of the reasons for low or no conflict in the area can be the low population densities of this primate in the area as evident from the encounter rates estimated. It is expected that the findings of the current study could assist wildlife managers and other stakeholders in effective conservation planning and making sound management decisions for this threatened primate.

Acknowledgements: We are thankful to The Mohamed bin Zayed Species Conservation Fund for providing financial support for this study. We thank the Department of Forests and the Department of Wildlife Protection for their support and cooperation in conducting this study in the Kashmir valley. We thank our field assistants and the field staff of the Forest and Wildlife Departments for helping us in their areas of posting. We thank all the interviewees for giving their precious time and valuable information.

References

Ahmed, K. B., Awanand, M. S., & Anwar, M. (1999). Status of major wildlife species in the Moji Game Reserve, Leepa valley Azad Kashmir. *Proceedings of the Pakistan Congress of Zoology, 19,* 173–182.

Anandam, M., Ahuja, P. V., & Shah, T. A. (2013). Conserving the Himalayan Grey Langur, *Semnopithecus ajax,* an Endangered, endemic species of primate. *Report submitted to Conservation Leadership Program,* 25.

Arekar, K., Sathya, S., & Karanth, K. P. (2019). Integrative taxonomy confirms the species status of the Himalayan langurs, Semnopithecus *schistaceus* Hodgson 1840.

Journal of Zoological Systematics and Evolutionary Research. http://doi.org/10.1101/602243.

Bagchi, S., Goyal, S. P., & Sankar, K. (2003). Prey abundance and prey selection by tigers (*Panthera tigris*) in a semi-arid, dry deciduous forest in western India. *Journal of Zoology, 260*(3), 285–290.

Biswas, S., & Sankar, K. (2002). Prey abundance and food habit of tigers (*Panthera tigris tigris*) in Pench National Park, Madhya Pradesh, India. *Journal of Zoology, 256*(3), 411–420.

Brandon-Jones, D. (2004). A taxonomic revision of the langurs and leaf monkeys (Primates: Colobinae) of South Asia. *Zoos' Print Journal, 19*(8), 1552–1594.

Groves, C. (2001). *Primate taxonomy.* Smithsonian Institution Press.

Habib, B., Gopi, G. V., Noor, A., & Mir, Z. R. (2014). Ecology of Leopard (*Panthera pardus*) in relation to prey abundance and land use pattern in Kashmir Valley. *Project completion report submitted to department of science and technology, Govt. of India.* Wildlife Institute of India, pp. 72.

Hill, W. C. (1939). An annotated systematic list of the leaf-monkeys. *Ceylon Journal of Science, 459*, XXI.

Iftikhar, N. (2006). Wildlife of Azad Kashmir (in Urdu). Al-sheikh Press, Muzaffarabad, AJK.

IUCN. (2019). *IUCN red list of threatened species.* Retrieved January 28, 2020, from http://www.iucnredlist.org/.

Meher-Homji, V. M. (1971). The climate of Srinagar and its variability. *Geographical Review of India, 33*(1), 1–14.

Minhas, R. A., Ahmed, K. B., Awan, M. S., Zaman, Q., Dar, N. I., & Ali, H. (2012). Distribution patterns and population status of the Himalayan Grey Langur (*Semnopithecus ajax*) in Machiara National Park, Azad Jammu and Kashmir, Pakistan. *Pakistan Journal of Zoology, 44*(3), 869–877.

Minhas, R. A., Ahmed, K. B., Awanand, M. S., & Dar, N. I. (2010). Habitat utilization and feeding biology of Himalayan Grey Langur (*Semnopithecus entellus ajax*) in Machiara National Park, Azad Kashmir, Pakistan. *Zoological Research, 31*, 1–13.

Minhas, R. A., Ali, U., Awan, M. S., Ahmed, K. B., Khan, M. N., Dar, N. I., Qamar, Z.Q., Ali, H., Grueter, C.C., & Tsuji, Y. (2013). Ranging and foraging of Himalayan Grey Langurs (*Semnopithecus ajax*) in Machiara National Park, Pakistan. *Primates, 54*(2), 147–152. https://doi.org/10.1007/s10329-013-0345-7.

Mir, Z. R. (2016). *Monitoring prey dynamics and diet fluctuations of Leopard (Panthera pardus) in Dachigam National Park, Srinagar, Jammu and Kashmir* [Doctoral Thesis]. Wildlife Institute of India, pp. 206.

Mir, Z. R., Noor, A., Habib, B., & Gopi, G. V. (2015). Seasonal population density and winter survival strategies of endangered Kashmir Grey Langur (*Semnopithecus ajax*) in Dachigam National Park, Kashmir, India. *Springer Plus, 4*(562), 1–8. https://doi.org/10.1186/s40064-015-1366-z.

Nowak, R. M. (1999). *Walker's mammals of the world* (6th ed.). The Johns Hopkins Press.

Roberts, T. J. (1997). *The mammals of Pakistan* (rev ed., pp. 1–525). Oxford University Press.

Rodgers, W. A., & Panwar, S. H. (1988). *Biogeographical classification of India.* New Forests.

Sayers, K., & Norconk, M. A. (2008). Himalayan Semnopithecus entellus at Langtang National Park, Nepal: Diet, activity patterns, and resources. *International Journal of Primatology, 29*(2), 509–530.

Sharma, N., & Ahmed, M. (2017). Distribution of endangered Kashmir Grey Langur (Semnopithecus ajax) in Bhaderwah, Jammu and Kashmir, India. *Journal of Wildlife Research*, 5(1), 1–5.

Sugiyama, Y. (1976). Characteristics of the ecology of the Himalayan langurs. *Journal of Human Evolution*, 5(3), 249–277.

Walker, S., & Molur, S. (2004). *Summary of the status of South Asian primates* (2nd ed.). CAMP 2003 Report. Zoo Out.

20 Forest Classification of Panna Tiger Reserve

Talat Parveen and Orus Ilyas

Introduction

Land use land cover classification has been widely used to analyse land cover using remote sensing (King, 2002). Such types of map are essential tools for managing natural terrestrial ecosystem for multiple purpose, especially biodiversity conservation. One of the most important terrestrial ecosystem are formed by forest. It provides a number of ecosystem goods and services, especially tropical forest. These services include the food, water, shelter, and nutrient cycling, carbon storage in regenerating tropical forest, and play a fundamental role in biodiversity conservation. These forests are influenced by natural and human activities including degradation, deforestation, wildfires, and conversion of forested area into non-forest use, particularly agriculture. Agriculture expansion is considered to be most significant driver of deforestation, primarily occurring in the tropics, and due to this, 70–95% of the forests are lost.

To understanding the environmental change dynamics, a detailed and accurate information of forest cover is important to be used to ensure sustainable development of forest resources (Gomez et al., 2016). The mapping of the forest cover has been done by remote sensing and conventional ground-based method. But the remote sensing domain are superior from conventional ground-based method of vegetation mapping (Jenning et al., 1999; Roy et al., 2001; Rautiainen et al., 2005; Panta et al., 2008; Paletto & Tosi, 2009). A number of earlier studies have been documented forest types using remote sensing from local to global level (Woodcock et al., 2001; Patenaude et al., 2005; D'Annunzio et al., 2014; Huang et al., 2010; Kennedy et al., 2010; Tateishi et al., 2014; Zhu, 2013; Addo-Fordjour et al., 2017), and a number of image classification approaches have been used to estimate forest canopy density from satellite data (Rikimaru, 1996; Roy et al., 1996; Nandy et al., 2003; Chandrashekhar et al., 2005; Joshi et al., 2006; Nangendo et al., 2007; Panta et al., 2008; Mon et al., 2012) for monitoring and assessing deforestation and forest degradation as well (Food and Agriculture Organization (FAO), 2000, 2005; Chandrashekhar et al., 2005; Joshi et al., 2006; Panta et al., 2008; and Mon et al., 2010). Moreover, Lidar remote sensing has been widely used to measure canopy cover, particularly in the vertical dimension and species habitat suitability. By using Lidar data, Mason et al. (2003) measured

DOI: 10.4324/9781003321422-23

habitat predictor variable over large geographical areas and facilitated Skylark (*Alauda arvensis*) habitat model in Oxfordshire, UK. Hyde et al. (2005) assessed the ability of Lidar to measure canopy structure over the diverse Montane forest of the Sierra Nevada. While using the Lidar derived data, Martinuzzi et al. (2009), and Graf et al. (2009) evaluated habitat suitability map for avian species and an endangered forest grouse, capercillie (*Tetro urogallus*), respectively. That's how Psomas et al. (2011) assessed the potential of hyperspectral remote sensing for mapping aboveground biomass in grassland habitats along a dry mesic gradient in the central part of the Swiss plateau near the city of Aarau. However, for the mapping of shrubs and tree cover on grassland habitat in Denmark, Hellesen and Matikainem (2013) used Lidar data and colour-infrared ortho image.

There is a vast volume of literature on vegetation mapping and deforestation using different satellite data such as Landsat (Morain and Klankamsorn, 1978; Lanley, 1982; Woodwell et al., 1984; Nelson and Holben, 1986; Tardin and Cunha, 1990; Niemann, 1993; Garcia, 1994; Elumnoh and Shrestha, 2000; Manandhar et al., 2009; Pasquarella et al., 2016), WiFS (Singh et al., 1999; Joshi et al., 2004a), and NOAA/AVHRR (Achard and Biasco, 1990; Cross et al., 1991; Amaral, 1992). The concept of remote sensing data and its use in wildlife management that provide spatial data for wildlife habitat inventory, evaluation, and wildlife census have been widely studied across the world. In forested areas, Davis and Goetz (1990) predicted the distribution of coast live oak (*Quercus agrifolia*) near Lampac, California. For forest successional stages, Fiorella (1992) used Landsat thematic data to analyse the structure of both old and young conifer forest at H.J. Andrew's experimental forest in the central Cascade Mountain of Oregon. With reference to elk's (*Cervus elaphus nelsoni*) potential habitat, where elk and human interact and compete for space and other resources, Huber (1992–1993) used a combination of different data from home range algorithm, satellite image classification, and GIS. Tiger (*Panthera tigris*) distribution in Nepal in relation to habitat quality was reported by Smith et al. (1998) through the combination of digital thematic mapper satellite data, interviews with local people, and information from previous studies as well. In 1997, An ecological assessment of juvenile queen conch (*Strombus gigas*) was done through RS and GIS data in the Bahamas (Jones and Stoner, 1997). According to habitat categorisation and species distribution patterns in the Greater Yellowstone ecosystem, Debinski et al. (1999) used remotely sensed data and GIS. Similarly, Debeljak et al. 82001), Osborne et al. (2001), Danks and Klein (2002), Wang (2003), and Sharma et al. (2004) used remotely sensed data with regards to evaluate potential habitat for red deer (*Cervus elaphus* L.), great bustard muskoxen (*Ovibos moschatus*), giant panda (*Ailuropoda melanoleuca*), and Tibetan wild ass (*Equus Kiang*) in south-central Slovenia, central Spain, northern Alaska, and China and Nepal respectively. The spatial and temporal habitat use of impala (*Aepyceros melampus*) was studied in Bostwana Okavango Delta by Bommel et al. (2006). In order to develop a Vector Ruggedness Measure (VRM), Sappington et al. (2007) used GIS and examined the terrain relationship of VRM to slope, distance to water, and springtime for modelling adult female of big-horn sheep's (*Ovis canadensis nelsoni*) habitat in Mojave Desert. For mapping the spatial distribution of giant panda's (*Ailuropoda melanoleuca*) habitat, Vina et al. (2008) compared two models using data from different satellite sensors in Wolong Nature Reserve China. The abundance model of non-native sambar (*Cervus unicolor*) in south-eastern Australia was developed by Forysth et al. (2009). In contrast, Pringle et al. (2009) estimated habitat availability for Australia's most endangered broad-headed snakes (*Hoplocephalus bungaroides*) using pixel and object-based remote sensing in Morton National Park. According to Jones et al. (2009), land use and land cover change, which was determined by remote sensing and GIS analysis, and in surrounding areas may influence the natural resource within parks. That's how by using remotely sensed data, Goetz et al. (2009) identified the core habitat of the north-eastern and mid-Atlantic USA. While using structural complexity indices, Estes et al. (2010) described spatial distribution model of rare mountain bongo antelopes (*Tragelaphus eurycerus*isaaci). On the other hand, by using eco-logical data in combination with spectral absorption data, Knox et al. (2012) was able to evaluate spatial and temporal variation in forage quality in grassland-savannah ecosystems that support high biomass of both ungulate and domestic livestock. In favour of winter habitat selection of roe deer (*Capreolus capreolus*) in montane forest, Ewald et al. (2014) used Lidar remotely sensed data and GPS telemetry data. Recently, Schubler et al. (2020) analysed past forest loss and pre-dicted future deforestation probabilities to define crucial stepping stone habitats for biodiversity conservation in north-eastern Madagascar using remote sensing and comparative predictive modelling. In India, Pant et al. (1999), Kushwaha et al. (2000), Singh et al. (2009), Imam and Kushwaha (2013), Alam et al. (2014), and Prajapati et al. (2015) used remote sensing and GIS modelling to evaluate the habitat for sambar (*Cervus unicolor*) in Corbett National Park, goral (*Nemorhaedus goral*) in Rajaji National Park, one-horned rhino (*Rhinoceros uni-cornis*) in Kaziranga National Park, gaur (*Bos gaurus*) in Chandoli Tiger Reserve, striped hyena (*Hyaena hyaena*) in Gir National Park and Sanctuary, and chi-tal (*Axis axis*) in Panna Tiger Reserve respectively. Moreover, the availability of remotely sensed data has been used for habitat evaluation of tiger (*Panthera tigris*), the largest predator in Asia (Weber and Rabinowitz, 1996) by Singh et al. (2009, 2013) and Kumar et al. (2013) in Corbett and Panna Tiger Reserves, respectively.

Study Area

Panna Tiger Reserve (PTR) was declared as the 22nd tiger reserve of India in the year 1994. It lies between coordinates 24°15'–24°20' latitude N and 80°00'–80°15' longitude E. PTR covers an area of 1578.12 km^2 and its 77.42% lies in Panna district whereas, remaining in Chatarpur district of Madhya Pradesh. The area falls in Vindhaya range within biogeographic province 6A "Deccan Peninsula-Central Highland" (Rodgers and Panwar, 1988).

The total area of the reserve is about 1578.12 km^2 out of which 576.12 km^2 is managed as the core area. The remaining 1002 km^2 is managed as buffer area which is constituted by reserve forest, protected forest and revenue land. Ken river, one of the 16 perennial rivers of Madhya Pradesh flows through the western part of Panna Tiger Reserve. The terrain of the reserve is characterised by extensive plateau and gorges, viz. Upper Talgaon Plateau which lies in Panna Range, Middle Hinauta Plateau which lies in Hinauta Range, and the River Valley that lies in Madla and Chandranagar range. The altitude of the reserve varies from 212 to 538 M.S.L (Mean Sea Level). Climatically, the area has four seasons. The summer from March to June, monsoon (July to August), a short post-Monsoon from September to October, and winter from mid-November to February (Figure 20.1).

Flora and Fauna

Ecologically, Panna Tiger Reserve is classified as tropical dry deciduous forest. Floristically, it can be classified according to Champion and Seth (1968) as: (1) southern tropical dry deciduous teak forest, (2) northern tropical dry deciduous mixed forest, (3) dry deciduous scrub forest, (4) anogeissus pendula forest, (5) boswellia forest, and (6) dry bamboo brakes.

About 35 species of mammals are observed in the Panna Tiger Reserve and a sizeable number of are of endangered status. The carnivore fauna is represented by the tiger (*Panthera tigris*), leopard (*Panthera pardus*), dhole (*Cuonalpinus*), jungle cat (*Felis chaus*), Asiatic wildcat (*Felis silvestrisorneta*) and Indian fox (*Vulpes benghalensis*). Wolves (*Canis lupus*) occur on the fringes and outside the reserve limits. Striped hyena (*Hyaena hyaena*), sloth bear (*Melursus ursinus*), and jackal (*Canis aureus*) make up the rest of the carnivore fauna of the reserve. Chital (*Axis axis*), sambar (*Cervus unicolor*), nilgai (*Boselaphustragocamelus*), wild pig (*Sus scrofa*), chinkara (*Gazella bennetti*), and chousingha (*Tetraceros quadricornis*) are the wild ungulate species found in the study area. The common langur (*Semnopithecus entellus*) and rhesus macaque (*Macaca mulatta*) represent the primate fauna of the area. The Indian porcupine (*Hystrix indica*), honey badger (*Mellivora capensis*), and black-naped hare (*Lepus nigricollis nigricollis*) also occur in this tiger reserve. The protected area even supports over ten species of reptiles, and some are of endangered status, such as Indian cobra (*Naja naja*), Indian rock python (*Python molurus*), rat snake (*Ptyas mucosus*), and common monitor lizard (*Varanus bengalensis*). The Ken river provides shelter for a variety of aquatic fauna, and a total of 14 species of fis are found in the protected area. Species like tengra (*Mystus tengara*), catfish (*Wallago attu*), mahseer (*Tor tor*), katai (*Mistris singhala*), and hilsa (*Hilsa ilisha*) are of rare/endangered status. Avifauna diversity in the tiger reserve is quite remarkable and supports several species of vultures, including white-rumped vulture (*Gyps bengalensis*), red-headed vulture (*Sarcogpys calvus*), and Indian vulture (*Gyps indicus*).

Methodology

The study was completed in three phases. In the first data was collected from satellite and topographic maps, while in the second phase a systematic field survey

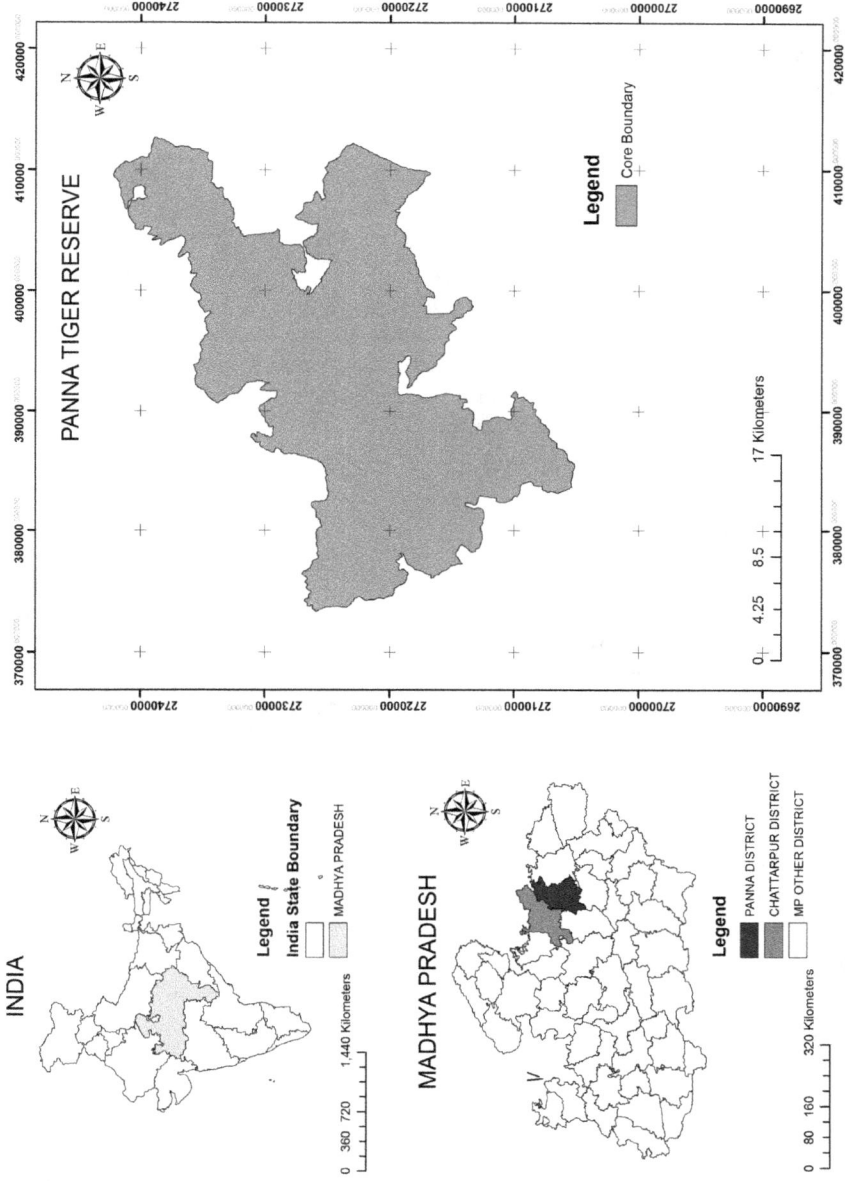

Figure 20.1 Map of the study area.

was conducted for ground trothing. The third phase included preparation of a Land Use Land Cover map and forest crown density map.

Data Collection and Data Processing

For this purpose, two types of data were used: primary and secondary data. The primary data was taken by field survey. During the field work observation on habitat, vegetation types, and their associated species and all ecological features were recorded using GPS. Various other sources like National Park maps, topographic maps, and satellite images were used as secondary data.

Satellite Data

The satellite images of Landsat 8 OLI/TIRS were acquired from USGS earth explorer on 24 April 2017, Path 144, Row 43. Data processing, interpretation, and analysis were done in ERDAS IMAGINE 2018 software. These data are going to be used in conjunction with other spatial data so that it is important to georeference the distorted data to a coordinate system. Since Landsat data is available in georeferenced form, this exercise was not required for the present satellite data. A False Color Composite (FCC) was created with combination of three bands (6,5,4). The printout of the FCC was taken on A-3 size paper, and this map was taken to the field for ground truthing.

Collateral Data

The Survey of India (SOI, 1975) top sheets (54 P/14, 54 P/15, 63 d/1, 63 d/2) at scale of 1: 50,000 were used to prepare map of the study area. All these top sheets were scanned and exported to ERDAS IMAGINE 2018 in image (.img) format for geo-referencing. The values of latitude/longitude of grid intersection were used as ground control points. These coordinates were entered as longitude of topographic map in X field and latitude of topographic map in the Y field in degree/minutes/seconds (DD MM SS) format that automatically converted into decimal format (Imam, 2011). After this step, the map was resampled by nearest neighbourhood method and re-projected into UTM- WGS 84 projection (Lillesand and Kiefer, 1994). Then all top sheets were mosaicked into one image that was overlaid on rectified Landsat 8 Image, and the top sheet features like roads, railway lines, canal were used to check the accuracy of mosaiced map. From this mosaiced topo sheet image, a subset of area of interest (AOI) was extracted to take information on various aspects of park boundary, land cover, human habitation, etc.

Field Survey

The field survey was conducted during March to April 2018. A systematic survey of accessible area was done and it was traversed on foot. Using GPS, the pattern, texture, shape, and size of the particular topographic feature were recorded on paper printout of FCC map.

Post Classification

Preparation of Land Use Land Cover Map

The satellite data was used for preparing FCC that served as the basic to develop Land Use Land Cover and Forest Crown Density Maps. The LULC map was prepared using unsupervised classification in ERDAS IMAGINE 2018. This was developed through field survey of the study area. In this method, the study area was classified in 200 classes. Later on, these classes were merged on the basis of similar pixel values and similar land cover types in seven major classes. Resulted in, the entire study area was classified into seven classes (Table 20.1, Figure 20.2)

Preparation of Forest Density Map

On the basis of NDVI values, the forest density map was prepared that was categorised into four canopy density classes, viz. < (less than) 10%, 10–40%, 40–70%, and > (more than) 70%. NDVI is a very simple method used for measuring and mapping the density of vegetation. It is calculated by the formula: $NDVI = (IR-R)/(IR+R)$, where R = red light and IR = Infrared light (Lellesand and Kiefer, 1994). The group of pixels having NDVI values from 0.06 to 0.20 were categorised under the crown density class of < 10%, 0.20 to 0.29 as under the crown density class of 10–40%, and 0.29 to 0.35 were categorised as 40–70%. The group of pixels having NDVI value 0.35 to 0.49 were kept under the crown density class of more than 70% (Table 20.2, Figure 20.3)

Result and Discussion

Knowledge of habitat type plays a vital role in species conservation and restoration activities. For the existence of any individual species is determined by presence of suitable habitat type. A suitable habitat consists of a source of food,

Table 20.1 Area covered by different land cover types, Panna Tiger Reserve, India (2017)

S. No.	Land cover	Area in km²	Percentage area
1	Dense forest	230.78	42.27%
2	Fairly dense forest	93.67	16.20%
3	Water	9.53	1.26%
4	Open forest	85.3	13.01%
5	Scrubland	79.82	14.07%
6	Grassland with scattered trees	43.21	7.96%
7	Fallow land/blank area	33.81	5.23%
Total		576.12	100%

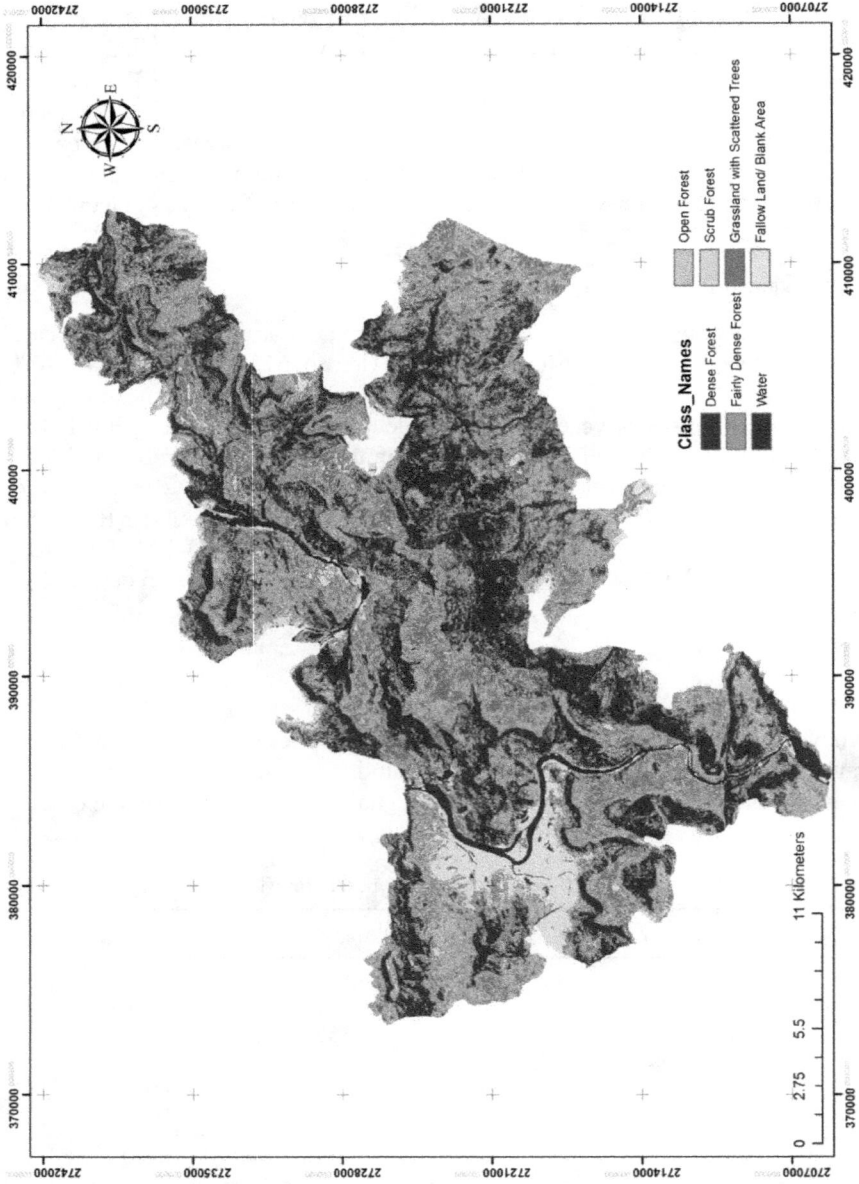

Figure 20.2 Land Use Land Cover map of Panna Tiger Reserve, India (2017).

Table 20.2 Area of different forest crown density classes, Panna Tiger
        Reserve (2017)

S. No.	Forest crown density class	Area in km²
1	Less than 10%	28.86
2	10–40%	103.46
3	40–70%	194.94
4	More than 70%	215.67
Total		576.12

protective cover, space, and water, henceforth an update of the Land Use Land
Cover map of Panna Tiger Reserve has been developed. The Land Cover map of
PTR depicted that out of 576.12 Km², 230.78 Km² area are covered as dense forest followed by fairly dense forest (93.67 Km²), open forest (85.3 Km²), scrub land
(79.82 Km²), grassland with scattered trees (43.21 Km²), fallow land (33.81 Km²),
and water bodies (9.53 Km²) (Table 20.1, Figure 20.2). The tropical dry deciduous
forest of Panna has several discrete cover patches, which supports ungulate population that have evolved in different environmental conditions. In such a way,
sambar originated in Oriental regions and has adapted in forested environments
(Corbet & Hill, 1992; Schaller, 1967). Chital and nilgai are autochthonous, for
the former is adapted to forest edges and ecotones (Eisenberg 1981); Schaller,
1967). Whereas chinkara has evolved in open country, it is closely related to
African gazelles (Corbet & Hill, 1992; Prater, 1971). Dense forest was confined
to southern side of tiger reserve where teak species was dominant. Other associated species are mainly dhaba (*Anogeissus Latifolia*), ghont (*Zizyphus xylopyra*),
seja (*Legerstromia parviflora*), and saj (*Terminalia tomentosa*). This type of forest
was found at many places in Hinauta and Madla range, especially around Badar,
Bilhata, Bundelababa, Katari, and Mahua Mod beats. The area around Badar
serves as ecotone, and has water and well protected forest patch, which have tall
teak trees with a variety of grass on gentle terrain. Chital usually prefer to be near
water and drink once a day but more frequently in summer (Laura, 2016). That's
the reason these forests support good populations of chital and nilgai. Nilgai density is influenced by water availability in their native range in India and Nepal
because individuals frequently drink water (Bagchi et al., 2008). Nilgai are intermediate feeders, and they can switch their diet from browse to grasses (Hines,
2016). Similarly, Chakrabarty (1991) and Bhat (1993) concluded that chital prefer forest with higher tree density (50–150 tree/ha) class, grass cover of (25–50%)
class, herb cover of (50–75%) class, and sparse shrubs during summer. During the
winter season, it shows preference for the tree density of (50–100 tree/ha) class,
grass cover of (50–75%) class, and herb cover of (25–50%) class. There are many
palatable grass and tree species that are consumed by chital. Existing literature
reports about 160–190 plant species from the species range. Chital is mainly a
grazer and feeds on short sprouting grasses. During the winter season; more fallen

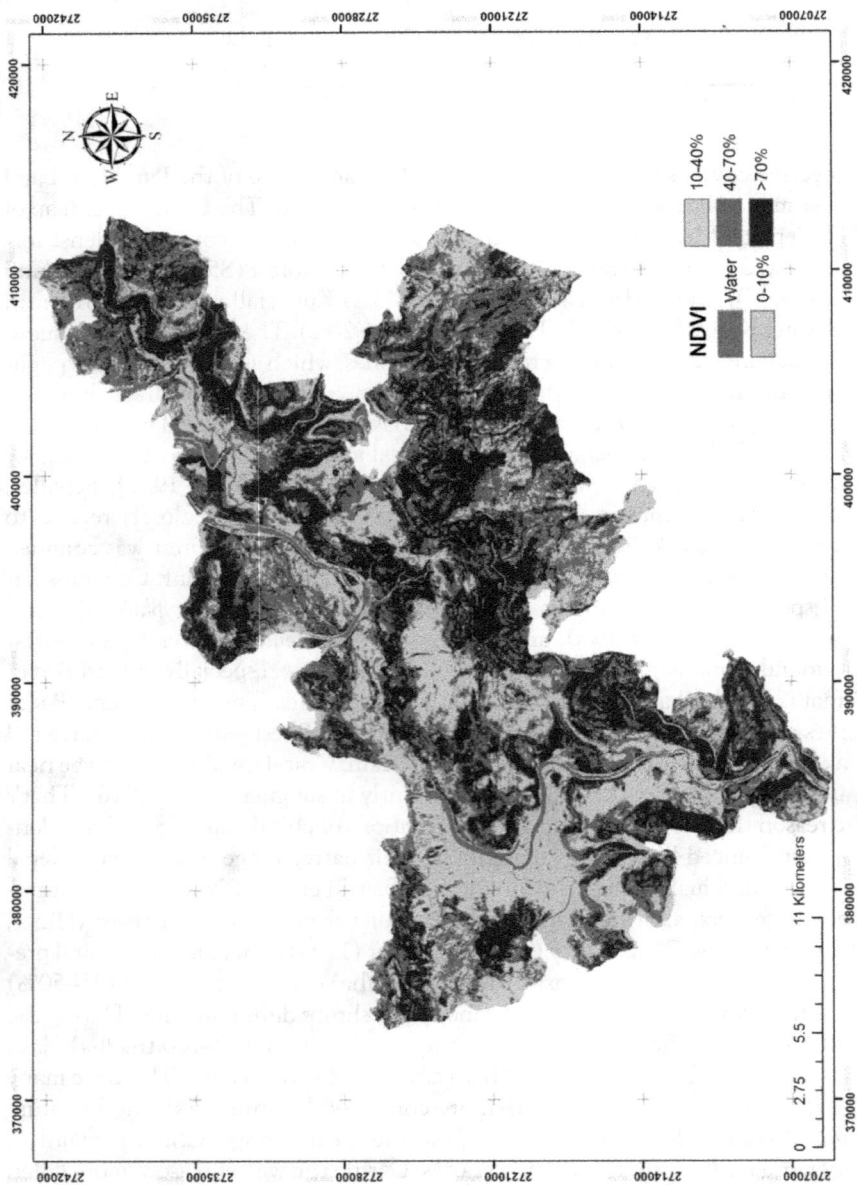

Figure 20.3 Forest crown density map of Panna Tiger Reserve, India (2017).

leaves, flowers, and fruits that are high in nutrient (Sharma & Chalise, 2014) eaten by chital.

Open forest was found at northern side of tiger reserve where mixed type of forest patch was found and characterised by open grass-shrub association with sparse tree species. In such a way, these forest patches support nilgai, chinkara and four-horned antelope species. It was mainly present in North Hinauta, West Hinauta, and many places in Bargadi beat in Hinauta range. Previous studies showed that nilgai and four-horned antelope prefer open areas, avoiding very dense forest and preferring scrubland with low tree and shrub densities (Dunbar Brander, 1923; Prater, 1971; Sankhala, 1977). Moreover, availability of water is important to the four-horned antelope (Jerdon, 1874; Blanford, 1888; Prater, 1971). These areas have several artificial and man-made waterholes.

Scrub forest was found mainly in Kishangarh and Chandranagar Range, and near the edge of river in Pipartola and Nararan beat and in some places of Hinauta range, viz. Gangau beat which was dominated by *Zizyphus* and acacia species. These species provide palatable leaves and fruits to herbivores like chital and nilgai. Hofmann (1985) classified chital as an intermediate or mixed feeder while Rodger (1988) classified chital as a generalist feeder with a diet consisting of grasses, forbs, and leaves of woody plants. On the other hand, Chakrabarty (1991) studied habitat use pattern for chital in Sariska Tiger Reserve and found that *Zizyphus* mixed vegetation and acacia mixed scrub were preferred by chital during winter and summer seasons, respectively.

Fairly dense kardhai (*Anogeissus pendula*) forest was found in many patches within Hinauta, Madla, and Gehrighat ranges, and some small patches in and around Bargadi near Kardhai Nala, while large contiguous patches was found from Pipartola to Gangau A and Gangau B beats. These forest patches were present on highly undulating terrain which was not preferred by chital and replaced by sambar and nilgai. It has been widely reported that chital prefer flat and gentle undulating terrain in both seasons (Chakrabarty, 1991; Bhat, 1993; Kumar, 2010).

Fairly dense salai (*Boswellia serrata*) forest was found in very small patches between Khamariya and Gangau B boundary and some in Badar beat, and associated with teak species. These forest patches are utilised by mainly sambar, nilgai, and four-horned antelope. Varman and Sukumar (1993) found that the preference of dense forest may be an anti-predator strategy, and sambar is the main prey of tiger in their native range (Biswas and Sankar, 2002; Bagchi et al., 2003; O'Brien et al., 2003). Previous studies in tropical and temperate regions have shown that a variety of habitats are used by sambar at high and low altitudes, and they are able to utilise a wide variety of plant species (Schaller, 1967; Downes, 1983; Ngampongsai, 1987; Bentley, 1998). Dry bamboo brakes were occurred on mostly slopes and valley, viz. Phata Nala in Gehrighat range andDudan Ghati near Khamariya beat in Hinauta range. It occurs in clusters between deciduous trees likely in Pipartola beat. These bamboos have palatable leaves that are utilised by herbivores mainly sambar. Padmalal et al. (2003) described sambar as predominantly being grazers, while some studies have suggested sambal is an intermediate feeder that utilises a mixture of browse and herbs (Varman & Sukumar, 1993; Stafford, 1997), and others have categorised sambar as predominantly being a browser (Santiapillai et al., 1981; Burke, 1982; Ngampongsai, 1987; Shea et al., 1990; Semiadi et al., 1995). Previous studies have concluded that they have a tendency to switch their habitat preference with respect to seasonal food resource availability, forage quality, and abundance (Santiapillai et al., 1981; Varman and Sukumar, 1993; Davies et al., 2001).

Grassland with scattered trees was found mainly in Hinauta range. Originally, there were 13 villages present inside the core boundary of the tiger reserve (Kumar et al., 2013). The land that was evacuated by these villages has been transformed into successional grassland with scattered trees. The best known place of grassland with scattered trees is in Hinauta near Bargadi A, Bargadi B, Khamariya, and some small patches in north and west Hinauta beats. This habitat provides good foraging of fallen leaves and fruits from trees like bahera, mohla, gurar, karhar and kaitha and supports significant population of chinkara, nilgai, four-horned antelope, and chital. Dookia and Goyal (2007) observed that the chinkara selectively prefer low fibre and high crude protein-containing leaves. Despite the fact the young and sprouting shoots of grasses are consumed by chinkara, Roberts (1977) has observed such diet preference during monsoon season and noticed that chinkara changed their diet from browsing to grazing only in monsoon season for a limited time. One study from Bandipur National Park found that four-horned antelope prefers the tree-savanna deciduous habitat, characterised by relatively open habitat with lower tree density and high degree of deciduousness (Krishna et al., 2008). A similar result was found by Chundawat et al. (2001), who concluded the four-horned antelopes prefer open habitats characterised by lower canopy cover. The highest density of four-horned antelope was reported from Panna Tiger Reserve in Central India. According to Mallon (2003), habitat loss is one of the main reasons for the declining four-horned antelope population in the wild.

The assessment of forest crown density of the tiger reserve was done on the basis of NDVI values. The result indicates that the majority of the forest area was covered by crown density of more than 70% (215.67 Km^2) (Table 20.2, Figure 20.3). In the river valley and along the areas of seasonal drainage were areas of tall grass and closed woodland. The forest area with crown density class of 40–70% was mainly distributed by the side of dense forest covering 194.94 Km^2 of the tiger reserve. The forest (103.46 Km^2) area covered by crown density of 10–40% was mostly grassland with scattered trees. This is due to translocation of villages from the tiger reserve which lies mainly in undisturbed Bargadi, West Hinauta, North Hinauta, Khamariya, and Mahua Mod Beats. The rest of the tiger reserve, 28.86 Km^2, was covered by crown density of 0–10%, which was mostly agriculture land and fallow/blank land. This category is facing high disturbance due to the presence of villages near Chandranagar and Kishangarh Ranges. There are 79 villages within 10 km of the boundary outside the reserve. These villages partially or fully depend on the PTR resources. They exploit the forest resources through collection of minor forest produce (MFP), illegal timber collection, extraction of fodder, and grazing a large number of livestock. Apart from these disturbances, one of the major threats to wildlife, especially the prey species of tiger, is livestock grazing, which induces competition. Several studies have suggested that the distribution and abundance of wild herbivores have been affected by domestic livestock grazing (Schaller, 1977; Jackson and Ahlbom, 1987; Paudyal and Bauer, 1988; Bauer, 1990; Prins, 1992; Mishra, 2001; Bagchi et al., 2002; Raghavan, 2003). This study supports earlier findings (Singh et al., 2009, Singh et al., 2013; Kumar et al., 2013) that can provide better habitats for tigers and their prey. Recent studies (Eisenberg and Seidensticker, 1976; Karanth and Sunquist, 1992; Karanth and Nichols, 1998; Carbone and Gittleman, 2002) suggest that the densities of ungulate prey species is directly related to the abundance of tigers. Tigers require large suitable habitats to survive in their natural state due to their territorial nature and wide movement pattern (Seidensticker, 1976a). The main tiger prey species are medium to large sized ungulates (Schaller, 1967; Seidensticker, 1976b; Karanth and Sunquist, 1995; Tamang, 1979). It has been reported that the main prey species of tiger are chital and sambar, along with other common species such as wild pig, gaur, and nilgai in South Asia (Sedensticker, 1976a). If the habitat is adequately protected, then it will support a fairly high prey biomass that is necessary for a healthy tiger population. Therefore, these areas have been given more attention for conservation and their sustainable management from human interference and other anthropogenic activities.

Acknowledgements: This study was funded by the Council for Scientific and Industrial Research (CSIR), Government of India. The authors thank the Chairman, Department of Wildlife Sciences, AMU, Aligarh for providing all necessary facilities. Thanks are also due to PCCF Chief WLW and all the forest staff of MP for permitting us to work in Panna Tiger Reserve. The authors sincerely thank Dr. Ekwal Imam and Ms. Farah Akram of Dept. of Wildlife sciences for immense help and support.

References

Achard, F., & Blasco, F. (1990). Analysis of vegetation seasonal evolution and mapping of forest cover in West Africa with the use of NOAA AVHRR HRPT data. *Photogrammetric Engineering and Remote Sensing, 56,* 1359–1365.

Addo-Fourdjour, P., & Ankomah, F. (2017). Pattern and drivers of forest land cover changes in tropical semi-deciduous forest in Ghana. *Journal of Land Use Science, 12,* 71–86.

Alam, M. S., Khan, J. A., Kushwaha, S. P. S., Agrawal, R., Pathak, B. J., & Kumar, S. (2014). Assessment of suitable habitat of near threatened striped hyena (*Hyaena hyaena* Linnaeus, 1758) using remote sensing and geographic information system. *Asian Journal of Geoinformatics, 14,* 1–10.

Amaral, S. (1992). Deforestation estimates in AVHRR/NOAA and TM/Landsat images for a region in central Brazil. *International Archives of Photogrammetry and Remote Sensing, VII*(29), 764–767.

Bagchi, S., Goyal, S. P., & Sankar, K. (2003). Prey abundance and prey selection by tigers (*Panthera tigris*) in a semi-arid, dry deciduous forest in western India. *Journal of Zoology*, 260(3), 285–290.

Bagchi, S., Goyal, S. P., & Shankar, K. (2008). Social organization and population structure of ungulates in a dry tropical forest in western India (Mammalia, Artiodactyla). *Mammalia*, 72, 44–49.

Bagchi, S., Mishra, C., Bhatnagar, Y. V., & McCarthy, T. (2002). Out of steppe? Pastoralism and ibex conservation in Spiti. CERC Technical Report No. 7. Nature Conservation Foundation, Mysore, Wildlife Institute of India, Dehradun and International Snow Leopard Trust, Seattle, USA.

Bauer, J. J. (1990). The analysis of plant-herbivore interactions between ungulates and vegetation on alpine grasslands in the Himalayan region of Nepal. *Vegetation*, 90(1), 15–34.

Bentley, A. (1998). *An introduction to the deer of Australia with special reference to Victoria* (3rd ed.). Australian Deer Research Foundation.

Bhat, S. D. (1993). *Habitat use by chital (Cervus axis) in Dhaulkhand, Rajaji national park, India* [M.Sc. Dissertation]. Saurashtra University.

Biswas, S., & Sankar, K. (2002). Prey abundance and food habit of tigers (Panthera tigris tigris) in Pench National Park, Madhya Pradesh, India. *Journal of Zoology*, 256(3), 411–420.

Blanford, W. T. (1888). *The fauna of British India, including Ceylon and Burma – Mammalia* (pp. 519–521). Taylor and Francis.

Bommel, F. P. J., Heitkonig, I. M. A., Epemat, G. F., Ringroset, S., Bonyongot, C., & Veenendaalt, E. M. (2006). Remotely sensed habitat indicators for predicting distribution of Impala (Aepyceros melampus) in the Okavango Delta, Botswana. *Journal of Tropical Ecology*, 22(1), 101–110.

Burke, P. (1982). Food plants utilized by sambar. *Australian Deer*, 7, 7–12.

Carbone, C., & Gittleman, J. L. (2002). A common rule for scaling of carnivore density. *Science*, 295(5563), 2273–2276.

Chakrabarty, B. (1991). *Habitat use by radio instrumented chital, sambar and nilgai in Sariska Tiger Reserve* [M.Sc. Dissertation]. Saurashtra University.

Champion, H. G., & Seth, S. K. (1968). *Revised survey of the forest types of India*. Manager, Publications.

Chandrashekhar, M. B., Saran, S., Raju, P. L. N., & Roy, P. S. (2005). Forest canopy density stratification: How relevant is biophysical spectral response modeling approach? *Geocarto International*, 20(1), 15–21.

Chundawat, R. S. (2001). *Strengthening tiger conservation through understanding predator-prey relationships in dry tropical forests of India* [Annual Report]. Center for Wildlife Studies.

Corbet, G. B., & Hill, J. E. (1992). *The mammals of the Indo-Malayan region: A systematic review*. Oxford University Press.

Cross, A. M., Settle, J., Drake, N. A., & Paivinen, R. T. M. (1991). Subpixel measurement of tropical forest cover using AVHRR data. *International Journal of Remote Sensing*, 12(5), 1119–1129.

D'Annunzio, R., Lindquist, E., & MacDicken, K. G. (2014). Global forest land-use change from 1990 to 2010: An update to a global remote sensing survey of forests. *Food and Agriculture Organization of the United Nations*. Report from FAO and European Commission Joint Research Centre.

Davies, G., Heydon, M., Leader-Williams, N., MacKinnon, J., & Newing, H. (2001). The effects of logging on tropical forest ungulates. In R. A. Fimbel, A. Grajal, & J.

G. Robinson (Eds.), *The cutting edge: Conserving wildlife in logged tropical forest* (pp. 91–104). Colombia University Press.

Davis, F. W., & Goetz, S. (1990). Modeling vegetation pattern using digital terrain data. *Landscape Ecology*, 4(1), 69–80.

Debeljak, M., Dzeroski, S., Jerina, K., Kobler, A., & Adamic, M. (2001). Habitat suitability modelling for red deer (Cervus elaphus L.) in South-central Slovenia with classification trees. *Ecological Modelling*, 138(1–3), 321–330.

Debinski, D. M., Kindscher, K., & Jakubauskas, M. E. (1999). A remote sensing and GIS based model of habitats and biodiversity in the greater yellowstone ecosystem. *International Journal of Remote Sensing*, 20(17), 3281–3291.

Danks, F. S. & Klein, D. R. (2002). Using GIS to predict potential wildlife habitat: A case study of muskoxen in northern Alaska. *International journal of Remote Sensing*, 23(21), 4611–4632.

Dookia, S., & Goyal, S. P. (2007). Chinkara or Indian gazelle. In K. Sankar & S. P. Goyal (Eds.), *Ungulates of peninsular India* (pp. 103–114). ENVIS bulletin, Wildlife Institute of India.

Downes, M. C. (1983). *The forest deer project 1982: A report to the forests commission Victoria*. Australian Deer Research Foundation.

Dunbar Brander, A. P. (1923). *Wild animals in central India. Reprinted in 1982* (pp. 241–246). Natraj Publishers.

Eisenberg, J. F. (1981). *The mammalian radiations: An analysis of trends in evolution, adaptation and behavior*. Athlone Press.

Eisenberg, J. F., & Seidensticker, J. (1976). Ungulates in Southern Asia: A consideration of biomass estimates for selected habitats. *Biological Conservation*, 10(4), 293–305.

Elumnoh, A., & Shrestha, R. P. (2000). Application of DEM data to Landsat image classification: Evaluation in a tropical wet - dry landscape of Thailand. *Photogrammetric Engineering and Remote Sensing*, 66, 297–304.

Estes, L. D., Reillo, P. R., Mwangi, A. G., Okin, G. S., & Shugart, H. H. (2010). Remote sensing of structural complexity indices for habitat and species distribution modeling. *Remote Sensing of Environment*, 114(4), 792–804.

Ewald, M., Dupke, C., Heurich, M., Müller, J., & Reineking, B. (2014). LiDAR remote sensing of forest structure and GPS telemetry data provide insights on winter habitat selection of European roe deer. *Forests*, 5(6), 1374–1390.

FAO. (2000). *Global forest resource assessment 2000*. Food and Agriculture Organization, the United Nations.

FAO. (2005). *Global forest resource assessment 2005*. Food and Agriculture Organization, the United Nations.

Fiorella, M. (1992). *Forest and wildlife habitat analysis using remote sensing and geographic information systems*. Master of Science, Oregon State University.

Forsyth, D. M., McLeod, S. R., Scroggie, M. P., & Whit, M. D. (2009). Modelling the abundance of wildlife using field surveys and GIS: Non-native sambar deer (Cervus unicolor) in the Yarra Ranges, South-Eastern Australia. *Wildlife Research*, 36(3), 231–241.

Garcia, M. C., & Alvarez, R. (1994). TM digital processing of a tropical forest region in Southeastern Mexico. *International Journal of Remote Sensing*, 15(8), 1611–1632.

Goetz, S. J., Jantz, P., & Jantz, C. A. (2009). Connectivity of core habitat in the Northeastern United States: Parks and protected areas in a landscape context. *Remote Sensing of Environment*, 113(7), 1421–1429.

Gómez, C., White, J. C., & Wulder, M. A. (2016). Optical remotely sensed time series data for land cover classification: A review. *ISPRS Journal of Photogrammetry and Remote Sensing*, 116, 55–72.

Graf, R. F., Mathys, L., & Bollmann, K. (2009). Habitat assessment for forest dwelling species using LiDAR remote sensing: Capercaillie in the Alps. *Forest Ecology and Management*, 257(1), 160–167.

Hellesen, T., & Matikainen, L. (2013). An object-based approach for mapping shrub and tree cover on grassland habitats by use of LiDAR and CIR orthoimages. *Remote Sensing*, 5(2), 558–583.

Hines, S. L. (2016). *Cattle, deer, and nilgai interactions* [Dissertation]. Texas A&M University of Kingsville.

Hofmann, R. R. (1985). The royal society of New Zealand. *Bull*, 22, 393–407.

Huang, C., Goward, S. N., Masek, J. G., Thomas, N., Zhu, Z., & Vogelmann, J. E. (2010). An automated approach for reconstructing recent forest disturbance history using dense Landsat time series stacks. *Remote Sensing of Environment*, 114(1), 183–198.

Huber, T. P. (1992–93). Integrated remote sensing and GIS techniques for elk habitat management. *Journal of Environmental Systems*, 22(4), 325–339.

Hyde, P., Dubayah, R., Peterson, B., Blair, J. B., Hofton, M., Hunsaker, C., Knox, R., & Walker, W. (2005). Mapping forest structure for wildlife habitat analysis using waveform lidar: Validation of montane ecosystems. *Remote Sensing of Environment*, 96(3–4), 427–437.

Imam, E. (2011). Mapping of landscape cover using remote sensing and geographic information system in Chandoli National Park, India. *Momona Ethiopian Journal of Science*, 3(2), 78–92.

Imam, E., & Kushwaha, S. P. S. (2013). Habitat suitability modelling for Gaur (Bos gaurus) using multiple logistic regression, remote sensing and GIS. *Journal of Applied Animal Research*, 41(2), 189–199.

Jackson, R., & Ahlborn, G. (1987). A high altitude wildlife survey of the Hongu valley with special emphasis on snow leopard [Report]. HMG, Department of National Parks and Wildlife Conservation (DNPWC), and King Mahendra Trust (KMTNC).

Jennings, S. B., Brown, N. D., & Sheil, D. (1999). Assessing forest canopies and understory illumination: Canopy closure, canopy cover and other measures. *Forestry*, 72(1), 59–74.

Jerdon, T. C. (1874). *A handbook of the mammals of India* (pp. 273–275). Reprinted by Mittal Publications.

Jones, D. A., Hansen, A. J., Bly, K., Doherty, K., Verschuyl, J. P., Paugh, J. I., Carle, R., & Story, S. J. (2009). Monitoring land use and cover around parks: A conceptual approach. *Remote Sensing of Environment*, 113(7), 1346–1356.

Jones, R. L., & Stoner, A. W. (1997). The integration of GIS and remote sensing in an ecological study of queen conch, strombus gigas, nursery habitats. *Proceeding of Gulf and Caribbean Fisheries Institute*, 49, 523–530.

Joshi, C., Leeuw, J. D., Skidmore, A. K., Duren, I. C. V., & Oosten, H. V. (2006). Remotely sensed estimation of forest canopy density: A comparison of the performance of four methods. *International Journal of Applied Earth Observation and Geoinformation*, 8(2), 84–95.

Joshi, P. K., Joshi, P. C., Singh, S., Agarwal, S., & Roy, P. S. (2004a). Tropical forest covers type characterization in central highlands of India, using multi-temporal IRS-1C WiFS data. *Indian Journal of Forestry*, 27, 157–168.

Karanth, K. U., & Nichols, J. D. (1998). Estimation of tiger densities using photographic captures and recaptures. *Ecology*, 79(8), 2852–2862.

Karanth, K. U., & Sunquist, M. E. (1992). Population structure, density and biomass of large herbivores in the tropical forests of Nagarahole, India. *Journal of Tropical Ecology*, 8(1), 21–35.

Karanth, K. U., & Sunquist, M. E. (1995). Prey selection by tiger, leopard and dhole in tropical forest. *Journal of Animal Ecology, 64*(4), 439–450.

Kennedy, R. E., Yang, Z., & Cohen, W. B. (2010). Detecting trends in forest disturbance and recovery using yearly Landsat time series: 1. LandTrendr – Temporal segmentation algorithms. *Remote Sensing of Environment* (in press).

King, R. B. (2002). Land cover mapping principles: A return to interpretation fundamentals. *International Journal of Remote Sensing, 23*(18), 3525–3546.

Knox, N. M., Skidmore, A. K., Prins, H. H. T., Heitkönig, I. M. A., Slotow, R., Waa, C. V. D., & de Boer, W. F. (2012). Remote sensing of forage nutrients: Combining ecological and spectral absorption feature data. *Journal of Photogrammetry and Remote Sensing, 72*, 27–35.

Krishna, C., Krishnaswamy, J., & Kumar, N. S. (2008). Habitat factors affecting site occupancy and relative abundance of four-horned antelope. *Journal of Zoology, 276*(1), 63–70.

Kumar, A. A., Sivakumar, R., & Ramesh, K. (2013). Assessing habitat sutability for tiger (*Panthera tigris*) in Panna Tiger Reserve, Madhya Pradesh, India: A geospatial approach. *Journal Scientific Transactions in Environment and Technovation, 7*, 77–81.

Kumar, N. S. (2010). *Assessment of distribution and abundance of ungulate prey using spatial models in Nagarahole and Bandipure Tiger Reserves of India* [Ph.D. Dissertation]. Manipal University.

Kushwaha, S. P. S., Roy, P. S., Azeem, A., Boruah, P., & Lahan, P. (2000). Land area change and Rhino habitat sutability analysis in Kaziranga National Park, Assam. *Tiger Paper, xxvii*, 9–17.

Lanley, J. P. (1982). Tropical forest resources. *F.A.O. Forestry Paper 30*, United Nations, Food and Agriculture Programme Publications, Rome.

Laura, C. (2016). *Activity pattern and interaction of European mouflon (Ovis musimon) and Axis deer (Axis axis) on Island of Rab, Croati* [Bachelor Thesis]. University Of Bologna.

Lillesand and Kiefer. (1994). *Remote sensing and image interpretation* (3rd ed.). Jhon Wiley and Sons, Inc.

Mallon, D. P. (2003). *Tetracerus quadricornis*. In *2006 IUCN Red list of threatened species*. Available at htpp:// www.iucnredlist.org.

Mallon, D. P., & Kingswood, S. C. (2001). *Antelopes: Part 4: North Africa, the middle East, and Asia.* Global Survey and Regional Action Plans. IUCN/SSC Antelope Specialist Group, pp. 248.

Manandhar, R., Odeh, I. O. A., & Ancev, T. (2009). Improving the accuracy of land use and land cover classification of Landsat data using post-classification enhancement. *Remote Sensing, 1*(3), 330–344.

Martinuzzi, S., Vierling, L. A., Gould, W. A., Falkowski, M. J., Evans, J. S., Hudak, A. T., & Vierling, K. T. (2009). *Mapping snags and understory shrubs for a LiDAR-based assessment of wildlife habitat suitability.* USDA Forest Service/UNL Faculty Publications, p. 213.

Mason, D. C., Anderson, G. Q. A., Bradbury, R. B., Cobby, D. M., Davenport, I. J., Vandepoll, M., & Wilson, J. D. (2003). Measurement of habitat predictor variables for organism–habitat models using remote sensing and image segmentation. *International Journal of Remote Sensing, 24*(12), 2515–2532.

Mishra, C. (2001). *High altitude survival: Conflicts between pastoralism and wildlife in the Trans Himalaya* [Ph.D. Thesis]. Wageningen University.

Mon, M. S., Kajisa, T., Mizoue, N., & Yoshida, S. (2010). Monitoring deforestation and forest degradation using FCD Mapper in Bago Mountain areas, Myanmar. *Journal of Forest Planning, 15*(2), 63–72.

Mon, M. S., Mizaue, N., Htun, N. Z., Kajisa, T., & Yoshida, S. (2012). Estimating forest canopy density of tropical mixed deciduous vegetation using Landsat data: A comparison of three classification approaches. *International Journal of Remote Sensing*, *33*(4), 1042–1057.

Morain, S. A., & Klankamsorn, B. (1978). Forest mapping and inventory techniques through visual analysis of Landsat imagery: Examples from Thailand. In *Proceedings of the 12th International Symposium on Remote Sensing of Environment* (pp. 417–426). University of Michigan.

Nandy, S., Joshi, P. S., & Das., K. K. (2003). Forest canopy density stratification using biophysical modelling. *Journal of the Indian Society of Remote Sensing*, *31*(4), 291–297.

Nangendo, G., Skidmore, A. K., & Oosten, H. V. (2007). Mapping East Africa tropical forests and woodlands – A comparison of classifiers. *Journal of Photogrammetry and Remote Sensing*, *61*(6), 393–404.

Nelson, R., & Holben, B. N. (1986). Identifying deforestation in Brazil using multi-resolution satellite data. *International Journal of Remote Sensing*, *7*(3), 429–438.

Ngampongsai, C. (1987). Habitat use by the sambar (Cervus unicolor) in Thailand: A case study for Khao-Yai National Park. In C. Wemmer (Ed.), *Biology and management of the Cervidae: A conference held at the conservation and research center, National Zoological Park, Smithsonian Institution* (pp. 289–298). Smithsonian Institution.

Niemann, O. (1993). Automated forest cover mapping using thematic mapper images and ancillary data. *Applied Geography*, *13*(1), 86–95.

O'Brien, T. G., Kinnaird, M. F., & Wibisono, H. T. (2003). Crouching tigers, hidden prey: Sumatran tiger and prey populations in a tropical forest landscape. *Animal Conservation*, *6*(2), 131–139.

Osborne, P. E., Alonso, J. C., & Bryant, R. G. (2001). Modelling landscape-scale habitat use using GIS and remote sensing: A case study with great bustards. *Journal of Applied Ecology*, *8*(2), 458–471.

Padmalal, U. K. G. K., Takatsuki, S., & Jayasekara, P. (2003). Food habits of sambar Cervus unicolor at the Horton Plains National Park, Sri Lanka. *Ecological Research*, *18*(6), 775–782.

Paletto, A., & Tosi, V. (2009). Forest canopy cover and canopy closure: Comparison of assessment techniques. *European Journal of Forest Research*, *128*(3), 265–272.

Pant, A., Chavan, S. G., Roy, P. S., & Das, K. K. (1999). Habitat analysis for sambar in Corbett National Park using remote sensing and Gis. *Journal of the Indian Society of Remote Sensing*, *27*(3), 134–138.

Panta, M., Kim, K., & Joshi, C. (2008). Temporal mapping of deforestation and forest degradation in Nepal: Applications to forest conservation. *Forest Ecology and Management*, *256*(9), 1587–1595.

Pasquarella, V. J., Holden, C. E., Kaufman, L., & Woodcock, C. E. (2016). From imagery to ecology: Leveraging time series of all available Landsat observation to map and monitor ecosystem state and dynamics. *Remote Sensing in Ecology and Conservation*, 152–170. https://doi.org/10.1002/rse2.24.

Patenaude, G., Milne, R., & Dawson, T. P. (2005). Synthesis of remote sensing approaches for forest carbon estimation: Reporting to the Kyoto Protocol. *Environmental Science and Policy*, *8*(2), 161–178.

Paudyal, D., & Bauer, J. J. (1988). A survey of wildlife, grasslands and pastoral systems of the upper Hinku and Hongu valleys. Field Doc. No. 15 FAO Project NEP/85/011. Department of National Parks and Wildlife Conservation and Woodlands Mountain Institute.

Prajapati, R. K., Triptathi, S., & Mishra, R. M. (2015). Habitat sutability analysis for Chital (Axis axis) using geo-spatial technology of Panna National Park (M.P.) India. *International Journal of Advanced Research in Science and Technology, 4*, 427–434.

Prater, S. H. (1971). *The book of Indian animals* (3rd ed.). Bombay Natural History Society.

Pringle, R. M., Syfert, M., Webb, J. K., & Shine, R. (2009). Quantifying historical changes in habitat availability for endangered species: Use of pixel- and object-based remote sensing. *Journal of Applied Ecology, 46*(3), 544–553.

Prins, H. H. T. (1992). The pastoral road to extinction: Competition between wildlife and traditional pastoralism in East Africa. *Environmental Conservation, 19*(2), 117–123.

Psomas, A., Kneubühler, M., Huber, S., Itten, K., & Zimmermann, N. E. (2011). Hyperspectral remote sensing for estimating aboveground biomass and for exploring species richness patterns of grassland habitats. *International Journal of Remote Sensing, 32*(24), 9007–9031.

Raghavan, B. (2003). *Interaction between livestock and Ladakh Urial (Ovis vignei vignei)* [M.Sc. Dissertation]. Saurashtra University.

Rautiainen, M., Stenberg, P., & Nilson, T. (2005). Estimating canopy cover in Scots pine stands. *Silva Fennica, 39*, 137–142.

Rikimaru, A. (1996). Landsat TM data processing guide for forest canopy density mapping and monitoring model. In *ITTO workshop proceedings on utilization of remote sensing in site assessment and planning for rehabilitation of logged-over forest*, 30 July–1 August (pp. 1–8).

Roberts, T. J. (1977). *The mammals of Pakistan*. Ernest Benn. Ltd., pp. 361.

Rodgers, W. A. (1988). The wild grazing ungulates of India: An ecological review. In P. Singh & P. S. Pathak (Eds.), *Rangelands - Resource and management. Proceedings of the national rangeland symposium* (pp. 404–419). IGFRI, November 9–12, 1987.

Rodgers, W. A., & Panwar, H. S. (1988). *Planning a wildlife protected area network in India: Volumes I & II*. Wildlife Institute of India.

Roy, P. S., Sharma, K. P., & Jain, A. (1996). Stratification of density in dry deciduous forest using satellite remote sensing digital data – An approach based on spectral indices. *Journal of Biosciences, 21*(5), 723–734.

Roy, P. S., Singh, S., Agarwal, S., & Joshi, P. K. (2001). Biome level classification of vegetation in Western India—An application of wide field sensor (WiFS). In *IGBP in India 2000—A status report on projects* (pp. 371–387). Indian National Science Academy.

Sankhala, K. (1977). *Tiger! The story of the Indian tiger*. Rupa and Co., Collins, pp. 220.

Santiapillai, C., Chambers, M. R., & Jayawardene, C. (1981). Observations of sambar Cervus unicolor Kerr, 1792 (Mammalia: Cervidae) in the Ruhuna National Park, Sri Lanka. *Ceylon Journal of Science (Biological Sciences), 14*, 193–205.

Sappington, J. M., Longshore, D. K. M., & Thompson, D. B. (2007). Quantifying landscape ruggedness for animal habitat analysis: A case study using bighorn sheep in the Mojave Desert. *Journal of Wildlife Management, 71*(5), 1419–1426.

Schaller, G. B. (1967). *The deer and the tiger: A study of wildlife in India*. University of Chicago Press.

Schaller, G. B. (1977). *Mountain monarchs: Wild sheep and goats of the Himalaya*. University of Chicago Press, p. 425.

Schüßler, D., Mantilla-Contreras, J., Stadtmann, R., Ratsimbazafy, J. H., & Radespiel, U. (2020). Identification of crucial stepping stone habitats for biodiversity conservation in northeastern Madagascar using remote sensing and comparative predictive modeling. *Biodiversity and Conservation, 29*(7), 2161–2184.

Sedeinsticker, J. (1976b). Ungulate populations in Chitwan valley, Nepal. *Biological Conservation, 10*(3), 183–210.

Seidensticker, J. (1976a). On the ecological separation between tigers and leopards. *Biotropica, 8*(4), 225–234.

Semiadi, G., Barry, T. N., & Muir, P. D. (1995). Dietary preferences of sambar (Cervus unicolor) and red deer (Cervus elaphus) offered browse, legumes and grass species. *Journal of Agricultural Science, 125*(1), 99–107.

Sharma, D. B., Clevers, J., de Graaf, R., & Chapagain, N. R. (2004). Mapping *Equus kiang* (Tibetan Wild Ass) habitat in Surkhang, upper Mustang, Nepal. *Mountain Research and Development, 24*(2), 149–156.

Sharma, G., & Chalise, M. K. (2014). Habitat preference of spotted deer (Axis axis) in Ghailaghari Buffer Zone Community Forest, Chitwan, Nepal. Special issue DNPWC-2071.

Shea, S. M., Flynn, L. B., Marchinton, R. L., & Lewis, J. C. (1990). Part II: Social behaviour, movement ecology and food habits. In Tall Timbers Research Station (Ed.), *Ecology of sambar deer on St. Vincent National Wildlife Refuge, Florida* (pp. 13–62). Tall Timbers Research Station.

Singh, G., Velmurugan, A., & Dakhate, M. P. (2009). Geospatial approach for tiger habitat evaluation and distribution in Corbett Tiger Reserve, India. *Journal of the Indian Society of Remote Sensing, 37*(4), 573–585.

Singh, R., Kumar, A., & Rajpoot, P. S. (2013). Spatial analysis of tiger habitat sutability in Panna Tiger Reserve of Madhya Pradesh. *India Life Science Bulletin, 10*, 95–98.

Singh, S., Agarwal, S., Joshi, P. K., & Roy, P. S. (1999). Biome level classification of vegetation in Western India—An application of wide field view sensor (WiFS). In *Joint workshop of ISPRS working groups I/1, I/3 and IV/4: Sensors and mapping from space* (pp. 81–96). 27–30 September 1999.

Smith, J. L. D., Ahearn, S. C., & Mcdougal, C. (1998). Landscape analysis of tiger distribution and habitat quality in Nepal. *Conservation Biology, 12*, 1338–1346.

SOI. (1975). Topographic maps of study area; 54P/14, 54P/15, 63d/1 and 63d/2 at 1:50000 scale, survey of India, government of India.

Stafford, K. J. (1997). The diet and trace element status of sambar deer (Cervus unicolor) in Manawatu district, New Zealand. *New Zealand Journal of Zoology, 24*(4), 267–271.

Tamang, K. M. (1979). *The population characteristics of the tiger and its prey* [Ph.D. Thesis]. Michigan State University.

Tardin, A. T., & Cunha, R. (1990). Evaluation of deforestation in the legal Amazon using Landsat TM images. *INPE Publication 501—RPE/609.* INPE.

Tateishi, R., Hoan, N. T., Kobayashi, T., Alsaaideh, B., Tana, G., & Xuan Phong, D. (2014). Production of global land cover data-GLCNMO2008. *Journal of Geography and Geology, 6*(3), 99.

Varman, K. S., & Sukumar, R. (1993). Ecology of sambar in Mudumalai Sanctuary, southern India. In N. Ohtaishi & H. I. Sheng (Eds.), *Deer of China: biology and management: Proceedings of the international symposium on deer of China* (pp. 273–284). Elsevier.

Viña, A., Bearer, S., Zhang, H., Ouyang, Z., & Liu, J. (2008). Evaluating MODIS data for mapping wildlife habitat distribution. *Remote Sensing of Environment, 112*(5), 2160–2169.

Wang, T. (2003). *Habitat analysis for Giant Panda in aoxiancheng nature reserve in the Qinling Mountains China* [Master Dissertation]. International Institute for Geo-Information Science and Earth Observation Enschede.

Weber, W., & Rabinowitz, A. R. (1996). A global perspective on large carnivore conservation. *Conservation Biology, 10*(4), 1046–1054.

Woodcock, C. E., Macomber, S. A., Pax-Lenney, M., & Cohen, W. B. (2001). Large area monitoring of temperate forest change using Landsat data: Generalization across sensors, time and space. *Remote Sensing of Environment, 78*(1–2), 194–203.

Woodwell, G. M., Hobbie, J. E., Houghton, R. A., Melillo, J. M., Moore, B., Park, A. B., Peterson, B. J., & Shaver, G. R. (1984). Measurement of changes in vegetation of the earth by satellite imagery. In G. M. Woodwell (Ed.), *The role of terrestrial vegetation in the global carbon cycle* (pp. 221–240). SCOPE 23. Wiley.

Zhu, Z. (2013). *Continuous change detection and classification of land cover using all available Landsat data* [Ph.D. Thesis]. B.E. Wuhan University.

21 Wetland Inventory of Aligarh District

Mohd. Mukhtyar Hussain, Orus Ilyas, and Ekwal Imam

Introduction

Wetlands are defined as lands transitional between terrestrial and aquatic ecosystems where the water table is usually at or near the surface or the land is covered by shallow water (Mitch & Gosselink, 1986). Ramsar Convention defines wetlands as follows:

> Wetlands as areas of marsh or fen, peatland or water, whether artificial or natural, permanent or temporary, with the water that is static or flowing, fresh, brackish or salt including areas of marine water, the depth of which at low tide does not exceed 6 m. Mangroves, corals, estuaries, bays, small creeks, flood plains, sea grasses, lakes, etc. are all covered under this definition.
>
> (Ramsar, 2013).

Wetlands are important as resting sites for migratory birds. Aquatic vegetation is a valuable source of food, especially for waterfowl. Apart from this, wetlands have great socio-economic and cultural values. Therefore, it is important to conserve and manage them in a sustainable manner. Identification, mapping, and preparation of inventory are foremost tasks. Satellite remote sensing has several advantages for monitoring wetland resources and with the combination of a Geographic Information System (GIS), this technique can be used as effective tools for wetland conservation and management. The application encompasses water resource assessment, hydrological modelling, flood management, reservoir capacity surveys, assessment and monitoring of the environmental impacts of water resources projects, and water quality mapping and monitoring (Jonna, 1999; Garg, 2013). Therefore, it has been widely used by various researchers at an international level. Shaikh et al. (2001) used remote sensing to determine environmental flows for wetlands of the Lower Darling River, New South Wales in Australia. During the next year, Arzandeh and Wang (2002) did the mapping of wetlands by evaluating the texture of RADARSAT Imagery while Sugumaran (2004) used remote sensing to study wetland dynamics in Lowa, USA. Rosso and Hastings (2005) used hyper-spectral data for mapping of marshland vegetation of San Francisco Bay, California. In 2007, Grenier et al., mapped the wetland of Quebec (Canada) by using RADARSAT-1 and Landsat ETM Image.

DOI: 10.4324/9781003321422-24

Henderson, and Lewis (2008) critically reviewed the utility of Radar in detection of wetland Ecosystems in Walpole island of Canada. In the next year, Reif et al. (2009) used GIS and remote sensing methods for mapping of isolated wetlands in a Karst Landscape of north-central Florida, USA. Similarly, Landmann (2010) used MODIS for wide area wetland mapping in semi-arid Africa. Interestingly, Cserhalmi et al. (2011) used panchromatic aerial photographs of a period of almost 50 years (1956–2002) to study the changes in wetland ecosystem in Bereg Plain of north east Hungary. Corcoran et al. (2012) integrated aerial photographs, SAR imagery, and polarimetric decompositions and evaluated their suitability to map wetlands in a forested region of northern Minnesota (USA). In 2015, Whiteside and Bartolo used high spatial resolution multispectral satellite imagery for mapping of aquatic vegetation in a tropical wetland of Alligator rivers region of northern Australia. Recently, Amani et al. (2017) integrated multi source SAR and optical data for wetland classification in Newfoundland and Labrador, while Sahel et al. (2017) did a comprehensive review of use of remote sensing in wetland classification.

India exhibits a great physiographical diversity and accordingly wetlands exist in all geographical regions. Considering its importance, scientific inventory of wetlands was prepared first time in India using 1992/93 post-monsoon and pre-monsoon seasons IRS 1A/1B LISS-I/II data (Garg et al., 1998). Inventory for the major states was made at 1:250 000 scale, while for all north-eastern states and some other states it was done on 1:50 000 scale. Preparation of the second wetland inventory was carried out at a 1:250,000 scale by Space Applications Centre (ISRO), Ahmedabad in GIS environment using 2004–2005 resources at AWiFS (spatial resolution 58 m) digital data. However, some lacuna was left in these two wetland inventories, and so the third project on wetland inventory was started in 2007. It was needed, because at the time of commencement of the first scientific inventory of wetlands, GIS in India was at an infancy level and only an attribute database was prepared (Garg & Patel, 2007). However, this gap was filled to an extent in the second wetland inventory. In order to prepare a registry of wetlands and to have a seamless country-wide spatial database, a national project, "National Wetland Inventory and Assessment," was taken up in 2007 by Space Applications Centre, Ahmedabad at the behest of the Ministry of Environment and Forests (Garg, 2013). Scale of analysis/map preparation was 1:50,000 and accordingly, all wetlands larger than 2.25 ha were delineated. This work was completed in 2011 (Anon, 2011) and state-wise atlases were prepared that also included information on high altitude wetlands. Encouraged by above mentioned works, several scattered studies have been carried out by various Indian researchers at individual level where they also used geospatial technology. Jagtap et al. (2001) used LANDSAT satellite data of 1985–1986 for assessing the coastal wetland resources of central west coast of Maharashtra (India). Again in 2002, Nayak used IRS LISS II and LANDSAT TM data for providing baseline information on coastal habitat and associated shore land features along the entire Indian coast on 1:250,000 and 1:50,000 scale. Furthermore, Patel et al. (2003) developed Wetland Information System (Winsys) with the collaboration of Space Applications Centre (ISRO) after using IRS- LISS-II satellite imageries of 1992–1993. Sarkar and Jain (2008) also used remote sensing data to study wetland dynamics of Harike wetland. Mabwaoga et al. (2010) used IRS P6, LISS IV satellite data to assess the water quality parameters of the Harike wetland (Ramsar site). During 2013, Gupta used geospatial techniques to study the changing land-use of East Kolkata wetland. Garg (2014) critically reviewed the works carried out in India for wetland conservation and management using geospatial techniques. Considering it as a potential tool, the present study was carried out where remote sensing was used for identification and mapping of wetlands of Aligarh district.

Materials and Methods

Study Area

Aligarh district lies in the western part of Uttar Pradesh and situated about 140 km away from New Delhi on Delhi-Howrah rail route. The latitudinal extension of Aligarh is 27° 33' to 28° 11' E and longitudinal 77° 28' to 78° 35' E. The elevation at the centre of the city is 187.38 metres. Aligarh district is stretched in between Yamuna and Ganga rivers from west to east and covering a span of 67 km. Commonly this area is known as Doab (a strip of land between two rivers). The total geographical area of district is 3713 km². The district is divided in to five administrative units of tehsil namely Atrauli, Gabhana, Iglas, Khair, and Koil. From agricultural point of view, this area is one of the most important lands in the Gangetic plain of India. Rice and wheat are the primary crops of the district and cultivated during rainy season (June–October) and winter (November–February) respectively. Monsoonal rainfall in 2016 was 640 mm considered to be below normal (Figure 21.1).

Except for some social forestry, plantation and orchards, the entire district has been converted to non-woody agriculture with human population densities ranging from 500 to 3,000 people km². The important thing is that amid these croplands various wetlands are found, and the majority of them are either small or isolated.

Wetland Mapping

The study was carried out during January 2016 to June 2016 and completed in three phases. In the first phase satellite and collateral data were collected and processed, while during the second phase, field survey was conducted for ground truthing. In the third phase mapping of different wetlands was done by using ERDAS IMAGINE (8.7) and Arc view 3.2 computer softwares.

Data Collection and Data Processing

The primary data was collected by conducting field surveys while secondary data was obtained from various sources like topographic maps, satellite imageries, and literature survey.

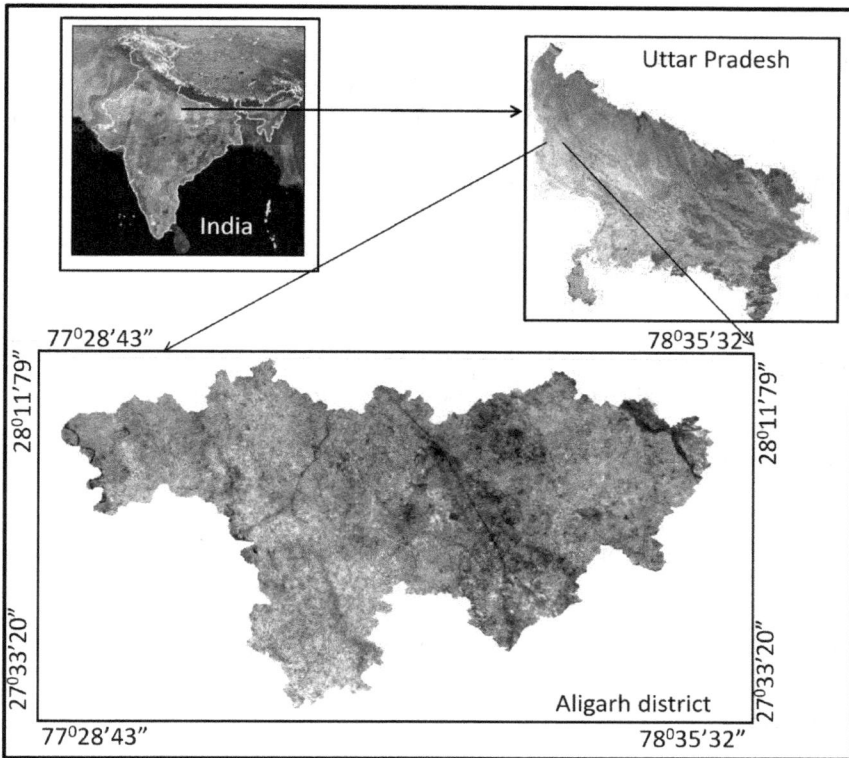

Figure 21.1 Map of the study area.

Collateral Data

The study area is covered by 14 different topographic sheets of 53H/08, 53H/12, 53H/16, 53L/04, 53L/08, 53L/12, 54E/09, 54E/13, 54E/14, 54I/01, 54I/02, 54I/05, 54I/06, 54I/09 (at 1: 50,000 scale). These topographic maps were acquired from Survey of India, Dehradun as well as Departments of Wildlife Sciences, Geography and Geology of Aligarh Muslim University Aligarh.

All the topographic sheets were scanned separately and exported to ERDAS IMAGINE 8.7 in image format (.img) for mosaicking. Before mosaicking, all scanned topographic maps were georeferenced to Geographic Lat/Long Projection to sub-pixel accuracy. The common uniformly distributed ground control points (GCP) were marked with root mean square error of one third of a pixel and images were re-sampled by nearest neighbour method. After geo-referencing, all topographic maps were mosaicked. Then this data was re-projected into UTM-WGS 84 projection for further analysis.

Satellite Data

The LANDSAT satellite data of April 2014 was acquired from the website http:// earthexplorer.usgs.gov. The satellite data was imported to ERDAS IMAGINE 8.7 software in an image format for stacking process. Since this satellite data is available in geo-referenced format, therefore there was no need to rectify it geometrically. However, radiometric correction was done when and ever needed for further use. In order to use this data in conjunction with other spatial data, it was re-projected in UTM-WGS 84 projection.

Field Survey

A field survey was carried out between 31 January and 30 April 2016. Ground truthing was done by matching the pattern, texture, association, shape, and size of the features from the FCC for a particular topographic feature using GPS locations. Before starting the extensive survey of the area, literature survey and preparation of questionnaire were done. The survey was conducted mainly along the different roads leading to different townships within the administrative boundaries of Aligarh district. These roads were considered as transects and covered with motorcycle with a low speed. Enquiries were made after every stoppage to know about wetland in nearby areas. To confirm their presence, the exact location was visited. These identified wetlands were revisited and relevant information collected. Out of total identified wetlands, five were selected for detailed study.

Assessment of Avian Diversity

Assessment of avian diversity was done for five selected wetlands during February and March 2016. Each selected wetland was visited atleast five times during morning hours on different days. Birds were observed from available vantage present on the periphery from where entire stretch of wetland area could be observed easily. For each wetland eight vantage points were selected randomly. It was taken care that each side of wetland must have atleast two vantage points. Researchers having good knowledge of bird identification were deputed for each vantage point. Observation for estimating the bird species were started simultaneously and terminated after four hours. Observers were asked to meander through wetland and flush the skulking species so that coverage of the entire wetland would be ensured. Inventory of bird species was made after combining the data collected by all eight observer groups. Binoculars (Olympus 10x50) and field guides (Ali, 2002; Ali & Ripley, 1995; Grimmet et al., 2011) were also used for identification of the birds.

Post-Field Work

Mapping of Wetland

To delineate the wetlands, it was decided to use supervised maximum likelihood classification technique. For this, satellite imagery was imported to ERDAS

IMAGINE (8.7) domain and modelled for supervised maximum likelihood classification. The image was categories into two classes of (i) wetland and (ii) non-wetland. Supervised classification is a procedure for identifying spectrally similar areas on an image by identifying "training" sites of known targets and then extrapolating those spectral signatures to other areas of unknown targets. Supervised classification relies on a priori knowledge of the location and identity of land cover types that are in the image. This can be achieved through field work, study of aerial photographs, or other independent sources of information. Training areas, usually small and discrete compared to the full images are used to "train" the classification algorithm to recognise land cover classes based on their spectral signatures, as found in the image. The maximum likelihood classifier (MLC) assumes that the training statistics for each class have a normal or "Gaussian" distribution. The classifier then uses the training statistics to compute a probability value of whether it belongs to a particular land cover category class. This allows for within-class spectral variance. In this, the image analyst uses a priori knowledge to weight the probability function. The MLC usually provides the highest classification accuracies (Imam et al., 2009).

First, we overlaid the wetland map with a 10mx10m grid corresponding roughly to 10 km x 10 km in ERDAS IMAGINE 8.7 domain. Identification of wetlands was done in each grid that helps in estimating the wetlands easily and avoiding recounting the same wetlands more than once. It also helps in estimating the density of wetland in each grid. Then different wetlands were arranged tehsil-wise. Geographical location was also recorded for 484 wetlands using GPS.

For detailed study, five wetlands namely Shekha Jheel Dabha-Dabhi, Barhad, Jamao, and Ahraula were selected. In selecting the wetlands, we were biased towards the size, and we chose the five largest wetlands. These wetlands were mapped separately and brought to Arc View 3.2 for further analysis.

Results and Discussion

A total of 929 wetlands were identified in Aligarh district. Out of which 165 were recorded from Atrauli tehsil, 121 from Gabhana, 163 from Iglas, 161 from Khair, and 319 from Koil tehsil (Table 21.1 and Figure 21.2)). Among wetlands selected for detailed study, Shekha Jheel, Dabha-Dabhi and Barhad wetland are present in Koil

Table 21.1 Tehsil-wise wetlands in Aligarh district (UP)

S. No.	Name of tehsil	Number of wetlands
1	Atrauli	165
2	Gabhana	121
3	Iglas	163
4	Khair	161
5	Koil	319
	Total	**929**

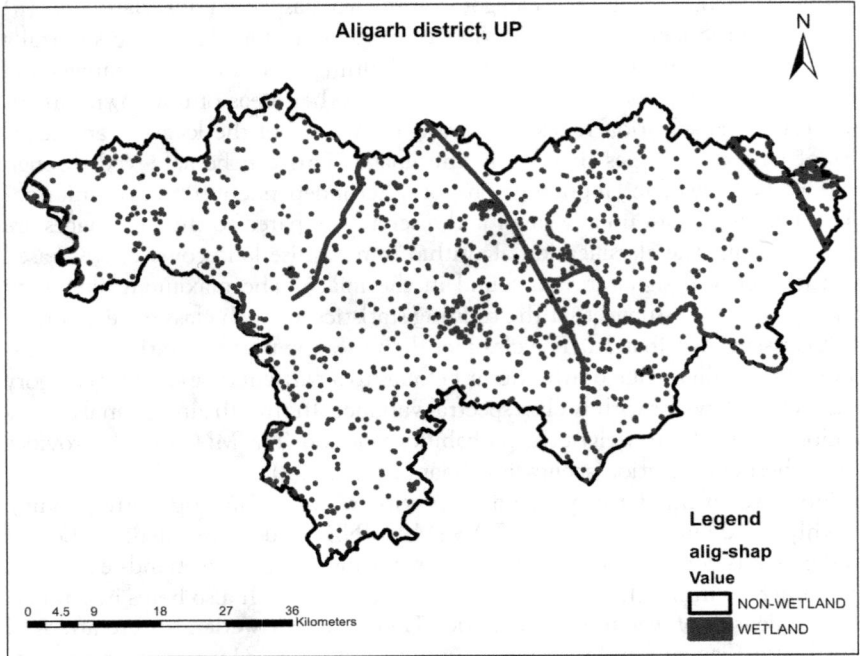

Figure 21.2 Distribution of wetlands in Aligarh district.

tehsil, while Jamao wetland lies in Iglas tehsil and Ahraula wetland in Khair tehsil. All these five wetlands are fresh water wetland and perennial in nature (Figure 21.3). Their source of water is either from upper Ganga canal or rainfall. Barhad, Jamao, and Ahraula wetlands also get water from tube well. Except for Shekha Jheel, the rests of the wetlands belong to Gram sabha (village council). Shekha Jheel is declared as a reserve wetland and has state government funding for conservation and protection.

Shekha Jheel

Shekha Jheel is spread over in an area of 26 hectares and located between 27° 51' 32" N to 78°13' 09" E. It is 18 km away from Aligarh city. Shekha Jheel is one of the most important wetlands of the district as it provides habitat for a large number of migratory and resident bird species. The number of bird species can be seen over here nesting, breeding, and feeding. Seventy-four species of birds were identified from Shekha Jheel, and out of these 20 were migratory. According to IUCN five bird species were near threatened, 69 least concern and one was endangered. Apart from birds, it also supports some fishes and amphibian biodiversity. Sheikha Jheel has also been developed as a picnic spot by district administration, which attracts large number of people from surrounding towns and villages (Figure 21.3).

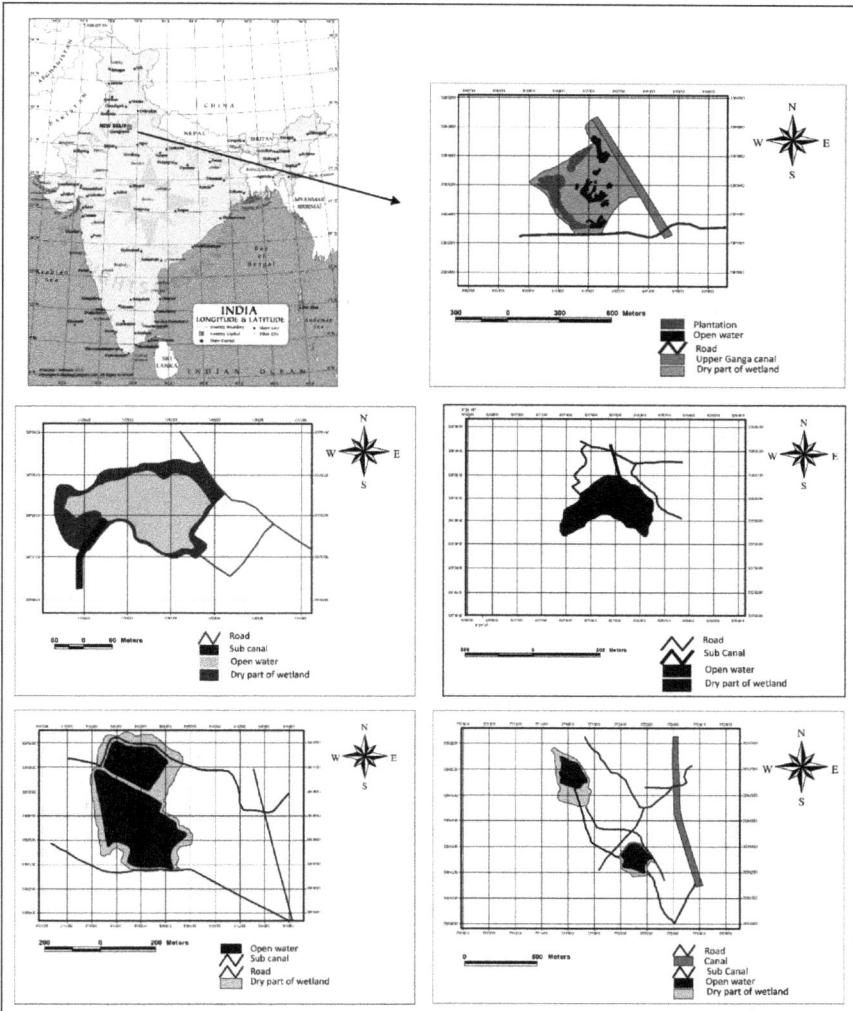

Figure 21.3 Selected wetlands of Aligarh district for the study.

The wetland is infested with water hyacinth (*Echhorni acrassipes*), which can be considered as one of the most potential threats. So far no serious attempt has been made by management to eradicate water hyacinth. Probably it may be due to financial constraints and government's lack of interest. Uncontrolled water hyacinth is not only resulting in eutrophication but also restricts the growth of food plants for birds.

Dabha-Dabhi Wetland

Dabha-Dabhi wetland is situated 30 km away from the Aligarh city. It lies under latitude of 27° 48' 14" N and longitude of 78° 19' 39" E. The total area of wetland is about 25 hectares and comes under Koil tehsil of Aligarh district. The primary source of water for this wetland is rain water. The wetland is also connected with upper Ganga canal and kali nadi by a sub-canal. However, sub-canal was found dry during the survey.

This wetland is habitat and breeding place for many bird species coming from different areas. Some of them are even migratory to India visiting from other countries while some of them are winter and summer visitors. Local birds are using this wetland for breeding and feeding purpose. About 50 species of birds were recorded from Dabha-Dabhi wetland. Out of them 36 were resident while 14 were migratory. Conservation status revealed that 44 species were of least concern, four near threatened, one vulnerable, and one was critically endangered. Dabha-Dabhi wetland is one of the ecologically important areas within the Aligarh district. The highest population of painted stork was reported during the survey probably due to availability of sufficient food. There were about 58 painted storks in a congregation and also 82 individuals of purple moorhen recorded in Dabha-Dabhi wetland (Figure 21.3).

Ahraula Wetland

The Ahraula wetland is situated 46 km away from Aligarh city in Khair tehsil of Aligarh district. It encompasses a total area of about nine hectares and lies between 27° 55' 30" N and 77°45' 33" E (Figure 21.3). Ahraula wetland consists of two ponds connected with a minor canal that is remotely joined with a tributary of upper Ganga canal. The size of these ponds is five hectares and four hectares respectively. Both wetlands were used for fish culture, and one of them was enclosed with thin iron wires just for protecting the fish from being eaten by birds. There are three sources of water for Ahraula wetland: Rainfall, tube well, and canal. However, we didn't see or hear water being released from this minor canal. From the point of avian diversity, Ahraula wetland attracts only 15 species of birds. They were all resident. According to IUCN status of conservation, 14 avian species were of least concern while only one was vulnerable (Figure 21.4 and Table 21.2).

Jamao Wetland

The Jamao wetland is situated about 44 km away from the Aligarh city and covering an area of five hectares. It lies between 27°43'43" N and 77°48'14" E in Iglas tehsil of Aligarh district.

Basically this wetland is owned by Gram samaj and generally used for fish culture. Main sources of water for this wetland are rainfall and tubewell. It is also connected with minor canal, but a practice of releasing water in this minor canal was stopped by irrigation department a year ago. Seventeen bird species were recorded from this wetland (Figure 21.3). Out of these, 12 species were resident while five

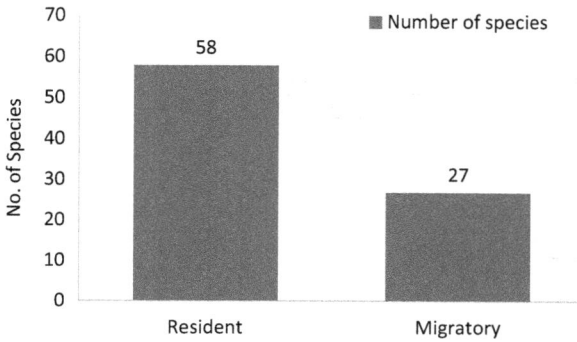

Figure 21.4 Resident and migratory status of the birds in five selected wetlands of Aligarh district (UP).

Table 21.2 Status of birds in different wetlands in Aligarh district (UP)

Name of wetland	IUCN status			
	LC	NT	EN	VUL
Shekha Jheel	68	5	1	0
Dhabi-Dhaba wetland	45	4	0	1
Ahraula wetland	14	0	0	1
Jamo wetland	15	1	0	1
Barhad wetland	11	0	0	0

were migrant. According to IUCN, 15 bird species were of least concern, one was near threatened, and one was vulnerable (Figure 21.4 and Tables 21.2 and 21.3).

Barhad Wetland

The Barhad wetland is situated about 38 km away from the Aligarh city in Koil tehsil of Aligarh district (27° 42' 31"N, 78° 12'25"E). It covers an area of about 14 hectares. The source of water for this wetland is rainfall and tubewell. The Barhad wetland is also used for fish culture and whenever there is scarcity of water, it is filled with water through private tubewell. Eleven species of birds were identified from Barhad wetland, categorised as of least concern by IUCN, and all were resident (Figure 21.3).

Threats to Wetlands

Wetlands of Aligarh district are facing severe anthropogenic pressures. The large number of wetlands in Aligarh district are affected due to harmful activities such

Table 21.3 Status of five wetlands selected for detail study in Aligarh district (UP)

S. No.	Name of wetland	Coordinates	Tehsil	Area (ha)	Wetland type	Source of water	Water quality	Conservation status
1	Shekha Jheel	27° 51' 25.9311"N to 78° 134'.7427"E	Koil	25.71	perennial	Upper Ganga Canal and rainfall	Fresh	Reserve wetland area
2	Dhabi-Dhaba wetland	27° 48' 17.5638"N to 78° 19' 34.9976"E	Koil	23.66	perennial	Rainfall	Fresh	Belongs to Gram Samaj
3	Barhad wetland	27° 42' 26.5197"N to 78° 12' 24.6867"E	Koil	14.11	perennial	Rainfall and tube well	Fresh	Belongs to Gram Samaj
4	Jamao wetland	27° 43' 42.5165"N to 77° 48' 15.8077"E	Iglas	5.5	perennial	Rainfall and tube well	Fresh	Belongs to Gram Samaj
5	Ahraula wetland A & B	27° 55' 31.3045"N to 77° 45' 33.2096"E & 27° 55' 9.9651"N to 77° 45' 50.6198"E	Khair	5.23 & 4.15	perennial	Rainfall and tube well	Fresh	Belongs to Gram Samaj

as encroachment, use of pesticides and fertilisers, livestock grazing, and weed infestation. Invasion by weeds could be considered as one of the most serious threats to some of the wetlands. Shekha Jheel, Chautal, and Kalidah are facing weed invasion by *Echhorni acrassipes* (water hyacinth). Most of the wetlands do not have any legal sanctity, therefore, eradication of weed is not done by government agencies. Even wetlands owned by gram panchayat (local administrations) do not have any budget for this purpose. Apart from this, some of the wetlands are facing eutrophication due to unwanted vegetation growth; however, its severity was not evaluated.

Most of the wetlands exist in rural areas, so grazing by cattle on the dried portion of land is obvious. Cattle grazing are not only responsible for habitat alteration but also disturb birds coming to that particular wetland. It has been observed that villagers visit Shekha Jheel for grass and fuel wood collection on a regular basis, which has a big negative impact on the wetland. During the survey it was recorded that some of the wetlands were used for fish culture, and to detract fish-eating birds, wetlands were enclosed with thin wire. Sound production activities were another means to protect the wetland. Erecting of enclosures may disturb other common birds and also cause death if strangled, while a high volume of sound may distract birds permanently.

As we know, Aligarh district is predominantly covered by agriculture, therefore almost all wetlands are surrounded by agricultural land. During rainfall, rain water drains into these wetlands carrying residuals of pesticides, vermicides, and fertilisers used by farmers for high yield of crops. These chemicals act as wetland pollutants and affect birds lives directly or indirectly.

Encroachment and reclamation of wetlands is another significant threat. The wetlands which are located near to village are partially encroached upon for construction of hutments for cattle shelter or dung cake storage. In most of the cases, surrounding land of wetlands is used for cultivation after water retreat. These practices may cause siltation of wetlands, disturbance to waterbirds, and permanent reclamation of land for agriculture purpose. During 2003 it was recorded that Kulwa wetland was a large size marshy land (19.2 hectare) attracting water birds (Ahmad, 2003). At present, several community toilets have been built by Gram sabha on one side of the wetland while other sides have been reclaimed as agricultural land by farmers who have their land adjacent to the wetland. Because of these activities, wetland has reduced to 1.86 hectares only. The Gular road wetland, being located in prime area of city, has attracted several land mafia, and its major portion has been grabbed illegally and sold for house construction. The satellite imagery shows that in 2001, the total area of wetland was 21.90 hectares, and after severe pressure and encroachment, it has reduced to 0.72 hectares only. According to a local newspaper, the district administration is trying to acquire back land of this wetland from illegal occupants. The matter is under litigation, but the government does not have total control over it. Chautal wetland, which is considered to be one of the major sink for drainage of Shamshad market area, has become a garbage dumping place for the municipality. Before this, large portions of the eastern side were filled during overhead bridge construction. A wetland located on the right side of

Aligarh-Kanpur railway track, known as Kalidah, is also facing severe problems of encroachment and land grabbing by local mafia. The satellite imagery revealed that during 2001, its total area was 18.28 hectares which reduced to 7.45 hectares in 2017. During the survey it was found that the land portion which has become permanently dry has been sold by land mafia for house construction to local people.

Wetlands are species-rich habitats performing valuable ecosystem services such as flood protection, water quality enhancement, food chain support, and carbon sequestration. In spite of that, wetlands have been drained to convert them into agricultural land or industrial and urban areas. A realistic estimate is that 50% of the world's wetlands have been lost. With the projected growth in human population during the next 25 years, there will be a challenge to increase the food production up to 50%.

The pressure to produce more food per area may lead to reclamation of more land for agriculture (FAO, 2003). All these developments together will inevitably lead to reclamation of natural or marginally used land for intensive crop production, and there will be increasing pressure to use wetlands for growing crops.

In India, where over 80% of rural households have marginal landholdings of less than one hectare and just 7% owe more than two hectares, and their per capita per day income is very low, ranging from Rs.15 to Rs. 31 for marginal farmers and small farmers respectively. Thus, all marginal farmers fall below the poverty line. Due to the inadequacy of agricultural income to meet their household expenditure, the small and marginal farmers have to devise livelihood strategies for their survival. Among other strategies, expansion of agricultural land through drainage and filling of wetlands (present on crop lands) is one of the efforts practiced by farmers, resulting in a reduction in large number of wetlands once present in the district.

The natural wetland ecosystems reclaimed in this way have lost much of their original character, leading to reduced biodiversity and reduced performance of functions (Hassan et al., 2005).

During the field survey, it has been recorded that most of the villagers, particularly farmers, know the socio-economic importance of wetlands but are not supportive of wetland restoration and conservation. It is felt that the awareness efforts about the wetlands made by government and non-government agencies are not adequate. Furthermore, no incentive has been offered to farmers for their role in conservation of wetlands. If the agriculture yield is optimised by providing new technologies and improved seed subsidised prices, the reclamation of wetlands can be reduced. Furthermore, new research (by agronomists and ecologists jointly) on improving traditional wetland agriculture systems and increasing the awareness among local communities and policy-makers of the importance of wetlands for provisioning food as well as other services (flood protection, water purification, biodiversity, etc.) may improve the situation.

Conclusion

Various works have been done in India on identification and mapping of wetlands, but more emphasis was given to those wetlands which are large in size or famous for various reasons in the country while small wetlands have been neglected up to now. It is true that whatever the size, wetlands may play their ecological, socio-economic, and cultural role. Garg (2013) also opined that a time may come when small wetlands will perish. Therefore, there is an urgent need to establish a mechanism for periodic monitoring of wetland resources in the country at micro level.

This can only be done if data on physical, biological, socio-cultural, structure, and functions of a wetland is available. Geospatial techniques have proved to play a critical and significant role in this direction. Remote sensing and GIS may help in preparation of wetland inventory, action plans, and doing monitoring that may ultimately help in conservation and management. Geospatial technology also has an important role in identification of suitable wetlands for waterfowl, habitat conservation, spatial modelling of physiological niche, and habitat suitability analysis. On the basis of these, a conservational priority list of wetlands can be prepared. The country like India, where funds are limited, all wetlands can't be managed simultaneously. Therefore, a wetland at a high level of priority may be managed first and so on. The present study is a small effort where various small and unknown wetlands of districts were identified and some of them mapped at 1:50,000 scale, trying to provide maps of wetlands at a micro level. Furthermore, it suggested to carry out inventories at a district level at larger scales (1:10000 or 1:5000).

Management Recommendations

On the basis of the present study, the following recommendations have been made for better management and conservation of wetlands present in Aligarh district:

Anthropogenic pressure should be reduced by providing alternatives for fuel and fodder to the local community. Local people are encroaching on the wetlands for agricultural practices due to which wetlands are shrinking, which is a major cause of biodiversity loss. Use of pesticides is a major issue, which should be checked, and water hyacinth should be removed so that space should be created for birds. However, the locals should also be provided with basic facilities like toilets in their villages. And they all should be made aware about the biodiversity values of wetlands, and local should also be involved in awareness campaigns and sensitisation programmes, otherwise all the above-mentioned recommendations will not be effective.

References

Ahmad, S. N. (2003). *A study on socio-economic importance of some wetlands of Aligarh.* Department of Geography, Aligarh Muslim University. Research Report No. 1.

Ali, S. (2002). *The book of Indian birds* (13th ed.). Bombay Natural History Society and Oxford University Press, p. 354.

Ali, S., & Ripley, S. D. (1995). *A pictorial guide to the birds of the Indian subcontinent* (2nd ed.). Bombay Natural History Society and Oxford University Press, p. 169.

Amani, M. B., Salehi, S., Mahdavi, J. G., & Brisco, S. D. (2017). Wetland classification in Newfoundland and Labrador using multi-source SAR and optical data integration. *GIScience and Remote Sensing, 54*(6), 779–796.

Anon. (2011). *National wetland Atlas.* Space Applications Centre (ISRO), p. 310.

Arzandeh, S., & Wang, J. (2002). Texture evaluation of RADARSAT imagery for wetland mapping. *Canadian Journal of Remote Sensing, 28*(5), 653–666.

Corcoran, J. J., Knight, B., Brisco, S., Kaya, A. C., & Murnaghan, K. (2012). The integration of optical, topographic, and radar data for wetland mapping in northern Minnesota. *Canadian Journal of Remote Sensing, 37*(5), 564–582.

Cserhalmi, D., Nagy, J., Kristof, D., & Neidert, D. (2011). Changes in a wetland ecosystem: A vegetation reconstruction study based on historical panchromatic aerial photographs and succession patterns. *Folia Geobotanica, 46*(4), 351–371.

FAO. (2003). Food energy – Methods of analysis and conversion factors: Report of a technical workshop. *FAO Food and Nutrition Paper, 77*, 18–37.

Garg, J. K. (2013). Wetland assessment, monitoring and management in India using geospatial techniques. *Journal of Environmental Management.* http://doi.org/10.1016/j .jenvman.2013.12.018.

Garg, J. K. (2014). Wetland assessment, monitoring and management in India using geospatial techniques. *Journal of Environmental Management, 148*, 112–123.

Garg, J. K., &Patel, J. G. (2007). National wetland inventory and assessment. Technical guideline and procedure manual. SAC/EOAM/AFEG/NWIA/01/07. Space applications Centre Ahemdabad, p. 102.

Garg, J. K., Singh, T. S., & Murthy, T. V. R. (1998). Wetlands of India project report. *Report No. RSAM/SAC/RESA/PR/01/98.* Space Applications Centre (ISRO), p. 239.

Grenier, M., Demers, A. M., Labrecque, S., Benoit, M., Fournier, R. A., & Drolet, B. (2007). An object-based method to map wetland using RADARSAT-1 and Landsat ETM Images: Test case on two sites in Quebec, Canada. *Canadian, 33*(Suppl 1), S28–S45.

Grimmet, R., Inskipp, C., & Inskipp, T. (2011). *Birds of Indian subcontinent.* Oxford University Press, p. 448.

Gupta, S. (2013). Changing land-use of East Kolkata wetland, India. *International Journal of Applied Research & Studies, 2*, 1–8.

Hassan, R., Scholes, R., & Ash, N. (2005). Ecosystems and human well-being: Current state and trends. In R. Hassan, R. Robert Scholes, & N. Neville Ash (Eds.), *Findings of the condition and trends working group of the millennium ecosystem assessment* (Vol. 1). Island Press, pp. 1–47.

Henderson, F. M., &Lewis, A. J. (2008). Radar detection of wetland ecosystems: A review. *International Journal of Remote Sensing, 29*(20), 5809–5835.

Imam, E., Kushwaha, S. P. S., & Singh, A. (2009). Evaluation of suitable tiger habitat in Chandoli National Park, India, using multiple logistic regression. *Ecological Modeling, 220*(24), 3621–3629.

Jagtap, T. G., Naik, S., & Nagle, V. L. (2001). Assessment of coastal wetland resources of central West Coast, India, using Landsat data. *Journal of Indian Society of Remote Sensing, 29*(3), 143–150.

Jonna, S. (1999). Remote sensing applications to water resources: Retrospective and Perspective. In S. Adiga (Ed.), *Proceedings of ISRS national symposium on remote sensing applications for natural resources.* ISRS, pp. 368–377.

Landmann, T. M., Schramm, R. R., Colditz, A. D., & Dech, S. (2010). Wide area wetland mapping in semi-arid Africa using 250-meter MODIS metrics and topographic variables. *Remote Sensing, 2*(7), 1751–1766.

Mabwaoga, S. O., Chawla, A., & Thukral, A. K. (2010). Assessment of water quality parameters of the Harike wetland in India, a Ramsar site, using IRS LISS IV satellite data. *Environmental Monitoring and Assessment, 170*(1–4), 117–128. https://doi.org/10.1007/s10661-009-1220-2.

Mitsch, W. J., & Gosselink, J. G. (1986). *Wetlands*. Van Nostr and Reinhold Company, p. 539.

Nayak, S. (2002). Use of satellite data in coastal mapping. *Indian Cartographer, 22*, 147–156.

Patel, J. G., Singh, T. S., Garg, J. K., Malik, A. K., & Bhattacharyya, S. M. K. (2003). Wetland information system (Winsys). *Space Applications Centre (ISRO)*. Project Report: SAC/RSAM/RESA/FLPG/WIS/ 01/2003. p. 82.

Ramsar. (2013). Ramsar convention, the Ramsar convention Bureau, *Rue Mauverney 28, CH-1196*, Gland, Switzerland. Retrieved December 13, 2013, from www.ramsar.org.

Reif, M. R., Frohn, C., Lane, C. R., & Autrey, B. (2009). Mapping isolated wetlands in a karst landscape: GIS and remote sensing methods. *GIS & Remote Sensing, 46*(2), 187–211.

Rosso, P. S. U., & Hastings, A. (2005). Mapping Marshland vegetation of San Francisco Bay, California, using hyperspectral data. *International Journal of Remote Sensing, 26*(23), 5169–5191.

Sahel, M., Bahram, S., Granger, J., Meisam, A. B. B., & Weimin, H. (2017). Remote sensing for wetland classification: A comprehensive review. *GIS & Remote Sensing, 55*. https://doi.org/10.1080/15481603.2017.1419602.

Sarkar, A., & Jain, S. K. (2008). Using remote sensing data to study wetland dynamics – A case study of Harike wetland. In M. Sengupta, & R. Dalwani (Eds.), *Proceedings of Taal 2007: The 12th World Lake Conference* (pp. 680–684).

Shaikh, M., Green, D., & Cross, H. (2001). A remote sensing approach to determine environmental flows for wetlands of the Lower Darling River, New South Wales, Australia. *International Journal of Remote Sensing, 22*(9), 1737–1751.

Sugumaran, R., Meyer, J., & Davis, J. (2004). A web-based environmental decision support system (WEDSS) for environmental planning and watershed management. *Journal of Geographical Systems, 6*(3), 1–16.

Whiteside, T. G., & Bartolo, R. E. (2015). Mapping aquatic vegetation in a tropical wetland using high spatial resolution multispectral satellite imagery. *Remote Sensing, 7*(9), 11664–11694.

22 Web of Socioeconomic Considerations for Nature Conservation in Manipur

Naveen Pandey, Mordecai Panmei, Jadumoni Goswami, Sumanta Kundu, Dibyajyoti Saikia, Sharad Kumar, and Kedar Gore

Introduction

"If we are to make real progress in conservation, we have to take the challenges of communication between different academic ways of understanding the world seriously" (Adams, 2007). There has been greater emphasis than ever on understanding social, political, cultural, and economic factors governing wildlife conservation efforts and outcomes in recent times. An interdisciplinary approach to conservation is being globally recognised and advocated to bridge the wide gap between ecologists and social scientists (Guha, 1997). The complex web of interactions between individuals, social organisations, cultural norms, and institutions constitute social processes that are often dynamic. The social processes, socioeconomic constraints, and opportunities largely shape the design and implementation of conservation initiatives, and inadequate attention to these are being recognised as causing failures in implementing conservation (Knight & Cowling, 2007). Spatial data on resource utilisation is crucial for conservation planning (Ban et al., 2013). In South Asian countries, where complex social structures, rampant poverty, ever-increasing populations, and rising expectations from developing economies characterise the landscapes that conservationists aspire to secure for the wild, social processes and resource utilisation take on far greater importance.

Following liberalisation and globalisation of the Indian economy, the Indian mainland has been subjected to rapid economic and technological changes in the last three decades. Insurmountable political pressure to meet the needs of growing populations necessitates land-use changes, increased exploitation of natural resources, and policy interventions challenging long-term conservation of nature and natural resources. The conservation challenges resulting from these forces are more pronounced in northeast India, where ethnic diversity comes into play. The northeast region of India, an ethnobotanical transition zone between India, China, Tibet, Burma, and Bangladesh (Ali & Das, 2003) is also a region of exemplary biodiversity. The northeast region is home to around 145 tribal communities (Aiyadurai, 2011), and the state of Manipur in northeast India alone has 38 recognised tribal groups (Devi & Salam, 2016). While nature worship, sacred

DOI: 10.4324/9781003321422-25

groves, and conservation of natural resources have been at the heart of many ethnic groups in northeast India (Khumbong Mayum et al., 2004), hunting as a tradition by the tribes in northeast India has been considered as a primary threat to wild populations by ecologists (Dutta, 2004, Mishra et al., 2006). Though the primary reason for the local community's dependence on hunting has been reported to be a lack of economic alternatives, socioeconomic and political motivations are said to have drastically changed hunting practices (Aiyadurai, 2011). Apart from hunting, changes in the traditional practice of Jhum cultivation have been adversely affecting soil nutrients and soil fertility due to a decline in the cycle of shifting cultivation (Ninan, 1992). An irregular supply of cooking gas and electricity has increased the demand for firewood (Yuhlung, 2014). Thus, there is a need for a better understanding of the socioeconomic considerations that shape the dependence of the local community on nature and natural resources. This paper aims to gain insight into the social and economic aspects

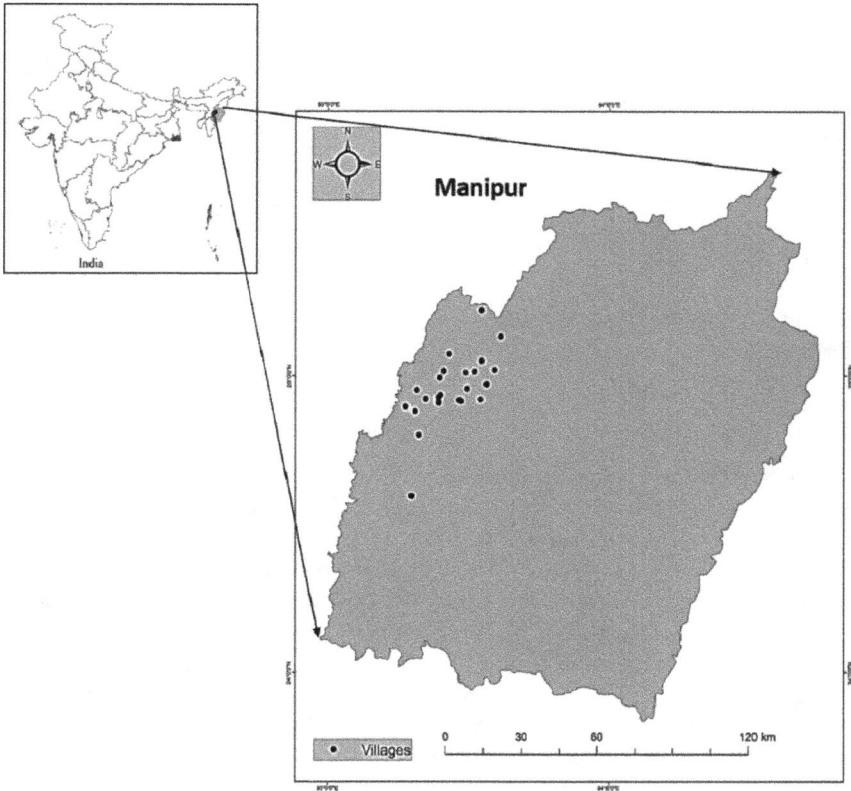

Figure 22.1 Location map of the study villages.

of the tribal populations in Manipur's Tamenglong district that are relevant to conservation management planning. Based on the findings of the study, recommendations have been put forward to accommodate the socioeconomic needs of the indigenous people.

Study Area

The study was conducted in 24 villages in the Tamenglong district of Manipur, which falls under the Indo-Burman (IBR) hotspot region. The villages are located in and around the Dailong Biodiversity Heritage Site and Community Reserve Forest of Azuram (Figure 22.1). Manipur, a small state in northeast India spreading over 22,327 sq km, is situated between 23° 50'N and 25° '41'N latitude and between 93° 2'E and 94° 47'E longitude (Vedaja, 1998). It is bounded by Nagaland on the north, Mizoram on the south, Myanmar on the east, and Assam on the west. Tamenglong, a hill district, is one of the 16 districts in Manipur state.

It is the farthest district from the state capital, Imphal. Barak, Irang, Makhru, Iring, Ijei, and Apah are the rivers that flow in this district. Most of the district is hilly, which makes it unsuitable for large-scale wet paddy cultivation. The study area is inhabited by tribal people only, and some of the notable tribes of the area are known as Paite, Hmar, Vaiphei, Anal, Thadou, Tangkhul, Kabui, Kom, Simte, Monsang, and Moyon, which are grouped into Naga, Kuki, Mizo, and others. The chief of the villages in the hill district owns the forests, and the simultaneous practice of customary laws and government's laws makes the situation very complex (Sitlhou, 2015). Dailong village, covering an area of 11.35 sq. km, has been declared as a Biodiversity Heritage Site under Section 37 (1) of the Biological Diversity Act, 2002. The Manipur State Government has declared the Azuram Community Reserve of 5.85 sq. km near Azuram village, a Biodiversity Heritage Site.

Methodology

The study aimed to enhance the understanding of the local socioeconomic considerations that governed conservation issues in Tamenglong and to recommend suitable conservation measures. The study was undertaken using a semi-structured questionnaire, a participatory risk mapping exercise, and by observing the fuel-wood collection pattern. Data was collected using a semi-structured questionnaire between December 2018 and August 2019. Questions were framed to collect information from each village on the demographic and socioeconomic background of the villagers, agricultural practices (types of crops, time of harvesting), intensity and pattern of wild animals raiding the crops, frequency and nature of fuelwood collection, mitigation measures adopted by the farmers, and the desired support for reducing dependence on forests.

Every fifth household was visited after randomly selecting the first house. Each household in a village was given a number starting from 1, which was noted down on a piece of paper. Randomly one piece of paper was drawn after the pieces of paper for each household with a designated number were mixed in a box. The household corresponding to the number on the paper was considered the first household to be visited for the survey in the village. This ensured the random selection of the households for interviewing. Only one member, over 18 years of age, from each selected household was interviewed. The "last birthday" method (also known as the Most Recent Birthday method) was used for the selection of the respondent in a family. A maximum of ten people were interviewed from a village, and a total of 225 people were interviewed in the study area. Each interview took around 15–20 minutes. Responses were noted on the response sheet. A Global Positioning System device (model Garmin Etrex 10) was used to record the location of every village visited under the study.

Participatory risk mapping meetings (Smith et al., 2000) were organised in three villages. Uncertain consequences and exposure to adverse situations were defined as risks (Smith et al., 2000). The first step in this approach included risk identification by the respondents. The respondents were asked to rank the risks in the second step. The respondents were selected from different socioeconomic backgrounds representing the community and included priests, teachers, farmers, artists, farm laborers, and representatives of local governing bodies. The severity index for each risk was calculated using the formula $Sj = 1 + (r-1)/(n-1)$, where r is the rank assigned by each respondent to each risk and n is the total number of risks listed by the respondents. Mean distribution was calculated for all the respondents, creating a score ranging from 1 to 2, wherein 1 indicated the most severe and 2 indicated the least severe. An incidence index (Ij) was created to mark the proportion of respondents mentioning a particular risk. 0 indicated not mentioned at all, and 1 indicated mentioned at all. A risk index (Rj) was obtained by dividing the incidence with severity. Every fifth respondent's household was observed for three months to note down the frequency of fuelwood collection from the forest.

Results and Discussion

Demographic and Socioeconomic Profile of Respondents

A total of 225 villagers were interviewed using a semi-structured questionnaire. Nearly a quarter of respondents were female (n=55) and three-quarters of them were male (n=170). The percentage of respondents who dropped out of school after their primary education was 40.4%, while 28% of the respondents had never been to school (Table 22.1). Farming (76.9%) is the major occupation of respondents followed by non-farming labour (6.7%), government job (4.9%), non-farming businesses (3.6%), and fishing (1.8%), while 6.2% of respondents were unemployed. All the respondents who practiced farming grew rice. While 77.5% of the farmers grew vegetables in addition to rice, nearly 65 % of the farmers adopted growing citrus fruits and bananas. In addition to rice, 42.8% of respondents grew king chilli (Figure 22.2). A total of 68 wild edible

Table 22.1 Demographic profile of respondents in the study

	Group	Number of respondents	% of total respondents
Age	18–30	25	11.1
	31–40	67	29.8
	41-50	48	21.3
	Above 50	85	37.8
Education	Primary school	91	40.4
	Illiterate	63	28.0
	High school	45	20.0
	Graduate	18	8.0
	Post-graduate	8	3.6

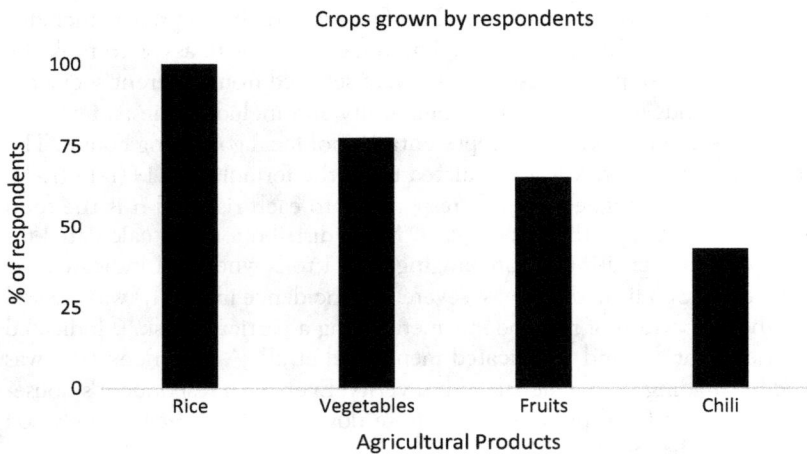

Figure 22.2 Crops grown by respondents.

vegetables belonging to 42 families were documented in Manipur in a previous study (Konsam et al., 2016), which calls for a need to invest in training, capacity building, and marketing of the vegetables grown by the farmers in Tamenglong.

Apart from consuming the agricultural products themselves, 82.4% of the farmers involved in the cultivation of king chilli were able to sell part of the produce in the local market. Only 6% of the rice cultivators were able to sell their crops after domestic consumption. All the respondents in the study reported the practice of jhum cultivation. The land is owned by the community through jhum cultivators, and the community land ownership system discourages development and investment in the agricultural land (Reimeingam, 2017).

The participatory risk mapping meetings helped identify perceived risk to the financial returns for the respondents (N=30). The respondents identified 11 such risks and ranked the threat from crop-damaging wildlife as the biggest threat

Table 22.2 Perceived risks for financial returns

Perceived risks	Incidence index	Severity index	Risk index
Wild animals	1	1.2	0.82
Difficult topography	0.9	1.3	0.71
Landslide	0.8	1.2	0.66
Poor market linkage	0.8	1.2	0.65
Poor road network	0.8	1.3	0.64
No industries around	0.8	1.4	0.59
Lack of veterinary care	0.9	1.6	0.56
Low population density	0.8	1.5	0.55
Poor healthcare	0.7	1.5	0.48
Illiteracy	0.7	1.5	0.47
Insects on crops	0.4	1.6	0.27

(perceived risk index 0.817) and the insects feeding on the crops as the smallest risk (perceived risk index 0.266) (Table 22.2). The other risks perceived were difficult topography, landslides, poor market linkages, poor road network, absence of industries, poor veterinary and healthcare, low population density, and illiteracy (Table 22.2). The role of king chilli in securing income for households is reflected in a high percentage of farmers being able to sell king chilli in the local markets. An earlier study showed that king chilli shared 27% of annual family income in the hills (Malangmeih & Sagolsem, 2015). As all the farmers reported only the local markets as the point for selling their additional agricultural produce, inadequate market linkage (perceived risk index of 0.649) seemed to be a risk for financial return. Another study has highlighted problems in marketing and unpredictable climatic conditions, leading to price and income instability (Malangmeih & Sagolsem, 2015). Greater engagement of the community in Tamenglong through alternative sources of income is necessary to discouraging their dependence on forests. Building up a market linkage to promote an improved chain of product flow will help conserve natural resources.

Wild boar was rated as the wild animal having the greatest risk (Table 22.3) for returns from the crops, with a perceived risk index of 0.808, followed by rats

Table 22.3 Perceived risk index of problematic species of wildlife as stated by the villagers during participatory risk mapping

Wild animals	Incidence index	Severity index	Risk index
Wild boar	0.87	1.08	0.81
Manipur bush rat	0.83	1.15	0.70
Barking deer	0.50	1.10	0.46
Parakeet	0.47	1.49	0.30
Macaque	0.33	1.53	0.22
Squirrel	0.30	1.65	0.18
Porcupine	0.20	1.72	0.12

Crop guarding tools by farmers

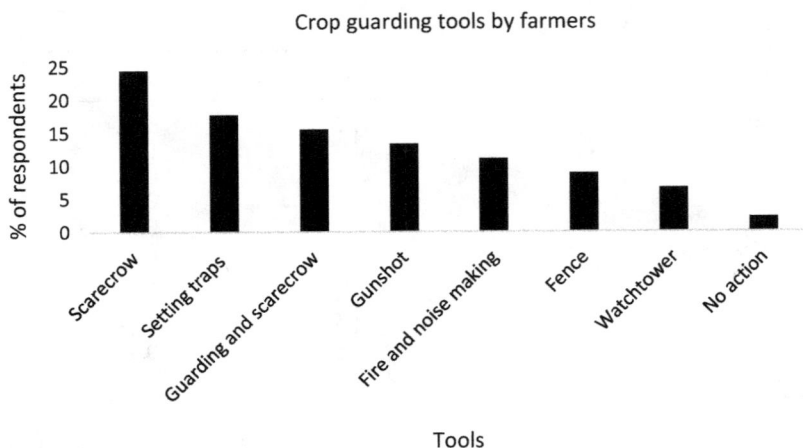

Figure 22.3 Crops guarding tools by farmers.

with a perceived risk index 0.705. While setting up a scarecrow is the most common crop guarding tool used by one-quarter of the respondents, nearly 31% of the respondents in the study set up traps in their fields or used gunshot to protect their crops (Figure 22.3).

The practice of trapping wild animals by nearly one-third of the respondents in this study should be seen along with traditional hunting practices that continue due to their linkage with local customs (Aiyadurai, 2011). Gaining insight into the existing hunting practices was beyond the scope of this study. Still, the respondents informally confirmed that the lack of sustainable economic alternatives for livelihoods had led to their continued dependence on the forest and hunting of wild animals. Lighting fires and fencing the farm were other tools used by the farmers for guard the crops.

Firewood Extraction from the Forest

Random observation of 45 villagers from different households indicated firewood extraction by all of them. Firewood collection was mainly a women's task, except in 3% of cases when men had to engage themselves in felling big trees. Their visits to forests for firewood collection averaged 11 times per month. One-third of the observed households had a Liquefied Petroleum Gas (LPG) connection. Firewood collection from the forest was not dependent on the availability of an LPG connection (ETA squared value 0.0718). During informal discussions, villagers indicated that the supply of LPG cylinders was not regular, and many of them did not use the LPG regularly despite owning them. The availability of LPG connections and the initial round of free cylinders are not enough to induce a behavioural change in favour of LPG cooking stoves.

Decadal growth of nearly 18% between 2001 and 2011 in Manipur is an important socioeconomic factor for an increase in deforestation and environmental degradation (Yuhlung, 2014).

Capacity-Building Areas for the Community

The respondents identified seven core areas for capacity building. Additional training for skills and knowledge in agriculture and animal husbandry was suggested jointly by 47% of the respondents. Training to meet ecotourism opportunities was listed by 13% of the respondents (Figure 22.4). The villagers were found to be aware of the progress made by Khonoma Forest Village in the neighbouring state of Nagaland. Khonoma village has attracted tourists, birders, nature enthusiasts, and researchers from all over the world since the declaration of the community-owned forest of Khonoma village as the Tragopan Sanctuary. The ecotourism initiatives in Khonoma have ensured people's participation in conservation, a complete ban on logging and hunting practices and formation of the Green Village Project (Chase & Singh, 2012). Beekeeping, carpentry, and road-repair skills were also recommended by the villagers to reduce their dependence on the forests. The findings of the study on farmers, emphasising the need for capacity building in agricultural practices, resonate with the observations made in another study where the lack of an improved method of cultivation, crop management, and extension services were recorded as the most critical constraints (Malangmeih & Sagolsem, 2015).

Conclusion and Recommendations

Socioeconomic considerations shape a community's perception of and interaction with nature and natural resources. Tamenglong in northeast India is predominantly occupied by various tribes who have depended on forests for generations. With a

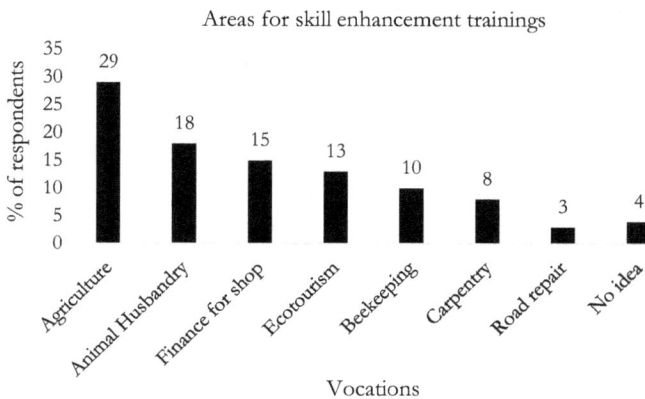

Figure 22.4 Areas for skill enhancement training and support.

growing population and increased extraction of natural resources, the age-old practice of sustainable harvesting of forest products has sharply come under threat. Shifting cultivation (jhum cultivation) is being increasingly practiced by the villagers in Tamenglong district. There is sufficient investment in improving soil conditions and employing modern mechanical tools in agriculture due to a lack incentive for developing agricultural land. Most of the agricultural produce is consumed, except some fruits, vegetables, and king chilli which farmers can sell in local markets.

On the one hand, wild animals are seen as a threat to financial return from agriculture due to damage to crops, poor market linkages, and bad roads. On the other hand, they make the price and income unstable. In the absence of sustainable socioeconomic alternatives, the dependence of villagers on the community forests continues to increase. The tools used for guarding crops by the farmers are not conservation-friendly. The traditional hunting practices and crop-protecting practices like setting up traps and gunshot pose threats to the conservation of existing wildlife in the community forests. The distribution of LPG stoves and cylinders has not discouraged villagers from collecting firewood from the forests. There is no socio-behavioural change in favour of utilisation of LPG because of irregular supply of filled in LPG cylinders and poor connectivity due to bad roads. The villagers continue to extract firewood irrespective of whether they have LPG stoves. This continuation of firewood collection from the forests further degrades the health of the forests.

Based on the observations made during the study, the following recommendations are made for securing wild habitats and their flora and fauna in the community forests of Tamenglong district:

Conduct systematic research to validate the perceived risks identified through this study.
Promote ecotourism based on a ban on logging and hunting to generate income from protected forests as practiced in Khonoma, Nagaland.
Improve the network of roads and market linkages to encourage investment and better practices in farming.
Establish a regular supply of filled LPG cylinders and educate villagers for behavioural change.
Invest in capacity building for skills and knowledge aimed at generating alternative livelihoods for the local community.

Acknowledgements: The authors are grateful to the Axis Bank Foundation for funding the project titled Systematic Rejuvenation of Indigenous Jungles and Nature (SRIJAN) in Tamenglong, Manipur. A very special note of thanks to Shri Arun R. S., IFS, Divisional Forest Officer, Tamenglong for the continuous support extended to the research team in the field.

References

Adams, W. M. (2007). Thinking like Ahuman: Social science and the two cultures problem. *Oryx, 41*(3), 275–276.

Aiyadurai, A. (2011). Wildlife hunting and conservation in Northeast India: A need for an interdisciplinary understanding. *International Journal of Galliformes Conservation, 2,* 61–73.

Ali, A. N., & Das, I. (2003). Tribal situation in North East India. *Studies on Tribes and Tribals, 1*(2), 141–148.

Ban, N. C., Mills, M., Tam, J., Hicks, C. C., Klain, S., Stoeckl, N., Bottrill, M.C., Levine, J., Pressey, R.L., Satterfield, T., & Chan, K. M. (2013). A social—Ecological approach to conservation planning: Embedding social considerations. *Frontiers in Ecology and the Environment, 11*(4), 194–202. Retrieved from http://www.jstor.org/stable/23470948.

Chase, P., & Singh, O. P. (2012). People's initiative for conservation of forests and natural resources: A success story of Khonoma village forest, Nagaland. *Nebio, 3*(3), 61–67.

Devi, M. R., & Salam, S. (2016). Wild edible plants used by the Monsang Naga tribe of Manipur, India. *Pleione, 10*(1), 90–96.

Dutta, B. (2004). *Biodiversity, livelihoods and the law: The case of Jogi-Nath snake charmers of India.* Wildlife Trust of India.

Guha, R. (1997). Social-ecological research in India: A 'status' report. *Economic and Political Weekly, 32*(7), 345–352.

Khumbongmayum, A. D., Khan, M. L., & Tripathi, R. S. (2004). Sacred grooves of Manipur-ideal centers for biodiversity conservation. *Current Science, 87*(4), 430–433.

Knight, A. T., & Cowling, R. M. (2007). Embracing opportunism in the selection of priority conservation areas. *Conservation Biology, 21*(4), 1124–1126.

Konsam, S., Thongam, B., & Handique, A. K. (2016). Assessment of wild leafy vegetables traditionally consumed by the ethnic communities of Manipur, Northeast India. *Journal of Ethnobiology and Ethnomedicine, 12*(9). https://doi.org/10.1186/s13002-016-0080-4.

Malangmeih, L., Dey, G., & Sagolsem, S. (2015). Rural livelihood system in Manipur with special reference to cultivation of King Chilli. *Journal Crop and Weed, 11*(1), 144–151.

Mishra, C., Madhusudan, M. D., & Datta, A. (2006). Mammals of the high altitudes of western Arunachal Pradesh, eastern Himalaya: An assessment of threats and conservation needs. *Oryx, 40*(1), 29–35.

Ninan, K. N. (1992). Economics of shifting cultivation in India. *Economic and Political Weekly, 27*(13), A2–A6.

Reimeingam, M. (2017). Shifting cultivation in Manipur: Land labor and development. *Journal of Rural Development, 36*(1), 97–119.

Sitlhou, H. (2015). Confronting the state land rights discourse in the hills of Manipur. *Economic and Political Weekly, 30,* 70–77.

Smith, K., Barrett, C. B., & Box, P. W. (2000). Participatory risk mapping for targeting research and assistance: With an example from East African pastoralists. *World Development, 28*(11), 1945–1959.

UN. (2009). Human development report. Overcoming barriers: Human mobility and development. United Nations Development Programme. Palgrave Macmillan.

Vedaja, S. (1998). *Manipur: Geography and regional development.* Rajesh Publications.

Yuhlung, C. C. (2014). Social ecology and rural development in Northeast India: Challenges and opportunities in Manipur. *Journal of Humanities and Social Science, 19*(8), 8–16.

Conclusion

Orus Ilyas and Sneha Singh

This book, *Case Studies on Wildlife Ecology and Conservation in India*, is one of the outcomes of the International Conservation Conference (ICC) held in 2019, organised by the Department of Wildlife Sciences, Aligarh Muslim University, India. It is a compilation of 21 articles based on various studies conducted by noted researchers in the different geographical regions ofIndia. These studies are solely based on the ecological concepts of wild animals and conservation under a single heading, *Case Studies of Wildlife Ecology and Conservation in India*, in the form of a book. The case studies discussed in the book are real-time, field-based studies from diverse backgrounds, drawing the attention of disparate readers. The data analysed for comprehending the ecological concepts were collected and collated by eminent field biologists and researchers from various study areas using various standard scientific statistical and analytical tools including, R-Software, SPSS, Remote Sensing, and GIS.

Resource selection in the protected and the non-protected regions, human-wildlife interactions, and landscape ecology and conservation are the three crucial elements of this book. The first section of the book discusses the role of resource availability in the distribution of species. In this section, eight studies were taken into account, elucidating the distribution pattern, the resource selection, and the movement patterns of the various species. Such types of information play a vital role in predicting the diet composition of species, local and migratory movement patterns of the species, and in the estimation of carrying capacities of the area for the species' survival and growth for designing more effective conservation strategies.

Resource availability, such as food, water, shelter, and mates, determines species distribution across any landscape. In short supply of these resources, the population tends to decline until resources rebound. However, the uneven distribution of natural resources plays a significant role in the allocation of resources among the species. Also, it has become an essential factor in regulating the population. It can be evident from the first chapter of the book, where the authors have put their efforts into conceptualising it in the real world through their review-based studies on the large predators' guild in Asia. Availability of prey is crucial for the carnivore species, just like the vegetation for herbivores for species' existence. This section raises concerns about conserving large predator species that could

DOI: 10.4324/9781003321422-26

be threatened due to a decline in the prey base and suggests ensuring the prey base, especially when a new protected area is declared Likewise, the study on the habitat preferences of leopards in the moist-temperate forest of Kashmir elucidates the ecological factors influencing their number and conservation measures to regulate their population. It is vital to understand the diet preferences of the species for the conservation of species at the landscape level. Similar studies have been described in the fourth chapter of the book about the golden jackal in the Sariska Tiger Reserve, Rajasthan using scat analysis as a tool in conservation management of the species. Sometimes, diet preferences get influenced by several climatic factors. The shifting diets of four-horned antelope in the Pench Tiger Reserve, Madhya Pradesh, and the Central Highlands of India illustrate the role of seasonal variations in the diet preferences of the species in the sixth chapter of the section.

Further, in this section of the book, the authors discussed the various resource requirements of the species, which vary with topography and the climatic conditions, especially in those species that have wide ranges of distribution. When two different species have similar requirements for resources, they divide those limited resources to avoid competition. Understanding resource partitioning among species may help anticipate how ongoing species declines would affect the functioning of ecosystems. The first two chapters of this section give insights into resource-partitioning through the spatial segregation of two sympatric species of sparrows (i.e., house sparrow and tree sparrow), indicating niche differentiation through their habitat utilisation pattern and illustrated using graphs and statistics. Simultaneously, the other study on the vocalisation of the Indian robin in the Chandrapur district of Maharashtra suggests the different types of songs, and territorial and mate acquisition in ten individuals, illustrating the various strophes and their structural variations, suggesting different resource requirements of the species at different spatial scales.

Environmental factors such as adequate food, shelter, water, and mates can determine the size of the population of a species. In other words, the availability of the resources directly influences the carrying capacities of the landscapes. This section also highlights species' population dynamics and variability across the landscape. One such study showed an increase in the population of blackbucks, a threatened species in the human-dominated landscape. The authors recommended further study to assess the level of human-wildlife conflict. Another study under the same theme provides a checklist of the total 140 species of avifauna in the Suru Valley of Ladakh.

A community is bound together by a matrix of influences that species have on one another. Resource sharing is one of such factors determining the ecological dynamics of the species. The second section of the book discusses human-wildlife interactions as negative interactions between humans and wild animals that have consequences for humans, wildlife, or both. However, a single resource may not be responsible for the human-wildlife conflict, as discussed in the third chapter of this section, where the food resource sharing between humans and sloth bears in Jessore Wildlife Sanctuary of Gujrat. Human-wildlife conflicts often occur when the requirements or behaviours of wildlife overlap with those of humans, resulting in adverse ramifications such as livestock depredations, crop damages, or even human casualties. The first two chapters of this section give insights into human-leopard conflict and its adverse consequences in the form of livestock depredation by leopards and tigers in the Corbett Tiger Reserve, Uttarkhand. It is essential to keep wildlife out of places with high human populations and agricultural areas to avoid human-wildlife conflicts. Some of the major causes responsible for the human-wildlife conflicts include the change inland use to meet the growing demands of expanding human populations, deforestation, urbanisation, or developmental activities, and over-grazing by livestock in the protected areas. The authors put their efforts into identifying the primary causes leading to such conflicts and, based on the data, recommendations were made for wildlife managers to improve the practice of livestock husbandry and provide possible alternatives for cattle grazing in the protected areas and to create awareness about the issues among local peoples.

Ecosystem function is critical for the existence and long-term survival of plant and animal communities. In the last few decades, human civilisation has undoubtedly resulted in the destruction, degradation, and fragmentation of forests, reducing the survival and reproductive rates of species. Moreover, the exploitation, pollution, and invasion of alien species further exacerbate the survival of species. These threats, however, do not influence all species equally. The fourth chapter of this section addresses the bird community structure of restored and unrestored ecosystems of the Aravalli Biodiversity Park in Delhi, serving as an excellent model for restoration, i.e., Aravalli Biodiversity Park, once a mining site, was restored and developed as a biodiversity park, certainly functioning as the lung of the most polluted capital, Delhi.

In the Indian scenario, the villages around the protected areas are remote areas, generally lacking basic amenities, consequently facing unemployment, and therefore, partially or completely dependent on the natural resources of the protected areas for their livelihood and livestock grazing. Generally, moderate grazing is considered to be beneficial for the vegetation. However, high grazing pressure influences the richness, abundance, cover, and biomass of plants as livestock utilise the vegetation for feed, especially graminoids and sedge species, thereby decreasing the biomass of plants above-ground, reducing the biodiversity of herbivores. Also, livestock grazing impacts over 60% of the world's agricultural lands. Therefore, livestock grazing is generally assumed to negatively affect wildlife through interference competition and changes in forage quality and quantity. Evaluation of the livestock grazing effects on more than one trophic level helps understand the effects of grazing on species. A study on the impact of cattle grazing in the Binsar Wildlife Sanctuary, Uttarakhand, suggests further research to provide an alternative solution to reduce the dependency of locals on the protected areas for livestock grazing.

Due to their sensitivity to the changing environmental conditions and their ecological peculiarities, insects are sometimes used as environmental bio-indicators to assess the level of change in the environment. The authors in the sixth chapter of this section provide a checklist of insect diversity and recommendations for long-term studies to evaluate the impact of climate change. The final chapter of this section discusses the role of dipterans as a forensic tool or indicator in identifying abuse through the presence of maggots in the anal region, wounds, and other uncleaned body parts, which speak of trauma caused to victims by, at times, caretakers' ignorance.

The idea of landscape ecology is to characterise the dynamic interactions between the ecological patterns and processes across various spatial scales. It is the scientific foundation for studying and managing landscapes and ecological systems. The first chapter of the third section was a review-based study on pheasants in Jammu and Kashmir. The study's primary goal was to develop science-based conservation breeding programmes to augment the existing populations of pheasants in the wild for their long-term survival. Besides this, the authors suggested undertaking status surveys to update the scientific knowledge on the ecology and the behaviour of Galliformes in general and pheasants in particular, for their effective management and long-term conservation in Jammu and Kashmir at the landscape level. Similarly, the next chapter of the section gives insights into the species diversity of birds and their conservation at Baba Ghulam Shah Badshah University in Jammu and Kashmir. It is a baseline study highlighting the importance of green space on university campuses to maintain ecological balance and sustainable urban development. Researchers will be able to better understand and analyse changes in species diversity in the future using this baseline data.

Landscape conservation is a holistic approach to landscape management that balances the competing goals of nature conservation and economic development. The challenge of safeguarding and conserving biodiversity in landscapes dominated by human land use is inextricably related to the widespread decline of ecosystems. During a study conducted on the endangered primate species of Kashmir grey langur, no direct threats were observed, presumably due to the low density of the species. However, the study came up with several conservation measures to protect the species at the landscape level.

In recent years, researchers and land-use planners have become highly concerned about the impact of urbanisation and industrialisation on biodiversity. Land use plays a crucial role in the long-term conservation of species at the landscape level by reducing the impacts of urbanisation on wildlife. However, due to the increasing need for the rapidly growing human population, a significant shift in land use patterns has been observed. Some of its adverse impacts are habitat loss, fragmentation of forests, and shifting of forests to agricultural land. A field-based study in Panna Tiger Reserve, Madhya Pradesh showcases the paradigm of land use in conservation using geospatial technology as a tool, as discussed in this section of the book.

Wetlands are one of the most important ecosystems on the planet, with crucial ecological functions in sustaining unique biodiversity and serving ecological services. To elucidate the importance of wetlands in the conservation of species at the landscape level, the authors in second last chapter of the book mapped the wetlands of Aligarh in Uttar Pradesh using remote sensing and GIS as a tool, suggesting the restoration of wetlands to recharge the water table of the area, as well as to attract aquatic and migratory bird species.

Many developing countries like India face challenges in achieving long-term socio-economic development in areas designated for ecological conservation due to conflict between ecological conservation and livelihood development. Most protected areas with limited access to basic amenities and ecological conservation practices restrict communities' ability to use resources. Tamenglong, in northeast India, is home to several tribes who have depended on the forest for generations. With a growing population and increased extraction of natural resources, the age-old practice of sustainable harvesting of forest products has sharply come under threat. Shifting cultivation (jhum cultivation) is increasingly practiced by the villagers in the Tamenglong district.

On the one hand, wild animals are considered as threat to agriculture's financial return due to crop damage, while farmers also competing with declining market relations and poor road conditions. On the other hand, the prices and income profits are volatile in the absence of sustainable alternatives, and villagers' increasing dependency on community forests The traditional hunting practices and crop-protecting practices like setting up traps and gunshots pose threats to the conservation of existing wildlife in the community forests. The authors made several recommendations based on their findings, including promoting ecotourism to generate income from protected forests, improving road networks to encourage investment and better farming practices, and investing in skill and knowledge development to generate an alternative source of livelihood among the region's local communities. The book's final chapter addresses the significance of the locals' socio-economic importance in conservation for achieving a more sustainable ecosystem. Hopefully this book will serve the purpose of researchers, academics, and scientists, and those who are interested in various aspects of wildlife studies. A copy of this book deserves a space on the shelf of every library.

Index

Note: Page numbers in *italic* indicate figures in **bold** indicate tables

For Product Safety Concerns and Information please contact our EU
representative GPSR@taylorandfrancis.com
Taylor & Francis Verlag GmbH, Kaufingerstraße 24, 80331 München, Germany

www.ingramcontent.com/pod-product-compliance
Lightning Source LLC
Chambersburg PA
CBHW052119230326
41598CB00080B/3885